RF Engineering for Wireless Networks

RF Engineering for Wireless Networks
Hardware, Antennas, and Propagation

Daniel M. Dobkin

AMSTERDAM • BOSTON • HEIDELBERG • LONDON
NEW YORK • OXFORD • PARIS • SAN DIEGO
SAN FRANCISCO • SINGAPORE • SYDNEY • TOKYO
Newnes is an imprint of Elsevier

ELSEVIER

Newnes

Newnes is an imprint of Elsevier.

30 Corporate Drive, Suite 400, Burlington, MA 01803, USA
525 B Street, Suite 1900, San Diego, California 92101-4495, USA
84 Theobald's Road, London WC1X 8RR, UK

⊗ This book is printed on acid-free paper.

Library of Congress Cataloging-in-Publication Data
Application submitted

British Library Cataloguing in Publication Data
A catalogue record for this book is available from the British Library

ISBN: 978-0-7506-7873-5

For all information on all Newnes/Elsevier Science publications visit our Web site at
www.books.elsevier.com

Transferred to Digital Printing in 2011

Contents

Introduction

1. The Beauty of Wires, the Inevitability of Wireless

The first gem of wisdom I ever acquired about consulting, obtained many years ago from a former schoolmate, was to ensure that everything is plugged in: no continuity, no data. Wires carry voltages and currents from one place to another. Their behavior is reasonably simple and predictable—at least for sufficiently low data rates and short lengths—and they can be seen, grabbed, traced, and tugged.

Wires are also an irritating and sometimes intolerable umbilical cord to the mother network. Look behind a typical personal computer or engineering workstation: you'll be hard put to find which plug goes with which cable. One person in a café with a laptop plugged into an Ethernet port on the wall is a curiosity; five, sitting at different tables with colorful category 5 cables snaking randomly along the floor between chairs and people, are an eyesore and a safety hazard. It's great to return your rental car and receive an instant printed receipt; the folks who provide the service would have a tough time dragging wires behind them down the long lines of cars, to say nothing of the angry businessmen and -women falling flat on the asphalt as they hurried to their flights. Sometimes wires won't do.

Accept that we have to cut the cords and break the old rules: no continuity, but lots of data. How is it done? What new complexities are introduced when the old metal conduit for current is removed? How are the resulting difficulties surmounted? What limits remain? It turns out that the way signals are modulated, transmitted, propagated, and received in a wireless link all change drastically from their wired counterparts. However, these changes can mostly be concealed from the data network: if things are done right, the network can't tell that the wires have been left behind.

This book is about the measures that must be taken, the obstacles that are encountered, and the limitations that result when data are to be moved wirelessly from place to place. The book is focused on local and personal area networks—*LANs* and *PANs*—although we will show how technologies developed for local communications can in some cases be deployed over long distances. We will concentrate on Institute of Electrical and Electronic Engineers (IEEE) 802.11–based wireless LANs—*Wi-Fi* networks—though we will discuss related technologies, and many of the lessons we will learn are broadly applicable.

Though we will touch on recent standards developments and provide examples of commercial practice, this book is not an attempt to provide an up-to-date snapshot or exhaustive survey of the state of the art; even in an age of rapid publication, any such summary will rapidly be rendered obsolete by the continual advancement of technology and industry. Rather, the reader should be armed upon completion of this book to understand why things work the way they do and thus distinguish between pretense and progress.

Most importantly, this book is about signals, not bits. We will touch upon the digital side of the problem of wireless networking, but we are primarily concerned with what happens between the point where bits are converted to voltages *here* and where the reverse operation occurs at a distant *there*. Some readers may be familiar with the *Open Systems Interconnect (OSI) reference model* for digital communications, a standard for imposing some hierarchical order on the various tasks required to communicate between two end-users within a network (Figure 1-1). In OSI terms, this book is about the *physical* layer of a wireless data link, with a few digressions into the *medium-access control* sublayer, but rarely higher. The reader whose interest extends also up the stack and sideways into network integration is referred to the many excellent texts that already touch upon the digital side of wireless LANs; a few examples can be found in the suggested reading at the end of this chapter.

2. What You Need to Proceed

The book is intended to be nearly self-contained: no previous acquaintance with radio technology is necessary. The reader ought to have some background in the physical sciences and an acquaintance with the basic electrical engineering concepts of voltage, current, resistance, capacitance, and inductance, along with their conventional schematic representations. Familiarity with the importance of harmonic (sinusoidal)

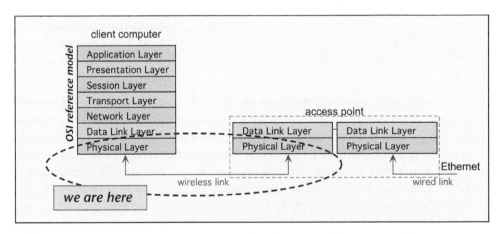

Figure 1-1: Location Relative to the OSI Reference Model

excitations is also assumed, though we provide a brief review of the use of complex exponentials to represent such signals. Though every effort is made to emphasize physical intuition before mathematical methodology, the reader ought to be comfortable with algebra and recognize the meaning of derivatives and integrals even if the mechanics of performing them is a bit rusty. Vectors and gradients will inevitably appear in the discussion of antennas and propagation. Certain aspects of Fourier transform theory, which play a vital role in wireless communications, will be briefly reviewed, though again a prior encounter with the notions of frequency and time representations of a signal is very helpful.

The more sophisticated reader may perhaps encounter a surprising dearth of certain expected vectorial complexities: I nowhere resort to Maxwell's equations, and nary a cross product is to be found. I do sink to taking a divergence of a vector field here and there, but the reader can bypass those scattered indulgences with little cost. Furthermore, though diffraction is examined in some detail, the estimable Mr. Huyghens and his principle are not in evidence, and though near-field and far-field regions are defined, the distinction is found to be mathematical rather than physical. In fact, the treatment of electromagnetism presented herein is exclusively in terms of the vector and scalar potentials, A and ϕ, acting on and created by actual currents and charges. In this attempt I have followed in part the prescriptions of Carver Mead in his book *Collective Electrodynamics*; I hope that the resulting presentation will provide a simpler and more appealing method of approaching a complex topic, though the results are equivalent to those obtained by conventional means, and readers with the requisite expertise can always fall back on more traditional approaches.

3. An Overview of What Is to Come

Figure 1-2 gives an overview of the remainder of the text.

We first cover some of the basics of wireless communications. We begin with the idea of multiplexing and then examine what happens when a high-frequency carrier is modulated to convey information, and why a trade-off between bandwidth, data rate, and noise tolerance must inevitably arise. We survey the typical modulation schemes used in digital data transmission, including a pair of somewhat exotic beasts—orthogonal frequency-division multiplexing and pulsed ultrawideband—that play an increasing role in the modern wireless world. We introduce the idea of a wireless link and the specialized terminology used in the radio world to describe voltage, power, and noise.

In the unavoidable exception to prove the rule, we digress into the digital domain to cover some basics of wireless local area networks, emphasizing the IEEE 802.11 committee's alphabet soup but not forgetting the unexpected radio legacy of King Harald "Bluetooth" Blätand of ancient Denmark, or some of the interesting activities of other task groups within the IEEE. While emphasizing the radio-related aspects of the standards, we also touch gingerly upon the domains of coding and encryption, by

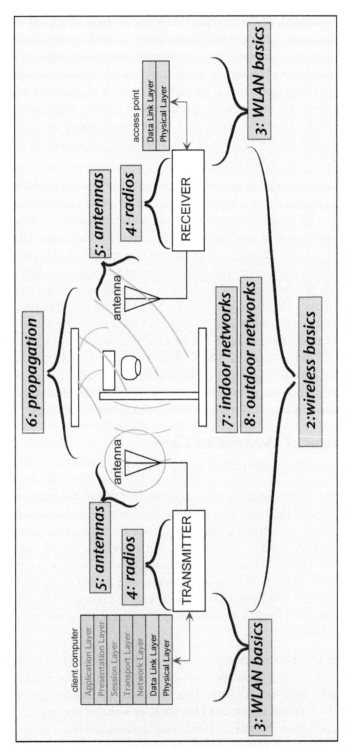

Figure 1-2: Overview of the Remainder of the Text

which the digital designers seek to compensate for the inadequacies of the uncertain wireless link.

We then plunge into the workings of digital radios: amplification, noise, and distortion; frequency conversion and spurious signals; frequency synthesis and phase noise; filters and filtering. We summarize by trying to design a radio chain to achieve a desired performance and examining how the requirements translate into the cost and complexity of the resulting chipset. The chapter ends with some examples of radio chips for wireless local area network and wireless personal area network applications and a quick look at the other pieces of the puzzle: the bit of PC board and external components that together with the integrated chips make a low-cost digital radio.

Having examined the creation of a signal, we examine its transmission and reception by means of antennas, exposing an antenna as nothing but a means to avoid cancellation of the effects of its currents. We show how antenna directivity allows us to estimate the amount of the transmitted power that arrives at the receiver given the antenna characteristics, at least for the ideal environment we know we will never encounter in this life. We briefly survey the operating principles behind a number of different types of commonly encountered antennas, with some emphasis on the tools for understanding how arrays of antennas can be used to produce interesting results not achievable from a single radiator.

We next abandon the pretense of ideality and treat propagation in some small fraction of the unmanageable complexity it deserves. Our poor transmitted wave finds itself reflected, absorbed, diffracted, and refracted by all manner of obstacles, on scales from the microscopic to the global, leading to the bugaboos of the radio world: attenuation, fading, and multipath. We examine how fading is surmounted with the aid of diversity antennas, appropriate modulations, and more sophisticated adaptive antenna arrays.

Our theoretical apparatus as complete as it is going to be, we plunge into the empirical realities of real indoor and outdoor networks. We review construction practices in sufficient detail to provide insight into existing buildings but not to obtain a general contractor's license and combine them with published information on microwave absorption and reflection to obtain some idea of where waves ought to go and where they ought not. We provide a survey of software tools to automate the tasks of getting the network to work, and some samples of their output. Our roving eye then moves into the wild outdoors, where we examine how buildings confine signals through reflection and diffraction, and how trees play a key role in limiting outdoor microwave links. We examine area coverage networks, point-to-multipoint networks for providing Internet access, and point-to-point links between fixed locations. The curvature of the earth, both real and effective, reminds us that we've gone as far as short-range radios, and this book, are going to take us. A brief review of safety precautions for outdoor work brings us to the end of the journey, save for the unusually determined reader, for whom appendices on regulatory and measurement considerations, as well as a few derivations mercifully avoided in the text, are provided.

4. Acknowledgments

The book you're about to read has benefited from the time, thought, and experience of many helpful people. Special thanks go to Jay Kruse of Tropos Networks, who reviewed many incarnations of the materials in detail and provided helpful insight and suggestions. Jay's colleagues Cyrus Behroozi, Tom Blaze, and Malik Audeh also provided support and allowed me to tag along on surveying expeditions. The folks at WJ Communications, including Ron Buswell, Don McLean, Mike O'Neal, Rich Woodburn, Mark Bringuel, John Tobias, Kevin Li, Ray Allan, and Steve Weigand, gave me access to test equipment and facilities for many of the measurements reported here as well as encouragement and microwave wisdom. Jim Mravca and Nathan Iyer were particularly helpful in providing insight on system and integrated circuit design. Bill McFarland and David Su of Atheros and Kevin Wang at Silicon Wave helped correct the descriptions of RF chipsets. Greg des Brisay and Bob Arasmith shared their practical knowledge of outdoor network installations; Pat McCaffrey of Hidden Villa kindly provided access to the network site described in Chapter 8. Rajeev Krishnamoorthy shared his experiences in the early development of wireless LAN technology. Miki Genossar provided guidance on the ultrawideband work of the IEEE standards bodies. Markus Moisio and Jussi Kiviniemi of Ekahau and Baris Dandar of ABP Systems provided demonstration software and guidance in surveying. Skip Crilly clarified the operation of the Vivato access point. Thanks also to Mark Andrews, Martin Chaplin, Rob Martin, Ana Bakas, William Stone, Franz Chen, W.R. Vincent, Richard Adler, David Freer, Simon Perras, Luc Bouchard, and Vinko Erceg. Harry Helms, my editor at Elsevier, was always entertaining as well as encouraging. Last but not by any means least, my long-suffering spouse, Nina, not only put up with innumerable weekend absences but also provided expert linguistic support on short notice to navigate the Chinese Ministry of Information Industries website.

Further Reading

General Introductions to Networking

Understanding Data Communications (6th Edition), Gilbert Held, New Riders, 1999: *Simple and accessible, though some coverage areas are already rather dated.*

Communication Networks, Alberto Leon-Garcia and Indra Widjaja, McGraw-Hill, 2003: *An academic text with modern emphasis on Internet Protocol.*

Introductions to Wireless LANs

802.11 Networks: The Definitive Guide, Matthew Gast, O'Reilly, 2002: *It is, at least on the digital side. A thorough review of standards and tools.*

Jeff Duntemann's Drive-By Wi-Fi Guide, Paraglyph Press, 2003: *Accessible introduction and hands-on guide to setting up common equipment.*

Wireless LANs, Jim Geier, Macmillan, 1999: *The technical information here is already somewhat old, but this book has a nice discussion of how to plan and manage network installation projects.*

Basics of Wireless Communications

1. Harmonic Signals and Exponentials

Before we begin to talk about wireless, we briefly remind the reader of a previous acquaintance with three concepts that are ubiquitous in radio engineering: sinusoidal signals, complex numbers, and imaginary exponentials. The reader who is familiar with such matters can skip this section without harm.

Almost everything in radio is done by making tiny changes—modulations—of a signal that is periodic in time. The archetype of a smooth periodic signal is the sinusoid (Figure 2-1), typically written as the product of the angular frequency ω and time t.

Both of these functions alternate between a maximum value of 1 and minimum value of −1; cosine starts at +1, and sine starts at 0, when the argument is zero. We can see that cosines and sines are identical except for an offset in the argument (the *phase*):

$$\cos(\omega t) = \sin\left(\omega t + \frac{\pi}{2}\right) \qquad\qquad [2.1]$$

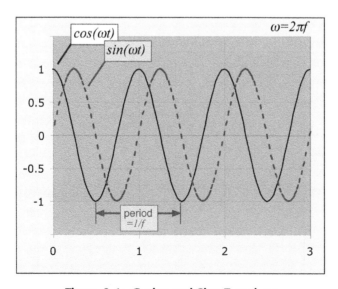

Figure 2-1: Cosine and Sine Functions

9

We say that the sine lags the cosine by 90 degrees. (Note that here, following common practice, we write angles in radians but often speak of them in degrees.) The cosine and sine are periodic with a period $= (1/f)$, where $f = \omega/2\pi$ is the frequency in cycles per second or *Hertz*.

Let us now digress briefly to discuss complex numbers, for reasons that will become clear in a page or two. *Imaginary* numbers, the reader will recall, are introduced to provide square roots of negative reals; the unit is $i = \sqrt{(-1)}$. A *complex* number is the sum of a real number and an imaginary number, often written as, for example, $z = a + bi$. Electrical engineers often use j instead of i, so as to use i to represent an AC; we shall, however, adhere to the convention used in physics and mathematics. The complex conjugate z^* is found by changing the sign of the imaginary part: $z^* = a - bi$.

Complex numbers can be depicted in a plane by using the real part as the coordinate on the x- (real) axis, and the imaginary part for the y- (imaginary) axis (Figure 2-2). Operations on complex numbers proceed more or less the same way as they do in algebra, save that one must remember to keep track of the real and imaginary parts. Thus, the sum of two complex numbers can be constructed algebraically by

$$(a + bi) + (c + di) = [a + c] + [b + d]i \qquad [2.2]$$

and geometrically by regarding the two numbers as vectors forming two sides of a parallelogram, the diagonal of which is their sum (Figure 2-3).

Multiplication can be treated in a similar fashion, but it is much simpler to envision if we first define the length (also known as the *modulus*) and angle of a complex number. We define a complex number of length 1 and angle θ to be equal to an exponential with an imaginary argument equal to the angle (Figure 2-4). Any complex number (e.g., b in

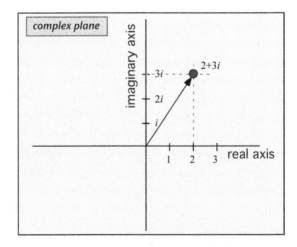

Figure 2-2: Complex Number Depicted as a Vector in the Plane

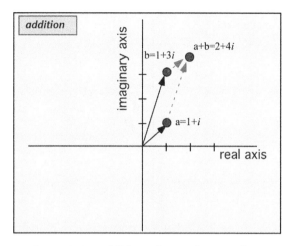

Figure 2-3: Addition of Complex Numbers

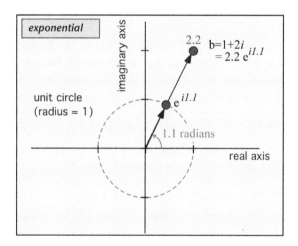

Figure 2-4: Imaginary Exponentials and Complex Numbers

Figure 2-4) can then be represented as the product of the modulus and an imaginary exponential whose argument is equal to the angle of the complex number in radians.

By writing a complex number as an exponential, multiplication of complex numbers becomes simple, once we recall that the product of two exponentials is an exponential with the sum of the arguments:

$$(e^a) \cdot (e^b) = e^{[a+b]} \qquad [2.3]$$

The product of two complex numbers is then constructed by multiplying their moduli and adding their angles (Figure 2-5).

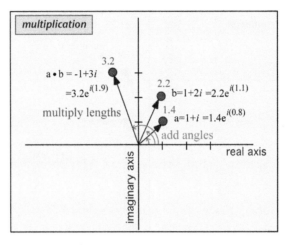

Figure 2-5: Multiplication of Complex Numbers

$$\left(\rho_1 e^{i\theta_1}\right) \cdot \left(\rho_2 e^{i\theta_2}\right) = [\rho_1\rho_2]e^{i[\theta_1+\theta_2]} \qquad [2.4]$$

We took the trouble to introduce all these unreal quantities because they provide a particularly convenient way to represent harmonic signals. Because the *x*- and *y*- components of a unit vector at angle θ are just the cosine and sine, respectively, of the angle, our definition of an exponential with imaginary argument implies

$$e^{i\theta} = \cos(\theta) + i\sin(\theta) \qquad [2.5]$$

Thus, if we use for the angle a linear function of time, we obtain a very general but simultaneously compact expression for a harmonic signal:

$$\begin{aligned} e^{i(\omega t+\phi)} &= \cos(\omega t+\phi) + i\sin(\omega t+\phi) \\ &= [\cos(\omega t) + i\sin(\omega t)] \cdot [\cos(\phi) + i\sin(\phi)] \end{aligned} \qquad [2.6]$$

In this notation, the signal may be imagined as a vector of constant length rotating in time, with its projections on the real and imaginary axes forming the familiar sines and cosines (Figure 2-6). The phase offset φ represents the angle of the vector at $t = 0$.

In some cases we wish to use an exponential as an intermediate calculation tool to simplify phase shifts and other operations, converting to a real-valued function at the end by either simply taking only the real part or adding together exponentials of positive and negative frequency. (The reader may wish to verify, using equations [2.5] and [2.6], that the sum of exponentials of positive and negative frequencies forms a purely real or purely imaginary sinusoid.) However, in radio practice, a real harmonic signal $\cos(\omega t + \phi)$ may also be regarded as being the product of a real carrier $\cos(\omega t)$ and a complex number $I + iQ = [\cos(\phi) - i\sin(\phi)]/2$, where the imaginary part is obtained through multiplication with $\sin(\omega t)$ followed by filtering. (Here I and Q

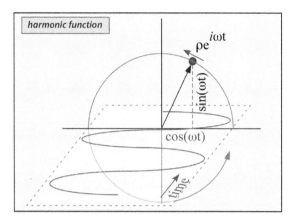

Figure 2-6: An Imaginary Exponential Can Represent Sinusoidal Voltages or Currents

denote "in-phase" and "quadrature," that is, 90 degrees out of phase, respectively.) We'll have more to say about the uses of such decompositions when we discuss radios in Chapter 4.

Finally, we note one other uniquely convenient feature of exponentials: differentiation and integration of an exponential with a linear argument simply multiply the original function by the constant slope of the argument:

$$\frac{d}{dx}(e^{ax}) = ae^{ax} \quad \int e^{ax}dx = \frac{1}{a}e^{ax} \qquad [2.7]$$

2. Electromagnetic Waves and Multiplexing

Now that we are armed with the requisite tools, let us turn our attention to the main topic of our discussion: the use of electromagnetic waves to carry information. An electric current element **J** at some location [1] induces a potential **A** at other remote locations, such as [2]. If the current is harmonic in time, the induced potential is as well. The situation is depicted in Figure 2-7.

The magnitude of the induced potential falls inversely as the distance and shifts in phase relative to the phase of the current. (The reader may wish to verify that the time dependence of **A** is equivalent to a delay by r/c.) The induced potential in turn may affect the flow of electric current at position [2], so that by changing a current **J**[1] we create a delayed and attenuated but still detectable change in current **J**[2]: we can potentially communicate between remote locations by using the effects of the electromagnetic disturbance **A**.

In principle, every current induces a potential at every location. It is this universality of electromagnetic induction that leads to a major problem in using electromagnetic waves in communications. The potential at our receiver, **A**, can be regarded as a medium of

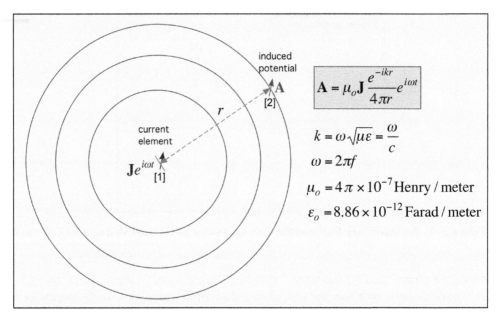

induced
potential
A
[2]

$$A = \mu_o J \frac{e^{-ikr}}{4\pi r} e^{i\omega t}$$

r

current
element

$$k = \omega\sqrt{\mu\varepsilon} = \frac{\omega}{c}$$

$$\omega = 2\pi f$$

$Je^{i\omega t}$
[1]

$$\mu_o = 4\pi \times 10^{-7} \text{Henry / meter}$$

$$\varepsilon_o = 8.86 \times 10^{-12} \text{Farad / meter}$$

Figure 2-7: A Harmonic Current at [1] Induces a Harmonic Potential at [2]

communications that is shared by every possible transmitter **J**. How do we detect only the signal we are interested in?

The sharing of a communications channel by multiple users is known as *multiplexing*. There are a number of methods to successfully locate the signals we wish to receive and reject others. A few important examples are the following:

- *Frequency-division multiplexing*: only receive signals with a given periodicity and shape (sinusoidal, of course).

- *Spatial multiplexing*: limit signals to a specific geographical area. Recall that induced potentials fall off as (1/distance) in the ideal case, and in practice attenuation of a signal with distance is often more rapid due to obstacles of various kinds. Thus, by appropriate choice of signal power, location, and sensitivity, one can arrange to receive only nearby signals.

- *Time-division multiplexing*: limit signals to a specific set of time slots. By appropriate coordination of transmitter and receiver, only the contents of the desired time slot will be received.

- *Directional multiplexing*: only listen to signals arriving from a specific angle. This trick may be managed with the aid of antennas of high directivity, to be discussed in more detail in Chapter 5.

- *Code-division multiplexing*: only listen to signals multiplied by specific code. Rather in the fashion that we can listen to a friend's remarks even in a crowded and noisy room, in code-division multiplexing we select a signal by the pattern it obeys. In practice, just as in conversation, to play such a trick it is necessary that the desired signal is at least approximately equal to other undesired signals in amplitude or power, so that it is not drowned out before we have a chance to apply our pattern-matching template.

In real communications systems, some or all of these techniques may be simultaneously used, but almost every modern wireless system begins with frequency-division multiplexing by transmitting its signals only within a certain frequency band. (We briefly examine the major exception to this rule, ultrawideband communications, in section 5.) We are so accustomed to this approach that we often forget how remarkable it is: the radio antenna that provides us with music or sports commentary at 105 MHz is also exposed to AM signals at hundreds to around a thousand kHz, broadcast television at various frequencies between 50 and 800 MHz, aeronautical communications at 108–136 MHz, public safety communications at 450 MHz, cellular telephony at 880 and 1940 MHz, and cordless telephones, wireless local area networks (WLANs), and microwave ovens in the 2400-MHz band, to name just a few.

All these signals can coexist harmoniously because different frequencies are *orthogonal*. That is, let us choose a particular frequency, say ω_c, that we wish to receive. To extract only the part of an incoming signal that is at the desired frequency, we multiply the incoming unknown signal $s(t)$ by a sine or cosine (or more generally by an exponential) at the *wanted* frequency ω_c and add up the result for some time—that is, we integrate over a time interval T, presumed long compared with the periodicity $1/f$ (equation [2.8]). The reader may recognize in equation [2.8] the *Fourier cosine transform* of the signal s over a finite domain. A similar equation may be written for the sine, or the two can be combined using an imaginary exponential.

$$\tilde{S}(\omega_c) = \frac{1}{T} \int_0^T s(t) \cos(\omega_c t) dt \qquad [2.8]$$

If $s(t)$ is another signal at the same frequency, the integral will wiggle a bit over each cycle but accumulate over time (Figure 2-8).

On the other hand, if the unknown signal is at a different frequency, say $(\omega_c + \delta)$, the test and unknown signals may initially be in phase, producing a positive product, but over the course of some time they will drift out of phase, and the product will change signs (Figure 2-9). Thus, the integral will no longer accumulate monotonically, at least over times long compared with the difference period $(1/\delta)$ (Figure 2-10); when we divide by T and allow T to become large, the value of $S(\omega_c)$ will approach zero.

Any signal that is periodic in time can be regarded as being composed of sinusoids of differing frequencies: in more formal terms we can describe a signal either as a function of time or as a function of frequency by taking its Fourier transform (i.e., by performing

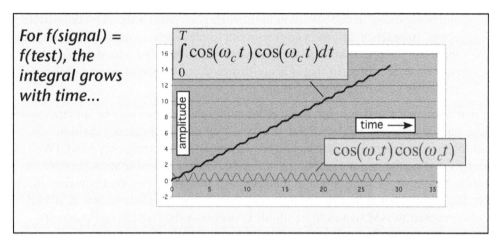

Figure 2-8: Unknown Signal at the Same Frequency as Wanted Signal

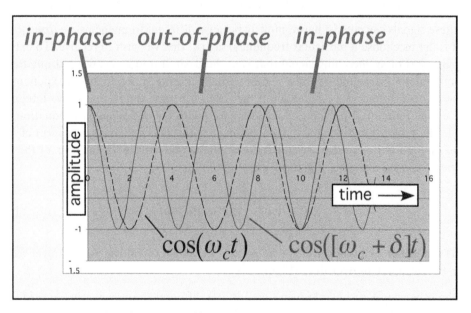

Figure 2-9: Two Signals at Different Frequencies Do Not Remain in Phase

the integration [2.8] for each frequency ω_c of interest.) The orthogonality of those differing frequencies makes it possible to extract the signal we want from a complex mess, even when the wanted signal is small compared with the other stuff. This operation is known generally as *filtering*. A simple example is shown in Figure 2-11. It is generally very easy when the frequencies are widely separated, as in Figure 2-11, but becomes more difficult when frequencies close to the wanted frequency must be

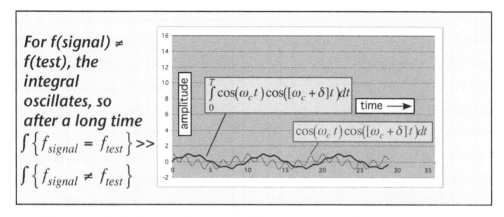

Figure 2-10: Unknown Signal at a Different Frequency from Wanted Signal

rejected. We examine some of the means to accomplish this task for WLAN radios in Chapter 4.

3. Modulation and Bandwidth

3.1. Simple Modulations

So far the situation appears to be quite rosy. It would appear that one could communicate successfully in the presence of an unlimited number of other signals merely by choosing the appropriate frequency. Not surprisingly, things are not so simple: a single-frequency signal that is always on at the same phase and amplitude conveys no information. To actually transmit data, some aspect of our sinusoidal signal must change with time: the signal must be *modulated*. We can often treat the modulation as a slowly varying function of time (slow being measured relative to the carrier frequency) multiplying the original signal.

"slowly" varying   sinusoidal vibration

modulation function   at carrier frequency   [2.9]

\\   /

$$f(t) = m(t)\cos(\omega_c t)$$

A simple example of a modulated signal may be obtained by turning the carrier on and off to denote, for example, 1 and 0, respectively: that is, $m(t) = 1$ or 0. This approach is known as *on–off keying* or OOK (Figure 2-12). OOK is no longer widely used in wireless communications, but this simple modulation technique is still common in fiber optic signaling.

A key consequence of imposing modulation on a signal at a frequency ω_c is the inevitable appearance of components of the signal at *different frequencies* from that of the original carrier. The perfect orthogonality of every unique frequency present in the

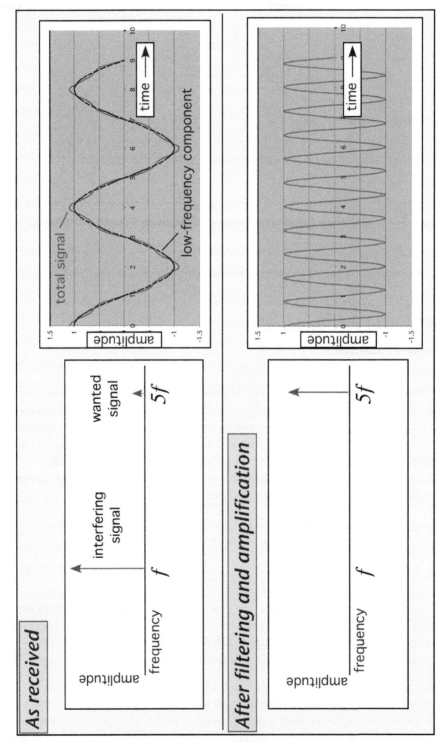

Figure 2-11: Extraction of a Wanted Signal in the Presence of a Large Unwanted Signal

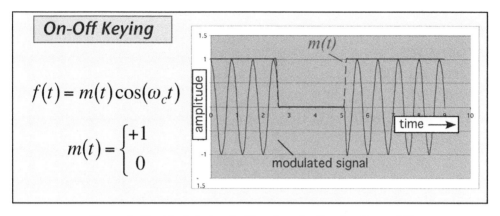

Figure 2-12: Modulation by Turning the Carrier On or Off

case of unmodulated signals is lost when we actually transmit data. Let us examine how this comes about for the particularly simple case of a sinusoidal modulation, $m = \cos(\omega_m t)$. Recall that the orthogonality of two different frequencies arose because contributions to the average from periods when the two signals are in phase are canceled by the periods when the signals are out of phase (Figure 2-9). However, the modulated signal is turned off during the periods when it is out of phase with the test signal at the different frequency ($\omega_c + \delta$), so the contribution from these periods no longer cancels the in-phase part (Figure 2-13). The modulated carrier at (ω_c) is now detected by a filter at frequency ($\omega_c + \delta$).

The astute reader will have observed that this frustration of cancellation will only occur when the frequency offset δ is chosen so as to ensure that only the out-of-phase periods are suppressed. In the case of a periodic modulation, the offset must obviously be chosen to coincide with the frequency of the modulation: $|\delta| = \omega_m$. In frequency space, a modulated carrier at frequency f_c acquires power at sidebands displaced from the carrier by the frequency of the modulation (Figure 2-14).

In the case of a general modulating signal $m(t)$, with Fourier transform $M(\omega)$, it can be shown that the effect of modulation is to translate the spectrum of the modulating or *baseband* signal up to the carrier frequency (Figure 2-15).

We can now see that data-carrying signals have a finite *bandwidth* around their nominal carrier frequency. It is apparent that to pursue our program of frequency-division multiplexing of signals, we shall need to allocate bands of spectrum to signals in proportion to the bandwidth those signals consume. Although the spectrum of a random sequence of bits might be rather more complex than that of a simple sinusoid, Figure 2-14 nevertheless leads us to suspect that the faster we modulate the carrier, the more bandwidth we will require to contain the resulting sidebands. More data require more bandwidth (Figure 2-16).

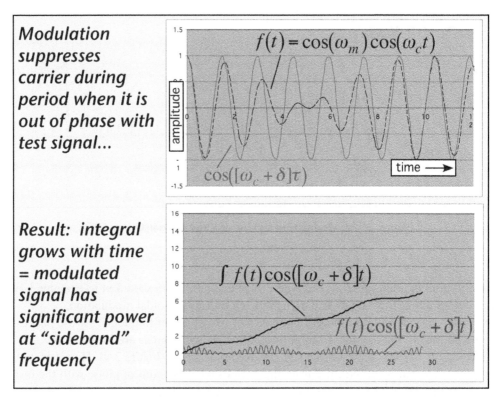

Modulation suppresses carrier during period when it is out of phase with test signal...

$$f(t) = \cos(\omega_m)\cos(\omega_c t)$$

$$\cos([\omega_c + \delta]\tau)$$

time →

Result: integral grows with time = modulated signal has significant power at "sideband" frequency

$$\int f(t)\cos([\omega_c + \delta]t)$$

$$f(t)\cos([\omega_c + \delta]t)$$

Figure 2-13: A Modulated Signal Is No Longer Orthogonal to All Other Frequencies

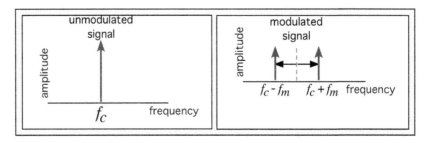

unmodulated signal

amplitude

f_C frequency

modulated signal

amplitude

$f_c - f_m$ $f_c + f_m$ frequency

Figure 2-14: Modulation Displaces Power From the Carrier to Sidebands

It would seem at first glance that the bandwidth required to transmit is proportional to the data rate we wish to transmit and that faster links always require more bandwidth. However, note that in Figure 2-16 we refer not to the bit rate but to the *symbol* rate of the transmitted signal. In the case shown, a symbol is one of three possible amplitudes, corresponding to a data value of 0, 1, or 2: this is an example of *amplitude-shift keying* (ASK), a generalization of OOK. (Note that in this and other

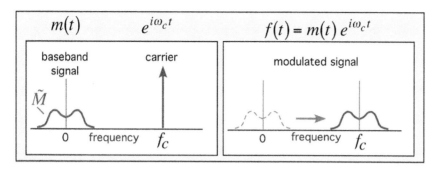

Figure 2-15: The Spectrum of a Carrier Modulated by a General Signal *m*(*t*)

examples we show abrupt transitions between different states of the carrier; in practice, the transitions are smoothed to minimize the added sidebands.) Each symbol might be said to contain 3/2 bit. The bit rate is thus 1.5 (symbol rate). More generally, we can envision a number of approaches to sending many bits in a single symbol. For example, we could use more amplitudes: if 8 amplitudes were allowed, one could transmit 3 bits in each symbol. Because the width of the spectrum of the modulating signal is mainly dependent on the rate at which transitions (symbols) occur rather than exactly what the transition is, it is clear that by varying the modulation scheme, we could send higher data rates without necessarily expanding the bandwidth consumed.

We can nevertheless guess that a trade-off might be involved. For example, the use of 8 distinct amplitudes means that the difference between (say) a "3" and a "4" is smaller than the difference between an OOK "1" and "0" for the same overall signal power. It seems likely that the more bits we try to squeeze into a symbol, the more vulnerable to noise our signal will become.

With these possibilities in mind, let us examine some of the modulation schemes commonly used in data communications. The first example, in Figure 2-17, is our familiar friend OOK. Here, in addition to showing the time-dependent signal, we have shown the allowed symbols as points in the phase/amplitude plane defined by the instantaneous phase and amplitude of the signal during a symbol. The error margin shows how much noise the receiver can tolerate before mistaking a 1 for a 0 (or vice versa).

Note that although we have shown the 1 symbol as a single point at a phase of 0 and amplitude 1, a symbol at any other phase—that is, any point on the circle *amplitude* = *1*—would do as well. OOK is relatively easy to implement because the transmitter doesn't need to maintain a constant phase but merely a constant power when transmitting a 1 and the receiver needs merely to detect the signal power, not the signal phase. On the down side, OOK only sends one bit with each symbol, so an OOK-modulated signal will consume a lot of bandwidth to transmit signals at a high rate.

As we mentioned previously, we might add more amplitudes to get more data: ASK (Figure 2-18). The particular example in Figure 2-18 has four allowed amplitudes and is

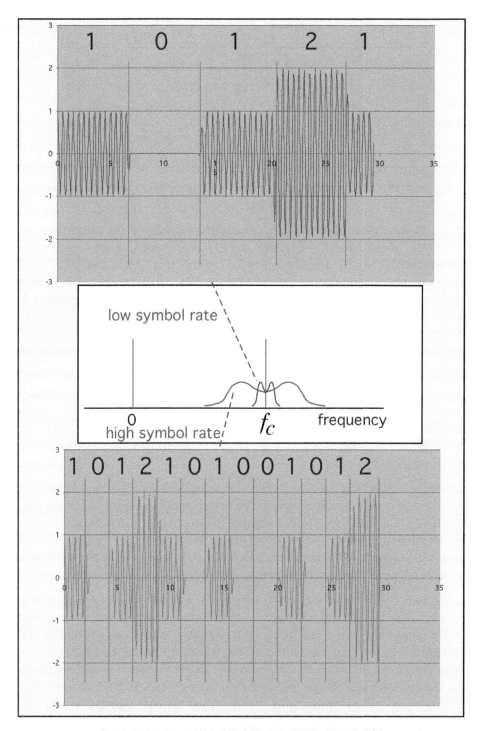

Figure 2-16: Faster Symbol Rate = More Bandwidth

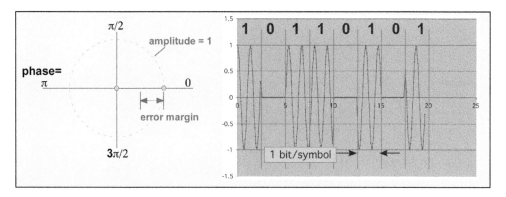

Figure 2-17: On–Off Keying (OOK)

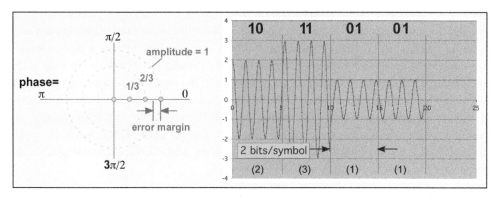

Figure 2-18: 4-Amplitude-Shift Keying (4ASK)

denoted *4ASK*. Once again we have collapsed the allowed states onto points for clarity but with the understanding that any point on, for example, the 2/3 circle will be received as (10), etc. 4ASK allows us to transmit 2 bits per symbol and would be expected to provide twice the data rate of OOK with the same bandwidth (or the same data rate at half the bandwidth). However, the margin available before errors in determining what symbol has been received (i.e., before *symbol errors* occur) is obviously much smaller than in the case of OOK. 4ASK cannot tolerate as much noise for a given signal power as can OOK.

Although it is obviously possible to keep adding amplitudes states to send more bits per symbol, it is equally apparent that the margin for error will decrease in inverse proportion to the number of amplitude states. A different approach to increasing the number of states per symbol might be useful: why not keep track of the phase of the signal?

The simplest modulation in which phase is used to distinguish symbols, *binary phase-shift keying* (BPSK), is depicted in Figure 2-19. The dots in the figure below the binary symbol

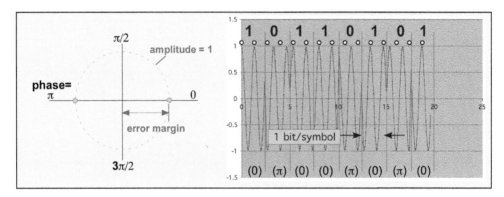

Figure 2-19: Binary Phase-Shift Keying (BPSK)

values are placed at constant intervals; a 1 is transmitted with the signal peaks coincident with the dots, whereas a 0 has its peaks between dots: 180 degrees or π radians out of phase. In phase-shift keying, the nominal symbols are points in the phase plane rather than circles: the group of points is known as a *signal constellation*. However, as long as the signal is large enough for its phase to be determined, the signal amplitude has no effect: that is, any received signal on the right half of the phase-amplitude plane is interpreted as a 1 and any signal on the left half is interpreted as 0. The error margin is thus equal to the symbol amplitude and is twice as large as the error margin in OOK for the same peak power. BPSK is a robust modulation, resistant to noise and interference; it is used in the lowest rate longest range states of 802-11 networks.

It is apparent that to get the best error margin for a given peak power, the constellation points ought to be spaced as far from one another as possible. To get 2 bits in one symbol, we ought to use four phases spaced uniformly around the amplitude = 1 circle: quaternary phase-shift keying (QPSK), shown in Figure 2-20. QPSK sends 2 bits per symbol while providing a noise margin larger than that of OOK at 1 bit per symbol for the same peak power. It is a very popular modulation scheme; we will encounter a variant of it in 802-11 WLANs.

Again, we could continue along this path, adding more phase states to provide more bits per symbol, in a sequence like 8PSK, 16PSK, and so on. However, as before such a progression sacrifices error margin. An alternative approach is to combine the two schemes that we have heretofore held separate: that is, to allow both amplitude and phase to vary. Such modulation schemes are known as *quadrature-amplitude-modulation* or QAM. An example, 16QAM, is depicted in Figure 2-21. Four bits can be sent with a single symbol, and the noise margin is still superior to that of 4ASK at 2 bits per symbol. More points could be added to the signal constellation; doubling the number of states in each axis for each step, we obtain 64QAM, 256QAM, and 1024QAM (the latter perhaps more aptly described as a clear night sky in the Sierra than a mere constellation).

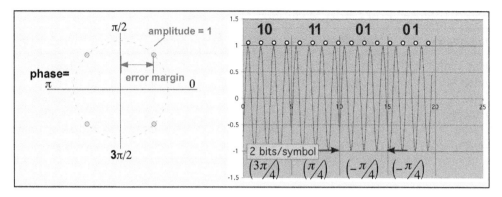

Figure 2-20: Quaternary Phase-Shift Keying (QPSK)

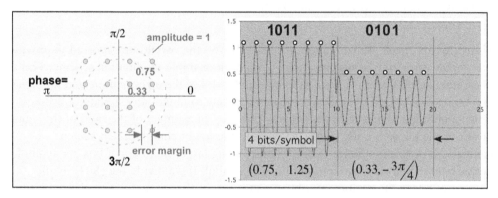

Figure 2-21: 16-Quadrature-Amplitude-Modulation (16QAM)

The various modulation schemes we examined are summarized in Table 2-1. Here the error margin is quoted in units where the peak signal amplitude is defined as 1. It is apparent that the choice of modulation is a trade-off between the target data rate, the bandwidth available to transmit it in, the noise immunity and therefore range of the transmission, and the complexity and cost of the resulting transmitter and receiver.

Although it seems likely that the noise susceptibility of a modulation scheme is increased when its error margin is reduced, it would clearly be helpful to make some quantitative statement about the relationship between a measure of error, such as the symbol or bit error rate, and a measure of noise, such as the ratio of the signal power to the noise power, the *signal-to-noise ratio* (*S/N*). Such relationships can be constructed, assuming the noise to act as a Gaussian (normally distributed) random variable in the phase plane, but the calculations are rather laborious. We can achieve almost the same result in a much simpler manner by exploiting the rule of thumb that almost all the area of a Gaussian distribution is within 3 standard deviations of the mean value. Assume the

Table 2-1: Summary of Carrier Modulation Approaches

Modulation	*Bits/Symbol*	*Error Margin*		*Complexity*
OOK	1	1/2	0.5	Low
4ASK	2	1/6	0.17	Low
BPSK	1	1	1	Medium
QPSK	2	$1/\sqrt{2}$	0.71	Medium
16QAM	4	$\sqrt{2}/6$	0.23	High
64QAM	6	$\sqrt{2}/14$	0.1	High

noise voltage is a Gaussian with standard deviation σ and thus average power proportional to σ^2. If the error margin of a modulation, measured as a fraction of the signal as in Table 2-1, is larger than 3σ, then the symbol error rate is likely to be very small (Figure 2-22).

Using this approach, we can obtain a value of (S/N) that would be expected to provide reasonably error-free communications using each example modulation scheme. The noise standard deviation is allowed to be equal to 1/3 of the error margin, and the signal amplitude is averaged over the constellation points to obtain an estimate of average signal amplitude; the ratio of the squares provides an estimate of the ratio of the signal power to the allowed noise power. However, one might complain that such a comparison does an injustice to the schemes using larger numbers of bits per symbol, because these deliver more information per symbol. If we divide the (S/N) by the number of bits in the symbol, we obtain a measure of the amount of signal required *per*

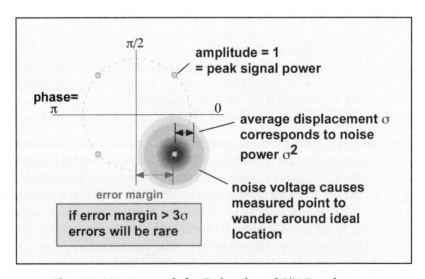

Figure 2-22: Approach for Estimation of S/N Requirement

bit of data, which seems a more appropriate way to compare differing modulation approaches, because it is after all bits, not symbols, that are the ultimate currency of data transfer. The quantity $(S/N)/(bits/symbol)$ can be reexpressed in a fashion that has an appealing interpretation:

$$\left\{\frac{S}{N}\right\}\bigg/\left\{\frac{bits}{symbol}\right\} = \frac{S}{N_o(BW)}\frac{symbols}{bit} = \frac{S}{N_o(1/T_{symbol})}\frac{symbols}{bit}$$
$$= \frac{ST_{symbol}}{N_o}\frac{symbols}{bit} = \frac{ST_{bit}}{N_o} = \boxed{\frac{E_{bit}}{N_o}}$$

[2.10]

In equation [2-10], we defined a quantity, N_o, that is the total noise divided by the signal bandwidth or the noise power in, for example, watts per Hertz. We also assumed that the bandwidth is inversely proportional to the symbol time (Figure 2-16). The product of symbol time and (symbols/bit) becomes an effective bit time; the product of bit time and signal power is the *energy per bit*, E_{bit}, often abbreviated E_b. The quantity (E_b/N_o) is thus a measure of the signal energy available per bit of data to be transmitted, relative to the fundamental amount of noise in the channel, and is a more appropriate basis for comparison of modulation approaches than raw (S/N).

The resulting extended comparison of modulation noise immunity is summarized in Table 2-2. The ratios (S/N) and (E_b/N_o) vary over a wide range and are conveniently expressed in *decibels* (dB): $(S/N)_{dB} = 10\log[(S/N)]$. The noise amplitude is simply 1/3 of the error margin. <Signal> is the average of the signal over the constellation points, normalized to the peak amplitude. This is a reasonable first estimate of the average signal power, though as we'll see in Chapter 4, it must also be corrected for the path the signal takes from one constellation point to another. The results are within 0.4 dB of the values of (E_b/N_o) required to achieve a bit error rate of 10^{-5} by a more accurate but much more laborious examination of the probability of error within the constellation point by point.

The results, particularly when expressed as the normalized quantity (E_b/N_o), confirm the initial impression that PSK and QAM modulations do a better job of delivering data in the presence of noise than corresponding ASK approaches. It is interesting to note that BPSK and QPSK have the same (E_b/N_o), even though QPSK delivers twice as

Table 2-2: Noise Immunity of Various Modulation Schemes

Modulation	*Noise Amplitude*	*<Signal>*	*(S/N) (dB)*	*(E$_b$/N$_o$) (dB)*
OOK	0.167	0.71	12.6	12.6
4-ASK	0.053	0.62	21.4	18.3
BPSK	0.333	1.00	9.5	9.5
QPSK	0.237	1.00	12.5	9.5
16QAM	0.077	0.74	19.7	13.7
64QAM	0.033	0.65	25.8	18.1

many bits per symbol. QPSK is an excellent compromise between complexity, data rate, and noise immunity, and we shall find it and its variants to be quite popular in wireless networking. Table 2-2 also reemphasizes the fact that higher rates require a higher S/N: faster means shorter range for the same bandwidth and power.

One more minor but by no means negligible aspect of modulating a signal must still be examined. The average value of the signal, normalized to a peak value of 1 for the constellation point most distant from the origin, is seen to decrease for more complex modulations: equivalently, the ratio of the peak to average power increases as more bits/symbol are transmitted. Other effects, having to do with both the details of implementation of the modulation and more subtle statistical issues for complex modulation schemes, give rise to further increases in peak-to-average power ratios. High peak-to-average ratios require that the transmitter and receiver must be designed for much higher instantaneous power levels than the average signal level would indicate, adding cost and complexity. A more detailed examination of the peak-to-average ratio is presented in the discussion of radios in Chapter 4.

We examined a number of common modulation schemes, exposing the trade-off between data rate, bandwidth, and noise. Any number of variants on these approaches could be imagined, including adjustments in the exact location of the constellation points, differing numbers of points, and different conventions on how to determine what value had been detected. Yet there certainly seems to be a trend, as shown in Table 2-1, that the more bits one transmits per symbol, the less noise can be tolerated. This observation suggests that perhaps the path of increasing symbol complexity cannot be continued indefinitely and that some upper limit might exist on the amount of data that can be sent in a given bandwidth in the presence of noise, no matter how much ingenuity is devoted to modulation and detection. In a series of seminal publications in the late 1940s, Claude Shannon of Bell Laboratories demonstrated that any communications channel has a finite data capacity determined by the bandwidth $[BW]$, signal power S, and noise power N in the channel:

$$\left\{ \frac{bits}{s} \right\} \leq [BW] \log_2 \left(1 + \frac{S}{N} \right) \qquad [2.11]$$

The ultimate capacity of a radio link, like any other channel, is determined by its bandwidth and the (S/N). To clarify the relevance of this limitation, let us look at an example. As explained in detail in Chapter 3, most of the 802.11-based (Wi-Fi) local area network protocols use a channel that is roughly 16 MHz wide. The (S/N) will vary widely depending on the separation between transmitter and receiver and many other variables, but for the present purpose let us adopt a modest and computationally convenient value of $(S/N) = 7{:}1$. We then have

$$\left\{ \frac{bits}{s} \right\} \leq [16 \times 10^6] \log_2 (8) = 4.8 \times 10^7 \qquad [2.12]$$

The capacity of an 802.11 channel is about 48 Mbps (megabits/second) at what we will find corresponds to a quite modest signal-to-noise requirement. If we assume that the channel bandwidth is approximately equal to the inverse of the symbol rate, this upper bound corresponds to about 3 bits/symbol. (The direct-sequence version of 802.11, and 802.11b, actually uses a symbol rate of 11 megasamples/second (Msps), so that 4 bits/symbol would be available.) We can thus infer that in this case, little or no advantage would result from using a modulation such as 64QAM (6 bits/symbol): the noise-created symbol errors could presumably be corrected by coding, but the overhead associated with correcting the mistakes would exceed the benefit gained. Note that if a higher S/N could be obtained, by turning up the transmit power, reducing the transmit-to-receive distance, or other means, the capacity of the channel would be increased, allowing for exploitation of more complex symbols, though the increase is logarithmic: an eightfold increase in the signal is required to achieve a doubling of the data rate. On the other hand, it is also apparent that there's a lot of room in the 802.11 channel for improvement over the 802.11b maximum data rate of 11 Mbps, even at quite modest (S/N).

A broader examination of the performance of some important communications channels relative to the Shannon limit is provided in Table 2-3. It is apparent that channels cannot normally make use of the whole theoretical capacity available and that in many cases less than half of the upper limit is achieved. Wi-Fi stands out as a notably inefficient user of its allotted bandwidth, which is a hint that more efficient modulations could be used (which has been done with the advent of 802.11g; see Chapter 3). Note also that a cable modem connection, being made over a wired link, can deliver a much larger (S/N) than is normally practical for a wireless link. A cable modem using 6 MHz can support actual data rates of 30 Mbps, notably larger than 802.11's 11 Mbps in 16 MHz, albeit a rather modest improvement given the huge increase in (S/N) necessary to achieve it.

Table 2-3: Actual and Theoretical Capacity of Some Communications Links

	EV-DO CDMA (cellphone)	*Cable Modem*	*Wi-Fi (802.11b)*
Bandwidth (MHz)	1.25	6	16
Configuration	Mobile, 2 km, LOS	Minimum FCC S/N	Indoors, 30 m, NLOS
(S/N)	6:1	2000:1	7:1
Maximum rate (Mbps)	3.6	65.8	48
Actual rate (Mbps)	2.5	30.3	11
Percent of maximum rate	70	46	23

LOS, line-of-sight from transmitter to receiver; NLOS, non–line-of-sight from transmitter to receiver.

3.2. Orthogonal Frequency-Division Multiplexing

The modulation approaches we discussed so far are called *single-carrier* modulations: at any given moment, a carrier with a particular phase and amplitude can be said to exist. Single-carrier modulations are versatile and relatively simple to implement. However, they have some fundamental limitations when used in a wireless link. To understand why, we need to introduce the concept of *multipath*.

Consider a typical real-world transmission between a transmitting and receiving antenna (Figure 2-23). Various sorts of obstacles may exist that reflect the signal, providing alternatives to the direct path. The time required to propagate along these alternative paths, t_2 and t_3, will in general differ from the time taken by the direct path, t_1.

If the delay associated with the other times is comparable with the symbol time and the signal strengths don't differ by too much, a serious problem is encountered at the receiver: the sum of all the received signals may not match the transmitted signal. In Figure 2-24, we depict the two delayed and attenuated replicas of the signal added to the directly transmitted version. The sum is garbled after the first bit, and there is no easy way to extract the transmitted symbols: *intersymbol interference* has wiped out the data even though there is ample signal power available.

We should note that in Figure 2-24, for simplicity we show the three signals as adding after demodulation, that is, after the carrier frequency has been removed. In fact, the signals add at the carrier frequency, which may give rise to the related but distinct problem of fading, a subject we deal with in more detail in Chapter 6.

When is this sort of distortion likely to be a problem? In general, multipath is important when propagation delays are comparable with symbol times. Using the convenient

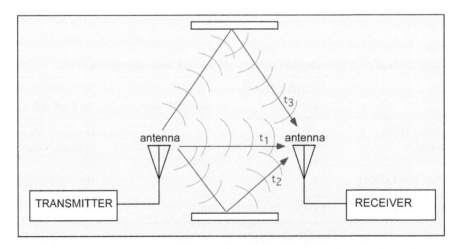

Figure 2-23: Multiple Paths May Exist Between Transmitter and Receiver

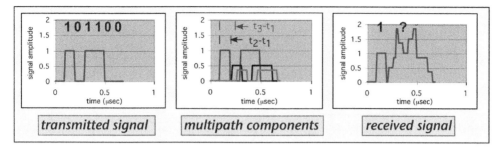

Figure 2-24: Multipath Results in Garbled Signals

approximation that the speed of light is about 3.3 nsec/m, Table 2-4 shows the typical regions that one can expect to cover at a given data rate before multipath rears its ugly head. We see that one can transmit a respectable 10 Msps over a sizable building with single-carrier modulations, but if we wish to achieve higher data rates over larger regions, some other approach is required.

An increasingly popular means of tackling multipath to allow higher rates over larger areas is the use of a specialized modulation, orthogonal frequency-division multiplexing (OFDM). OFDM is used in the WLAN standards 802.11a, 802.11g, and HiperLAN, as well as digital broadcast television in Europe and Asia (known as COFDM). OFDM has also been proposed for very-high-rate personal area networks to deliver high-resolution video within the home. In this section we'll take a general look at how OFDM works, in preparation for examining the specific OFDM modulations used in the WLAN standards described in Chapter 3.

OFDM uses three basic tricks to overcome multipath. The first is *parallelism*: sending one high-speed signal by splitting it into a number of lower speed signals sent in parallel (Figure 2-25). Serial-to-parallel conversion is very well known in the wired world: open up any desktop computer and you'll see numerous *ribbon cables*, composed of a large number of inexpensive wires carrying many simultaneous signals at a low rate, emulating the function of a more expensive and delicate coaxial cable carrying a much higher rate serial signal. In the case shown in Figure 2-25, the signal path is split into five

Table 2-4: Data Rate vs. Multipath-Free Region Size

Symbol Rate (Msps)	Symbol Time (μsec)	Path Distance (m)	Path Description
0.1	10	3300	City
1	1	330	Campus
10	0.1	33	Building
100	0.01	3	Room

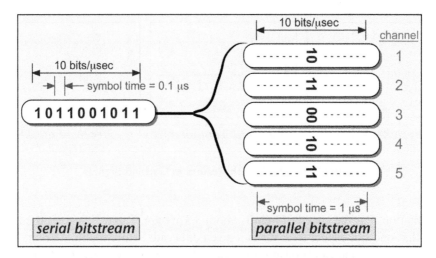

Figure 2-25: Split a High-Speed Serial Signal into Multiple Slow Parallel Signals

parallel paths, each carrying 2 bits/symbol and thus running at 1/10th of the rate of the original serial path, allowing a 10-fold increase in the transmission delay allowed before serious multipath distortion occurs. A serial symbol time of 0.1 μsec becomes a parallel symbol time of 1 μsec.

Where do we get these parallel channels? The obvious answer is to use separate carrier frequencies, one for each channel. Such an arrangement is shown in Figure 2-26. If implemented in a conventional fashion using separate transmitters and receivers with filters for each subcarrier, there are two problems: the cost of the equipment will increase linearly with the number of parallel channels or *subcarriers*, and the spectrum will be used wastefully due to the need for *guard bands* between the subcarriers to allow the receivers to filter their assigned subcarrier out of the mess.

To surmount these two obstacles, the two additional tricks at the heart of OFDM are necessary. Trick 2 is to exploit the *orthogonality* of different frequencies in a more subtle fashion than we have done heretofore. We already know that differing

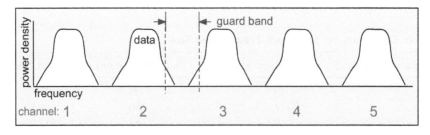

Figure 2-26: Conventional Frequency-Division Multiplexing for Subcarriers

frequencies are orthogonal when integrated over any long time period. However, we can be much more specific: two different frequencies are exactly orthogonal if the integration time contains an integral number of cycles of each frequency:

$$\int_0^T \cos\left(2\pi\left[\frac{n}{T}\right]t\right)\cos\left(2\pi\left[\frac{m}{T}\right]t\right)dt = 0 \;\; for\; m \neq n \qquad [2.13]$$

In equation [2.13], we can see that the two cosines undergo, respectively, $(n/T)*T = n$ and $(m/T)*T = m$ full cycles of oscillation in the time T. If we choose T as our symbol time, we can send symbols consisting of an amplitude and phase for each frequency in parallel at frequencies spaced by integers and still have every frequency orthogonal to every other frequency, as long as we are careful to integrate over the correct time interval. The situation is depicted graphically in Figure 2-27.

The result of such a scheme, as shown in Figure 2-28, is the elimination of the guard bands otherwise required between subcarriers, resulting in much more efficient use of a given slice of spectrum (contrast this image with that of Figure 2-26).

By thus using closely spaced orthogonal subcarriers, we can split a serial symbol stream into a number of slower parallel symbol streams. If we imagine that multipath delay was equal to the serial symbol time, the same delay will now constitute 1/5th of the symbol time of our parallel symbols. Although this is obviously an improvement over having

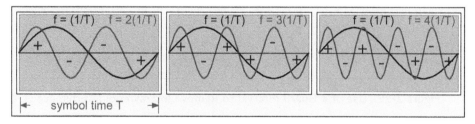

Figure 2-27: Products of Integer Frequencies Integrate to 0 Over an Integer Number of Cycles

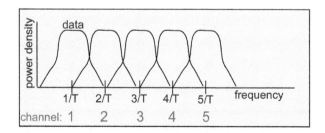

Figure 2-28: Orthogonal Subcarriers Use Spectrum Efficiently

the whole symbol garbled, we can do even better at the cost of some slight decrease in data rate. If the symbol time is extended by (in this case) an extra 20% and then we simply eliminate the beginning part where intersymbol interference takes place, we can perform our integration over the remainder of the symbol time where the received symbol is essentially ideal. Such a scheme is depicted schematically in Figure 2-29. The extra symbol time is called the *guard interval* and represents signal energy intentionally wasted to improve multipath resistance.

If one knew in advance exactly where the multipath boundary was, the guard interval could be implemented by simply turning the signal off during this time. However, in this case any error in determining the edge of the guard interval would result in integration of the signal over partial cycles: the orthogonality of the different subcarriers would be lost and interference between the parallel symbols would result. Instead, the guard interval is normally implemented by a *cyclic extension* of the original symbol, as shown in Figure 2-30. The portion of each subcarrier that occurs during the time period corresponding to the guard interval is simply repeated at the end of the base interval. Because all the subcarriers are periodic in T, this is equivalent to performing the same operation on the final symbol, which is the sum of all the subcarriers. The resulting

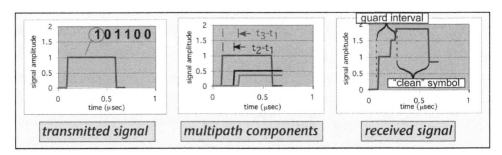

Figure 2-29: Use of a Guard Interval to Remove Intersymbol Interference

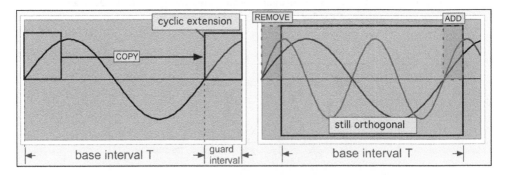

Figure 2-30: Cyclic Extension of an OFDM Symbol Makes Integration Insensitive to Small Time Offsets

cyclically extended symbol preserves the orthogonality of its subcarriers over the interval T even if the start of the interval is displaced slightly, because the portion that is removed from the beginning is just added at the end.

The combination of parallelism implemented with OFDM and a cyclically extended symbol to provide a guard interval gives us a method of transmitting a clean undistorted set of symbols at a high overall data rate in the presence of multipath distortion. Because the subcarriers are separated by an integration over a specific time interval rather than an analog filtering operation, only one radio is required to transmit and receive the signals. However, our putative radio still faces a significant challenge: to get a big improvement in multipath resistance, we need to use a lot of subcarriers. To extract each subcarrier, we must integrate the product of the received symbol and the desired frequency, and the number of points we must use in the integration is set by the highest frequency subcarrier we transmit. For example, in our five-subcarrier example, to extract the portion of a symbol at the lowest frequency $n = 1$, we must use enough points in our integration to accurately remove the part at the highest frequency, $n = 5(4 \times 5 = 20$ sample points in time, or 10 if we treat the data as complex, will do). For N subcarriers, it appears that we have to do on the order of N integrations with N points each, or roughly N^2 multiplications and additions. For small N, that's no big deal, but we get little benefit from using small values of N.

Let us consider the example of 802.11a/g, to be discussed in more detail in Chapter 3. In this case 64 subcarriers are defined (though not all are used): $N = 64$. To extract all the subcarriers, we must perform approximately 4096 multiplications and additions during each symbol time of 4 μsec. Each operation must be performed with enough resolution to preserve the original data: imagine that 8 bits is sufficient. If performed serially, this requires an operation every 49 nsec, or about 2 billion 8-bit adds and multiplies per second. Such an accomplishment is by no means impossible with modern digital hardware, but it is hardly inexpensive when one considers that the networking hardware is to be a small proportion of the cost of the device that makes use of it.

Fortunately, there is one trick left in the bag that immensely improves the situation: the *fast Fourier transform* (FFT). FFT algorithms achieve the necessary integration in roughly $N \log N$ operations. For large N, this represents a huge improvement over N^2: for 802.11a the problem is reduced to roughly 400 operations instead of 4000, and for schemes such as COFDM, which uses up to $N = 1024$, the improvement is on the order of 100-fold. We provide a very brief discussion of this important class of algorithms.

To see how the FFT works, we first examine in more detail the problem of extracting an approximation to the Fourier transform of a time signal sampled at discrete intervals τ. For simplicity we'll limit ourselves to an even number of points. The operation we wish to perform is shown schematically in Figure 2-31: we wish to convert N samples ($N = 16$ here) in time into an estimate of the Fourier transform of the signal at points $k\delta$, where $\delta = 1/N\tau$; k ranges from $-(N/2)$ to $(N/2)$.

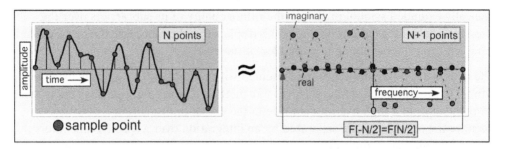

Figure 2-31: Discrete Fourier Transform of a Signal Sampled in Time

It is easy to see why the number of frequencies is limited to the number of samples. A sine or cosine with a frequency of $(N + 1)\delta$ has the same values at the sample points as one with frequency δ: $\cos[2\pi(N+1)\delta\tau] = \cos(2\pi(N+1)\tau/N\tau) = \cos(2\pi(1+1/N)) = \cos[2\pi + 2\pi\delta\tau] = \cos[2\pi\delta\tau]$. This phenomenon is known as *aliasing*. By the same argument, the values are the same at $k = \pm(N/2)$. In general, the signals could have complex values: in our example the time data are real, forcing the values in frequency space to be complex conjugates with respect to $f = 0$: that is, $F(-k\delta) = F^*(k\delta)$.

The basic mathematics are shown in equation [2.14]. We approximate the Fourier transform integral as a sum, where at the nth sample point the sampled value of the signal, $f(n)$, is multiplied by the value of the exponential for the frequency of interest, $k\delta$. Substituting for the frequency increment, we find that the sum is actually independent of both the sample increment τ and frequency increment δ and is determined only by the value of N and f for a given value of k. The part in brackets is often called the discrete Fourier transform, with the factor of τ added in later to set the time and frequency scales.

$$F[k\delta] = \tau \sum_n e^{-2\pi i n k \delta \tau} f(n) = \tau[f(0) + e^{-2\pi i k \delta \tau} f(1) + e^{-(2\pi)2ik\delta\tau} f(2) + \dots]$$

$$= \tau \sum_n e^{-2\pi i n k \left(\frac{1}{N\tau}\right)\tau} f(n) = \tau \sum_n e^{-2\pi i k \left(\frac{n}{N}\right)} f(n) \tag{2.14}$$

$$= \tau\left[f(0) + e^{-2\pi i k \left(\frac{1}{N}\right)} f(1) + e^{-2\pi i k \left(\frac{2}{N}\right)} f(2) + \dots\right]$$

To clarify this perhaps confusing expression, we depict the sums in tabular form in Figure 2-32 for the simple case of $N = 4$. A transform at normalized frequency k is obtained by summing all the terms in the kth row of the table.

The index k is conventionally taken to vary from 0 to $(N - 1)$; because of aliasing as noted above, the $k = 2$ row is the same as $k = -2$ and $k = 3$ is the same as $k = -1$.

Note that the sum of each row can be rewritten slightly by grouping the terms into those for which n is even and odd, respectively. An example of such a rearrangement is shown in Figure 2-33. Although so far we haven't actually saved any calculations, note that after the regrouping, both terms in brackets look like transforms with $N = 2$.

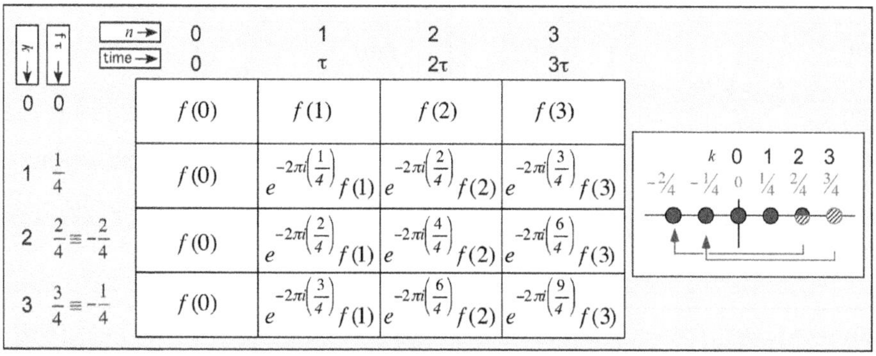

k		$n \rightarrow$ 0	1	2	3
		time $\rightarrow$ 0	τ	2τ	3τ
0	0	$f(0)$	$f(1)$	$f(2)$	$f(3)$
1	$\frac{1}{4}$	$f(0)$	$e^{-2\pi i\left(\frac{1}{4}\right)}f(1)$	$e^{-2\pi i\left(\frac{2}{4}\right)}f(2)$	$e^{-2\pi i\left(\frac{3}{4}\right)}f(3)$
2	$\frac{2}{4} \equiv -\frac{2}{4}$	$f(0)$	$e^{-2\pi i\left(\frac{2}{4}\right)}f(1)$	$e^{-2\pi i\left(\frac{4}{4}\right)}f(2)$	$e^{-2\pi i\left(\frac{6}{4}\right)}f(3)$
3	$\frac{3}{4} \equiv -\frac{1}{4}$	$f(0)$	$e^{-2\pi i\left(\frac{3}{4}\right)}f(1)$	$e^{-2\pi i\left(\frac{6}{4}\right)}f(2)$	$e^{-2\pi i\left(\frac{9}{4}\right)}f(3)$

k: 0 1 2 3
$-\frac{2}{4}$ $-\frac{1}{4}$ 0 $\frac{1}{4}$ $\frac{2}{4}$ $\frac{3}{4}$

Figure 2-32: Calculation of the Discrete Fourier Transform for $N = 4$

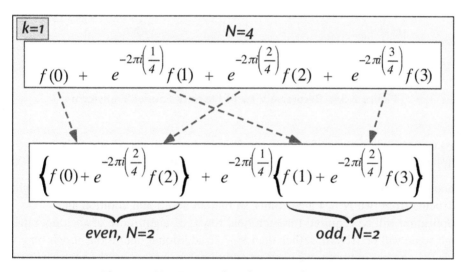

Figure 2-33: Regrouping the Sum Along a Row

When we now view the whole table in this light, we can see the utility of this rearrangement. In Figure 2-34 we show the sum for the odd rows $k = 1,\ 3$. In Figure 2-34(a), the sum of each row is written in recursive fashion as the sum of two transforms of order $N = 2$, denoted as 2E or 2O for even or odd, respectively. Each of these can be further regarded as the sum of two transforms of order $N = 1$ (which are just the time samples themselves). But we note that the transform $F_{2E,k=1}$ is exactly the same as the transform $F_{2E,k=3}$: only the multiplying factor is different. The same is true for the pair of odd transforms. Thus, as shown in Figure 2-34(b), only half as many calculations as we would naively suspect are actually needed. The same argument naturally can be used for the $k = 0,\ 2$ rows. To summarize, at first glance one would believe that to evaluate the table of Figure 2-32 would require nine multiplications (not counting the "×1"

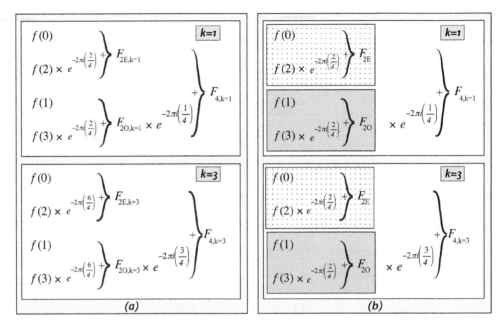

Figure 2-34: Recursive View of Discrete Fourier Transform

terms) and 16 additions, or 25 operations. When regarded recursively, only five multiplications and eight additions, or 13 operations, are required.

In general, because the approach is recursive, we see that an $N = 8$ transform will be constructed from two $N = 4$ algorithms, so that an additional addition and multiplication will be required for each final row (i.e., eight more of each operation, though some will be trivial), rather than $8^2 - 4^2$ additional operations. Each time we double the number of points, we incur an additional N operations: the complexity of the whole process is approximately $N \log_2 N$ instead of N^2.

A closer inspection will also discover that many of the multiplications are actually mere changes in sign: for example, in the $k = 1$ and $k = 3$ cases in Figure 2-34(b), the multiplying factors differ by $e^{-\pi i} = -1$. By some additional ingenuity, the multiplications can be reduced to a small number of nontrivial factors for any given N. More detailed treatments of the FFT are available in any number of texts and web locations; some are provided in the suggested reading at the end of the chapter.

We can now summarize the construction of an OFDM symbol, as depicted in Figure 2-35:

- We start with a serial data set, grouped as appropriate for the modulation to be used. For example, if QPSK were to be used on each subcarrier, the input data would be grouped two bits at a time.

- Each data set is converted into a complex number describing the amplitude and phase of the subcarrier (see Figure 2-20 for the QPSK constellation diagram) and that complex number becomes the complex amplitude of the corresponding subcarrier.

- We then take an *inverse* FFT to convert the frequency spectrum into a sequence of time samples. (The inverse FFT is essentially the same as the FFT except for a normalization factor.) This set of numbers is read out serially and assigned to successive time slots.

- The resulting complex numbers for signal versus time are converted into a pair of voltages by an *analog-to-digital converter* (ADC); the real part determines the in-phase or I channel and the imaginary part determines the quadrature or Q channel.

- The I and Q voltages are multiplied, respectively, by a cosine and sine at the carrier frequency, and the result is added (after filtering) to produce a real voltage versus time centered around the carrier frequency.

The received symbol is demodulated in a similar fashion: after extraction of the I and Q parts by multiplication by a cosine and sine, a digital-to-analog converter produces a serial sequence of complex time samples, which after removal of the guard interval are converted by an FFT to (hopefully) the original complex amplitude at each frequency.

Notice in Figure 2-35 that the behavior of the resulting OFDM symbol is quite complex, even for a modest number of subcarriers. In particular, large excursions in voltage occur in a superficially unpredictable fashion: the ratio of peak power to average power is much larger than for, for example, the QPSK signal contained in each subcarrier (refer to the right-hand side of Figure 2-20). The absolute peak power relative to the average power grows linearly with the number of subcarriers, because the peak power is the square of the sum of the voltages (N^2), whereas the average power is the sum of the individual average powers (N). Fortunately, as the number of subcarriers grows large, the few symbols with all the subcarriers in phase grow increasingly rare as a percentage of the possible symbols, but for practical signals, such as those encountered in 802.11a/g, peak-to-average power ratios nearing 10 dB are encountered. We discuss the consequences of high peak-to-average ratios in more detail in Chapter 4.

In Figure 2-35, almost all the heavy lifting associated with this specialized modulation is performed in the digital domain: analog processing is very similar to that used in a more conventional radio, though as one might guess from the discussion of the previous paragraph, the radio specifications are more demanding in some respects. This is characteristic of advanced communications systems: because digital functions continue to decrease in both power and cost as long as the integrated circuit industry can keep scaling its products according to Moore's law, whereas analog component cost and size do not scale with feature size, it is generally advantageous to add as much value as possible in digital signal processing. It is interesting to note that significant

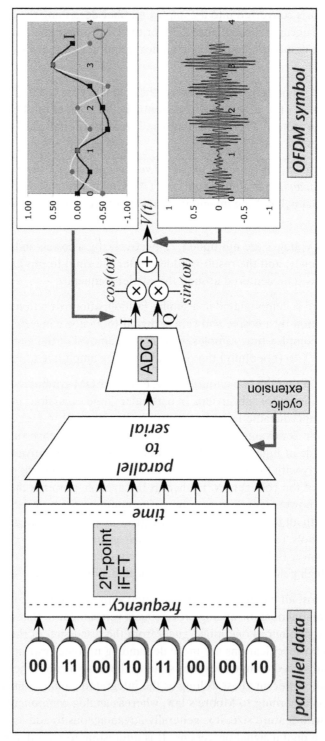

Figure 2-35: Schematic Depiction of OFDM Symbol Assembly for N = 8 Subcarriers

obstacles are being encountered in scaling digital circuits as metal-oxide-semiconductor field-effect transistor (MOSFET) gate oxides become comparable in thickness with a single monolayer of SiO_2. It remains to be seen whether a truly viable substitute for thermal silicon dioxide can be found or whether Moore's law will slowly fade as a driving force in electronic design until a new generation of clever technologies comes about.

3.3. Ultrawideband: A License to Interfere (Sort of)

In section 2 we learned that to share the electromagnetic medium, simultaneous users agree to occupy different frequencies. In section 3 we found that the need to modulate forces those users to take up not infinitesimal slices but sizable swathes of spectrum, the extent being proportional to the amount of data they wish to transfer, subject to Shannon's law. The combination of these two facts means that there is a finite amount of data that can be sent in a given geographical region with a given chunk of usable spectrum without having the various users interfere with one another so that nothing gets through. This fact has been recognized in a general way since the early twentieth century, when such incidents as the sinking of the ocean liner *Titanic* dramatized the difficulties of uncoordinated sharing of the radio medium. The solution has traditionally been for regulatory bodies in each nation or supranational region to parcel out the available spectrum to specific users and uses. As one might expect, after a long time (nearly a century) under such a regimen, it becomes difficult to obtain large chunks of "new" spectrum for "new" uses.

Various innovative responses to this conundrum have arisen. Cellular telephony uses conventional licensed spectrum allocations, but the service providers limit the power and range of individual transmitters so that the same spectrum can be reused at a number of locations, increasing the number of total users on the system: this is an example of spatial multiplexing. WLANs and other users of unlicensed spectrum go further, by limiting transmit power and using interference-minimizing protocols to allow a sort of chaotic reuse of the same frequency bands with at least graceful degradation of performance. In the last 10 years or so, a more radical approach has been advocated: that new users simply reuse whole swathes of spectrum already allocated for other purposes but with appropriate precautions to minimize the disruption of existing users while providing new data communications and other services. This approach has come to be known as *ultrawideband* (UWB) communications.

The basic approach to making UWB signals tolerable to other users is to keep the transmitted power in any given conventional band sufficiently small. As we discuss in more detail in section 4, any radio at a finite temperature has a certain noise threshold due to both intrinsic thermal noise and limitations of the radio's components: signals smaller than this noise level will have essentially no effect. The approach taken by UWB is to transmit a modest but appreciable power level but to spread it so thinly over the conventional bands that any conventional receiver reasonably far from the transmitter sees only a negligible addition to normal noise levels. The user of a special

UWB receiver combines all the UWB power back into a single signal, which is larger than the background noise and thus detectable. There are several approaches to performing this nifty trick, but in this section we take a brief look at *pulse UWB* radios. Pulse radios are quite different from the modulation approaches we discussed so far and harken back to the original spark-gap transmitters, like that of the *Titanic* and its contemporaries, used in the pre–Federal Communications Commission (FCC) era of radio communication.

A pulse radio, instead of using a modulated carrier of a single frequency, send a series of brief pulses in which each pulse consists of one or at most a few excursions of the signal voltage away from 0 (Figure 2-36). It is important to note that what is depicted in Figure 2-36 is not a baseband signal before modulation but the actual transmitted voltage. Typical pulses lengths are less than a nanosecond.

To see why such a signal can be considered as UWB, let's try to take the Fourier transform of a pulse following the procedure we used in section 2: we multiply the pulse by a sinusoid at the frequency of interest and integrate. This operation is shown in Figure 2-37.

It is easy to see that as long as the period of the sinusoid is long compared with the pulse, the integral of their product will be rather insensitive to ω. During the pulse, $\cos(\omega t) \approx$ a constant value or perhaps at most a linearly increasing or decreasing voltage. If the pulse is, for example, 0.3 nsec long, one would expect that the resulting frequency spectrum would be roughly flat up to a frequency of about $(1/0.3 \times 10^9)$ or 3.3 GHz (Figure 2-38). Because the total energy of the pulse is distributed over this broad band of frequencies, the amount of energy collected by a conventional narrowband receiver would be a small fraction of the total energy transmitted. The exact shape of the spectrum will depend on the detailed shape of the pulse, but the general features will be similar for any sufficiently short pulse.

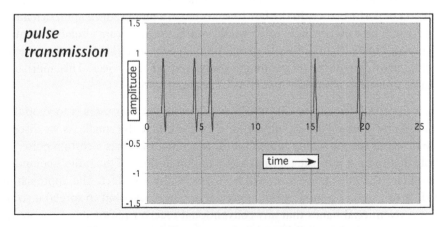

Figure 2-36: Pulse Transmission Voltage vs. Time

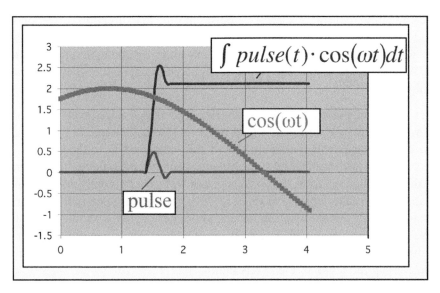

Figure 2-37: What Is the Component of a Short Pulse at Frequency ω?

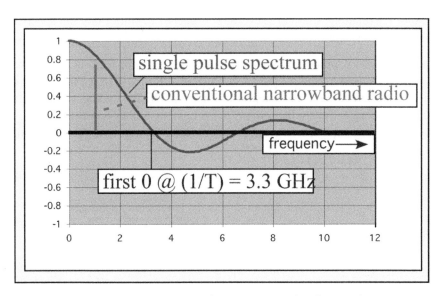

Figure 2-38: Representative Spectrum of a Short Pulse

We can easily envision a UWB communications scheme based on short pulses. A pulse could represent a 1 and the absence of a pulse 0; by implication, one would need to have defined time slots in which to look for the pulses. Such a scheme is known as *pulse-code modulation* and was used in what one could regard as the archetype of all digital communications systems, wired telegraphy using Morse code. A more sophisticated approach might be to displace successive pulses in time in a direction corresponding to the value of the applicable data bit: *pulse-position modulation*. However, any simplistic scheme of this nature runs into a serious problem when applied within the constraints of the UWB environment, that the power in any conventional radio should be small. Let us examine why.

Consider a periodic pulse train, that is, a series of identical pulses evenly spaced in time. If we multiply by a sinusoidal signal whose period is the same as that of the pulse train, successive pulses will contribute exactly the same amount to the integral (Figure 2-39(a)); if we multiply by some other sinusoid with a random incommensurate period, the contributions from the various pulses will vary randomly and sum to some small value. The result, depicted in Figure 2-39(b), is that the spectrum of a periodic pulse train is strongly peaked at frequencies whose periodicity is an integer multiple of the pulse train. Because any of these peaks may lie within a conventional radio's bandwidth, the overall pulse power must be reduced, or *backed off*, by the height of the peak, decreasing the range of the UWB radio.

Although a pulse-code-modulated stream is not exactly like the simple periodic pulse stream, because the presence or absence of a pulse varies randomly with the incoming data, it will still have a peaked frequency spectrum, limiting the transmit power of the UWB radio. Fortunately, there are several approaches to smoothing the spectrum to resemble that of a single pulse.

One simple method is called *biphase modulation*. In this approach, 1s and 0s are encoded with positive or negative pulses, respectively. It is easy to see that in the case of random or randomized data, the integral in Figure 2-39(a) would become 0 instead of accumulating, because the contribution of each pulse would cancel that of its predecessor. In general, if the data sequence is effectively random, the spectrum of the

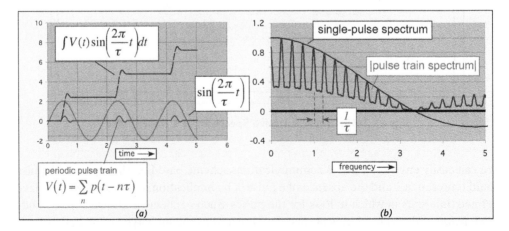

Figure 2-39: Spectrum of a Periodic Pulse Train

pulse train is found to be very smooth, approaching an ideal single-pulse spectrum in the limit of a perfectly random data stream. Biphase modulation is in some sense a generalization of BPSK and shares its nice properties of simplicity and noise robustness.

Biphase modulation can be regarded as the multiplication of each data bit by a simple randomizing sequence of either $+1$ or -1. A generalization of biphase modulation is the *pseudo-noise encoded biphase modulated wavelet*. This elaborate name merely means that instead of multiplying a data bit by a single value, each bit is multiplied by a longer sequence of pulses, where the sequences are chosen to "look" random, that is, they have the spectral characteristics of random noise even though, being found by some deterministic algorithm, they are not truly random. Such a scheme avoids the possibility of spectral lines resulting from inconvenient data (essentially long strings of 1s or 0s), while preserving the benefits of biphase modulation.

A different approach to eliminating spectral lines is to (pseudo-)randomly shift the position of the pulses with respect to their nominal periodic appearance, an approach known as *dithering*. As the magnitude of the average displacement increases, a dithered pulse stream's spectrum becomes smoother, approaching that of an ideal single pulse. However, as the amount of dithering increases, it also becomes more difficult to synchronize the receiver: how does the receiver figure out what the nominal pulse position is so that it can subtract the actual position from this nominal value to get the displacement?

By using very long sequence lengths either for pseudo-noise encoded biphase modulation or pseudo-random dithered sequences, extremely low signal powers can be demodulated. In formal terms, such techniques use a spread-spectrum approach and exploit spreading gain in the receiver. At heart, these are merely fancy averaging techniques, pseudo-random equivalents of sending the same signal again and again and taking the average of the result to improve the S/N. The benefit of such an approach—the spreading gain—is essentially proportional to the ratio of the bandwidth used by the transmission to the bandwidth actually required by the data. If we use a UWB radio, with a bandwidth of, for example, 1 GHz, to send a 10 kbps data stream, we have a spreading gain of $(10^9)/(10^4) = 100,000{:}1$. This is equivalent to saying that we get to transmit the same signal 100,000 times and take the average: noise and interference average to 0 and the wanted signal grows. By using spreading gain we can use power levels that are imperceptible to other radios and still obtain ample S/Ns.

However, note that if we increase the data rate, the spreading gain decreases; in the limit where the data rate is 1 Gbps, there is no spreading gain at all. Furthermore, just like conventional radios, a pulse radio can encounter multipath problems when the inverse of the data rate becomes comparable with the propagation delays in the environment. In an indoor environment, for example, where we expect propagation delays in the 10s of nanoseconds (Table 2-4), it is clearly easy to demodulate a data stream consisting of pulses at intervals of around a microsecond. Just as in conventional radio, we won't be bothered by a few echoes at, for example, 5 or 10 or 20 nsec, because we know that none of them could be a true data pulse. In fact, we might be able to use the echoes to help us decode the pulse: in such an approach, known as a *rake receiver*, the total received signal is correlated with a series of delayed versions of the desired signal, each multiplied by an appropriate empirically determined coefficient, and the various delayed versions summed to get a best guess at the transmitted signal. However, if we attempted to increase the data rate to 100 Mbps, requiring data pulses at rates of one every 10 nsec, we would anticipate serious problems distinguishing the transmitted pulse from its echoes.

Receivers for UWB pulse radio signals are somewhat different from the conventional radio components to be described in some detail in Chapter 4. A UWB receiver may be constructed from a bank of correlators combined with a very-low-resolution (1 or 2 bit) ADC. Essentially, the job of the kth correlator is to attempt to determine whether a pulse was present at each possible time offset t_k without any need to determine the size or detailed characteristics of the constituent pulse. UWB receivers of this type lend themselves to implementations on complementary silicon metal-oxide semiconductor MOS (CMOS) circuitry, because the receiver is based on a large number of parallel circuit elements, with each element required to act rapidly but not very accurately. In conventional radio terms, the correlators do not need to be very linear, and if a large spreading gain is used, they may not need to be very sensitive either.

In summary, pulse radios represent an interesting new approach to wireless communications. Although no new principles are required to understand them, pulse radios invert the normal relationships between frequency and time, which results in differences in optimal approaches for encoding, transmission, and detection.

It is also important to note that pulse radios are not the only possible approach to implementing UWB communications. As we discuss in more detail in Chapter 3, an alternative approach involves the use of very fast FFTs (FFFFFFTs?) and a wideband OFDM scheme.

4. Wireless Link Overview: Systems, Power, Noise, and Link Budgets

4.1. A Qualitative Look at a Link

We now know that a radio system must transmit a modulated signal centered around a known frequency and receive it with sufficient S/N to ensure that the demodulated signal is interpreted correctly most of the time. Let us use this knowledge to get a top-level view of how to design a link between the transmitter and receiver so that these requirements are met.

The basic elements of a radio link are depicted in Figure 2-40. A *transmitter* modulates a carrier according to the input signal and produces a sufficiently large output signal. The *transmitting antenna* radiates the output as a propagating wave, hopefully in the

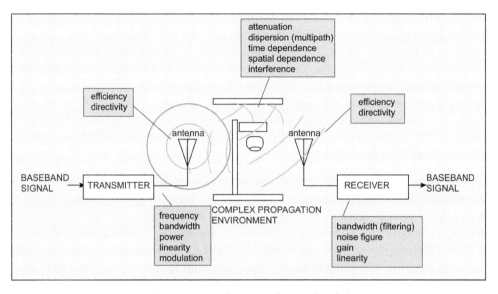

efficiency
directivity

attenuation
dispersion (multipath)
time dependence
spatial dependence
interference

efficiency
directivity

antenna

antenna

BASEBAND
SIGNAL

TRANSMITTER

RECEIVER

BASEBAND
SIGNAL

frequency
bandwidth
power
linearity
modulation

COMPLEX PROPAGATION
ENVIRONMENT

bandwidth (filtering)
noise figure
gain
linearity

Figure 2-40: Element of a Radio Link

direction of the receiver. The *environment* then makes a complete mess of the nice transmitted signal.

The receive antenna converts the propagating wave, and much of the other junk in the environment, into a voltage, which the receiver filters, amplifies, and demodulates to recover something bearing a resemblance to the original input signal. Let us briefly discuss each of these components of the link. The transmitter must be designed to operate at the desired frequency. High frequencies offer more absolute bandwidth (and thus more data) for the same percentage bandwidth and can use smaller antennas. However, more advanced and expensive components are needed. The choice of frequency is usually strongly constrained by regulatory requirements, cost, and mechanical limitations. The bandwidth of the transmitted signal must be large enough, in combination with the expected (S/N), to provide the desired data rate, but one is again typically constrained by the requirements of regulatory bodies and by the choices made in separating the available band into chunks available for specific users: *channelization*. For example, 802.11b operates within the ISM band from 2.4 to 2.483 GHz. The total bandwidth is thus 83 MHz. However, the available bandwidth is partitioned into three noninterfering 25-MHz slots to allow three multiple noninterfering networks in the same place. Within each channel, the actual signal energy is mostly contained in a region about 16 MHz wide and can carry up to 11 Mbps. The transmitter can be designed with various architectures, denoted by the nature of the *frequency conversion* operations used to get the data onto the carrier. The modulation of the carrier by the data can take place in a single step (a *direct conversion* transmitter) or multiple steps (a *superheterodyne* transmitter). The modulated signal must then be amplified to the desired transmit power. More power provides better (S/N) and thus higher data rates or longer range (or both), but the components are more expensive and consume more DC power, which impacts battery life of portable devices. Total output power is also often constrained by regulation. The transmitter must also transmit a reasonably undistorted version of the modulated signal: it needs good *linearity*. Distorted signals will contain components at undesired frequencies, possibly outside the channel or band assigned to the transmitter, resulting in interference with other radios. The choice of modulation is a trade-off between data rate, linearity, and tolerance to multipath.

After spending the power and money to create a transmitted signal, it is obviously desirable to have that signal converted into a radio wave with good *efficiency*. Antenna efficiency is generally close to 1 for large antennas but degrades for antennas that are very small compared with a wavelength of the transmitted or received signal. However, the size of the antenna may be constrained by the requirements of the application, with little respect for the wavelength: for example, antennas for a self-contained PC-card radio must fit within the PC-card form factor even if larger antennas would work better.

Antennas do not radiate equally in all directions: they have *directivity*. If you don't know what direction the receiver is in, a nearly isotropic antenna is best. If you do know exactly where you need the waves to go, a highly directional antenna can be used—but directional antennas are physically large and may be inconvenient. Adaptive or "smart" antennas can exploit directivity to minimize the power needed to reach the receiver at the cost of additional complexity.

Antennas much accept signal voltages from the transmitter without reflecting the input power back into the connection: they need to be *matched* to the transmitter impedance. This matching typically can only be achieved over a finite bandwidth, which must at least extend over the band the antenna is to be used for. Antennas need to be physically robust enough to survive in service: PC-card antennas will get whacked and banged, and outdoor installations must survive heat, cold, wind, rain, and snow. Finally, esthetics must also be considered: for example, Yagi antennas (see Chapter 5), although cheap and efficient, are ugly and are typically covered in plastic shrouds.

Radio performance is strongly influenced by the *environment* in which the transmitter and receiver are embedded. In empty space between the stars, the received power will decrease as the inverse square of the distance between transmitter and receiver, as it ought to conserve total energy. Thus, the power received 1 km from the transmitter is 1 million times smaller than that received at 1 m. However, on earth things are more complex. Indoors, the wave will encounter reflection and absorption by walls, floors, ceilings, ducts, pipes, and people as well as scattering by small objects and interference from other radios and unintentional emitters such as microwave ovens. Outdoors, in addition to reflection and absorption by buildings, the earth's often craggy surface, and lakes, rivers, or oceans, the wave can be scattered by cars and trucks and absorbed by foliage or heavy rain. The net result is that the received wave may have been severely attenuated, delayed, replicated, and mixed with other interfering signals, all of which need to be straightened out in the receiver.

The *receiver* begins its task by *filtering* out the band of interest, in the process rejecting most of the interfering signals that weren't rejected by the limited bandwidth of the antenna. The signal is then amplified; the initial stages of amplification are responsible for determining how much noise the receiver adds to whatever was present in the signal to begin with and thus the *noise figure* of the receiver. Receivers can also use single-step *direct conversion* or multiple-step *superheterodyne* architectures. The gain of the receiver—the ratio between the received voltage and the voltage delivered to the ADC—must be large enough to allow a faithful digital reproduction of the signal given the resolution of the ADC. Overall gain is generally inexpensive and does not limit receiver performance, but note that the incoming signal strength can vary over many orders of magnitude, so the gain must be adjustable to ensure that the wildly varying input is not too big or too small but appropriately matched to the *dynamic range* of the ADC.

All the tasks required by a successful radio link must be performed under severe real-world constraints. To keep cost down, the radio will be fabricated if at all possible on inexpensive circuit board material (the laminate FR4 is common). As many components as possible will be integrated into CMOS circuitry, constantly trading off total cost versus the use of large but cheap external discrete components (resistors, capacitors, inductors, etc.). Low-cost radios will have only enough linearity to meet specifications and mediocre noise figure relative to the ideal receiver. Filtering may provide only modest rejection of nearby interferers.

Intentional emitters must be certified by the FCC in the United States, or appropriate regulatory bodies in other jurisdictions, before they can be offered for sale. Regulations will normally constrain the frequency of transmission, the total power, and the amount of radiation emitted outside the allowed frequency bands. In unlicensed bands, constraints are typically imposed on the modulation approaches and other etiquettes that must be used to minimize interference with other users of the band.

For many uses, size is highly constrained. Add-on radios for laptop computers must fit within a PC-card (PCMCIA) form factor, often at the cost of using inefficient antennas; radios for personal digital assistants are constrained to the still-smaller compact-flash size. Radios for consumers must be simple to use and safe. Portable devices require long battery life, limiting transmit power, and receiver sensitivity. Though there is no evidence that radio waves in the ordinary ambient represent a safety concern, near the transmitter power levels are higher, and consideration must be given to the health of users. Antenna designs that minimize power delivered to the user may be desirable. The resulting product must dissipate the heat generated by its operations and survive thermal, mechanical, and electrical stress to perform reliably for years in the field. All these constraints must be met while minimizing the cost of every element for consumer products.

A key to constructing such high-performance low-cost radio links is to use as much digital intelligence as possible. Wireless links are complex, time varying, and unreliable. The digital hardware and protocols use modulation, coding, error correction, and adaptive operation to compensate for the limitations of the medium, ideally delivering performance comparable with a wired link without the wires.

4.2. Putting the Numbers In

A system or radio designer needs to do more than exhort his or her components to be cheap and good: the designer must be able to calculate what will work and what won't. At the system level, the most fundamental calculation to be performed is an estimate of the *link budget*: the amount of loss allowed to propagation and inefficiency given the transmit power and receiver noise that will still meet the accuracy requirements of the protocol given the modulation used. To perform such a calculation, we first need to introduce a few definitions.

A single-frequency harmonic signal dissipates power P in a resistive load proportional to the square of the signal voltage, just as in the DC case, though the constant of proportionality is divided by 2.

$$V(t) = v_O \cos(\omega t) \rightarrow \langle P \rangle = \frac{v_O^2}{2R} \qquad [2.15]$$

Some folks define a root-mean-square or *RMS* voltage to avoid the extra factor of 2.

$$v_{rms} = v_O/\sqrt{2} \rightarrow \langle P \rangle = v_{rms}^2/R \qquad [2.16]$$

The nominal load resistance associated with a power level is usually $50\,\Omega$ in radio practice, and when the load impedance is not stated, one may safely use this value. Fifty ohms was historically chosen as a compromise value for the impedance of coaxial cables, about midway between the best power handling (roughly $30\,\Omega$) and the lowest loss (about $75\,\Omega$). Note, however, that in cable television systems power levels are low and loss in the cable is important, so cable television engineering has historically used $75\,\Omega$ impedances.

The power levels encountered in radio engineering vary over an absurdly huge range: the transmitted power could be several watts, whereas a received power might be as small as a few attowatts: that's 1/1,000,000,000,000,000,000 of a watt. Getting this tiny signal up to a power level appropriate for digital conversion may require system gains in the millions or billions. Logarithms aren't just nice, they are indispensable.

Gain is typically measured logarithmically in decibels or *dB* (the notation dates back to Alexander Graham Bell, inventor of the telephone):

$$G(dB) = 10 \cdot \log_{10}\left(\frac{P_2}{P_1}\right) \qquad [2.17]$$

Thus, 10 dB represent a gain of 10, and 3 dB is (very nearly) a factor of 2.

To measure power logarithmically, we must choose a reference power level to divide by, because the argument of a logarithm must be dimensionless. In radio practice it is very common to use a reference level of 1 mW. Logarithmic power referred to 1 mW is known as decibels from a milliwatt or *dBm*:

$$P(dBm) = 10 \cdot \log_{10}\left(\frac{P_2}{1mW}\right) \qquad [2.18]$$

Other units are occasionally encountered: dBmV is dB from a millivolt but is a measure of power not voltage; dBμv (referenced to a microvolt) is also used. One may convert from dBmV to dBm by subtracting 49 dB.

It is a (in the present author's view unfortunate) convention that voltages and voltage gains are defined with the respective logarithms multiplied by 20 instead of by 10. This definition allows a gain to be constant in decibels whether voltages or power are entered into the equation but results in a consequent ambiguity in the conversion of decibels to numerical values: one must know whether a voltage or power is being described to know whether 10 dB is a factor of 10 or 3.

Some examples of the use of logarithmic definitions for harmonic signals are shown in Table 2-5. It is noteworthy that a milliwatt, which seems like a tiny amount of power, represents a sizable and easily measured 0.32-V peak into a 50 Ω load.

We already alluded several times to the importance of *noise* but have so far said little about its origin or magnitude. There are many possible sources of electrical noise, but at the frequencies of interest for most digital communications, the first contributor to noise we must consider is the universally present *thermal noise* generated by any resistive load at a finite temperature. The physical origin of this noise is the same as that of the blackbody radiation emitted by any object not at absolute zero: by the equipartition theorem of statistical mechanics, all degrees of freedom of a classical system at finite temperature will contain an equal amount of energy, of order kT where k is Boltzmann's constant (1.38×10^{-23} joules/K, but not to fear: we will use this only once and forget it), and T is the absolute temperature in Kelvins, obtained by adding

Table 2-5: Examples of Power Levels and Corresponding Voltages

Power (W)	Power (dBm)	Peak Voltage (V)	RMS Voltage (V)
1	30	10	7.1
0.1	20	3.2	2.2
0.001	0	0.32	0.22
10^{-6}	−30	0.01	7.1×10^{-3}
10^{-12}	−90	10^{-5}	7.1×10^{-6}

273 to the temperature in centigrade. (Note that this is only true when the separation between energy levels is small compared with kT; otherwise Fermi or Bose-Einstein statistics must be used and life gets more complex. Thus, the statements we're going to make about electrical noise at microwave frequencies cannot be generalized to the noise in optical fiber systems at hundreds of teraHz.)

Because the resistor voltage is a degree of freedom of a resistor, we must expect that if the resistor is at a finite temperature, a tiny random thermal noise voltage will be present. At microwave frequencies, the amount of power is essentially independent of frequency (for the aficionado, this is the low-frequency limit of the Planck distribution), and so it is convenient to describe the noise in terms of the noise power per Hz of bandwidth. The mean-square voltage is

$$\langle v_n^2 \rangle \approx 4kT[BW]R \qquad [2.19]$$

The power delivered to a matched load (load resistance = source resistance R, so half the voltage appears on the load) is thus kT[BW] independent of the value of the resistor. This power level at room temperature (300 K—a very warm cozy room with a fire and candles) is equal to 4×10^{-21} W/Hz or $-174\,dBm/Hz$. This is a very important number and well worth memorizing: it represents the lowest noise entering any receiver at room temperature from any source with a finite impedance of 50 Ω, such as an antenna receiving a signal. The noise floor of any room-temperature receiver cannot be better than -174 dBm/Hz.

How much power does this represent for bandwidths of practical interest? In 1 KHz (the ballpark bandwidth used in some types of cellular phones) the thermal noise is -144 dBm. In 1 MHz (closer to the kind of channels used in WLANs) the corresponding power is -114 dBm. Generally speaking, a receiver must receive more power than this to have any chance of extracting useful information, though specialized techniques can be used to extract a signal at lower power than the noise at the cost of reduced data rates.

Real radios always add more than just the minimum amount of thermal noise. The excess noise is measured in terms of what it does to the S/N. The noise factor of a receiver is defined as the ratio of output to input (S/N) and represents the effect of excess noise. The logarithm of the noise factor is known as the *noise figure*; the great utility of the noise figure is that it can simply be added to the thermal noise floor to arrive at an effective noise floor for a receiver. Thus, if we wish to examine a 1-MHz-wide radio channel using a receiver with a 3-dB noise figure, the effective noise floor will be $(-114 + 3) = -111$ dBm.

We've noted previously that the transmitted signal will ideally decrease as the inverse square of the distance between transmitter and receiver. Let us spend a moment to quantify this *path loss* in terms of the distance and the receive antenna. Assume for the present that the transmitting antenna is perfectly isotropic, sending its energy uniformly in all directions. (Such an antenna cannot actually be built, but any antenna that is much smaller than a wavelength provides a fair approximation.) Further, let us assume the receiving antenna captures all the energy impinging on some effective *A* (Figure 2-41). The collected fraction of the radiated energy—the path loss—is then just the ratio of the area *A* to the total surface area of the sphere at distance *r*. We show in Chapter 5 that this area is related to the directive characteristics of the antenna, but for the moment we'll just state a plausible value. If the transmitting antenna is in fact directional (and pointed at the receiver), more power will be received. The directivity of the antenna is measured in dB relative to an isotropic antenna or dBi; the received power is the sum of the isotropic received power in dBm and the antenna directivity in dBi, for ideally efficient antennas. (In Chapter 5 we'll provide a more detailed description of directivity.)

We're now equipped to take a first crack at figuring out the *link budget* for a plausible radio link. The calculation is depicted graphically in Figure 2-42. We'll set the frequency to 2.4 GHz, though at this point the choice has no obvious effect. The transmit power will be 100 mW, or 20 dBm, which turns out to be a reasonable power for a high-quality WLAN client card or access point. For the present, we'll assume the transmitting and receiving antennas are 100% efficient and deliver all the power they take in either to or from the electromagnetic medium; real antennas may not reach this nice quality level. We'll allow the transmitting antenna to be slightly directional—3 dBi—so that the

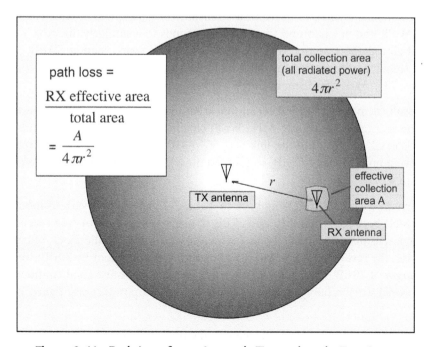

Figure 2-41: Path Loss for an Isotropic Transmitter in Free Space

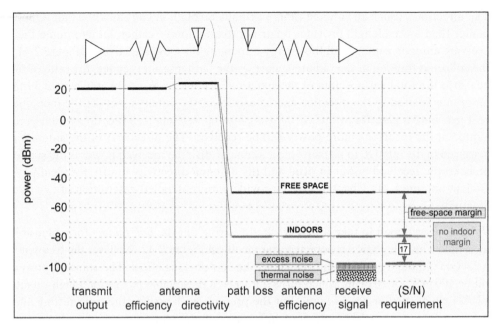

Figure 2-42: Link Budget Example

received signal should be increased by 3 dB over the nominal isotropic value. The receiving antenna, located 60 m from the transmitter, is assigned an effective area of 25 cm². (We'll find in Chapter 5 that this corresponds to a similarly directive receiver.) Putting it all together, we find that the path loss in free space is about 73 dB. The received signal, compensating for the directionality of the transmitter, is $(20\,\text{dBm} - 73 + 3) = -50\,\text{dBm}$, or 10 nW.

That doesn't seem like a lot, but recall that what matters is how it compares with the noise. The thermal noise in a 16-MHz Wi-Fi channel is about -102 dBm. Thus, if the receiver is ideal, the (S/N) of the link is $(-50 - (-102)) = 52$ dB. A realistic receiver might have a noise figure of 5 dB, so the real noise floor is -97 dBm and the (S/N) is reduced to 47 dB.

How much do we need? Referring back to Table 2-2, we see that various modulations require (S/N) between around 10 and 25 dB to achieve nearly error-free reception. We'll use a reasonable value of 17 dB (chosen to get round numbers out, of course). In this case, the received (S/N) of 47 dB exceeds the requirement by 30 dB: we have a *link margin* of 30 dB. With such a large margin, we could have good confidence that the link would work reliably, despite minor variations in power, noise figure, or distance.

Unfortunately, in a real indoor environment, the path loss can vary widely. It would not be unusual to have a path loss at 60 m indoors of as high as 103 dB (the exact value obviously chosen again to give a round number). In this case our received signal decreases to −80 dBm, the (S/N) is 17 dB, and we have no link margin at all. A link with no margin will fail to perform to specifications at least half the time. In practice, the radio would attempt to adapt to the circumstances by "falling back" to a modulation with less stringent (S/N) requirements and thus by implication to a lesser data rate.

The goal of the system designer is to ensure that the link budget for each link is adequate to reliably transport data at the desired rate, despite the constant efforts of nature and humanity to disrupt it.

5. Capsule Summary: Chapter 2

Let us briefly recap the main points of this chapter. The goal of a wireless communications system is to exploit propagating electromagnetic waves to convey information. The electromagnetic potential is fundamentally a shared communications medium, so measures must be taken to enable many users to coexist on the medium through multiplexing. The most basic and (almost) universal multiplexing approach is to confine radiation by frequency. However, to transmit information, waves must be modulated, resulting in a finite bandwidth for a given data rate and thus a finite amount of spectrum to share between differing uses and users. The modulation techniques that can be used differ in the amount of information they transmit per symbol, their resilience in the presence of noise, and their tolerance to multipath, but no matter what technique is used, the maximum amount of information that can be transmitted in a single noisy channel is less than Shannon's limit. We examined two specialized modulation techniques, OFDM and pulse radio, of increasing interest in the modern wireless world.

We gave a brief qualitative review of the considerations that go into creating a working wireless link. Once the modulation and data rate are known, the required (S/N) can be specified. A given radio link can then be examined to see whether this required signal quality can be delivered on the assigned frequency from the power available across the path of interest, despite inevitable uncontrolled variations in the equipment and the environment.

Further Reading

Modulations

Digital Modulation and Coding, Stephen Wilson, Prentice-Hall, 1996: *Includes a nice introduction to probability and information theory as applied to data communications.*

Information Transmission, Modulation, and Noise (4th Edition), Steven and Mischa Schwartz, McGraw-Hill, 1990: *Thorough and readable, but dated.*

OFDM

"Implementing OFDM in Wireless Designs," Steve Halford, Intersil; tutorial T204, Communications Design Conference, San Jose, CA, 2002

UWB/Pulse Radio

"Efficient Modulation Techniques for UWB Signals," Pierre Gandolfo, Wireless System Design Conference, San Jose, CA, 2003

Basics of Wireless Local Area Networks

1. Networks Large and Small

Although this book is primarily about voltages and waves, to make the text reasonably self-contained, we provide a brief introduction into the topic of networking in general and review the characteristics of several wireless local network technologies. In this review, the reader will encounter real examples of how the concepts introduced in Chapter 2 are put to use in getting data from one place to another wirelessly.

Our treatment is necessarily cursory and emphasizes aspects relevant to our main topic of wireless transmission and reception, giving short shrift to frame formats, management entities, and many other topics directed toward network implementation.

Data networks have existed in various forms since the invention of telegraphy in the nineteenth century. However, the use of data networks has expanded tremendously with the proliferation of computers and other digital devices. Today, networks extend everywhere, and the Internet—the network of networks—allows computers almost anywhere in the world to communicate with each other using these networks.

Data networks can be organized in a hierarchy based on physical extent. The physical reach of the network is a useful figure of merit both for technical and practical reasons. Information cannot move faster than the speed of light, so as the distance covered by a network grows, its *latency*—the minimum time needed to send a bit between stations—grows too. Latency has important consequences on how networks are designed and operated: if the transmitter has to wait a long time for a reply to a message, it makes sense to send large chunks (*packets*) of information at once, because otherwise the medium will simply be idle while awaiting confirmation. Long links also need low error rates, because the time required to correct an error is long. Shorter links can take a different approach, sending a packet and then waiting for a response, because the time required is not large. Small and large networks are also practically and commercially distinct: small networks are usually owned and operated by their users, whereas large networks are owned and operated by service providing companies who are not the primary users of their capacity.

With that brief preface, let us categorize data networks by size:

- *Personal area networks* (PANs): PANs extend a few meters and connect adjacent devices together. To speak formally, a data connection between a single transmitting station and a single receiving station is a link, not a network, so the connection between, for example, a desktop computer and an external modem (if there are any readers who remember such an impractical arrangement) is a data *link* rather than a data network. However, more sophisticated cabling systems such as the small computer system interface, which allow sharing of a single continuous cable bus between multiple stations, may be regarded as networks. More recently, wireless PANs (WPANs) have become available, with the ambition to replace the tangle of cabling that moves data between devices today. Wired PANs were purely dedicated to moving data, though some WPAN technologies (such as Bluetooth, discussed in section 5-1) also support voice traffic.

- *Local area networks* (LANs): LANs were invented to connect computers within a single facility, originally defined as a room or small building and later extended to larger facilities and multisite campuses. LANs extend a few hundred meters to a few kilometers. They are generally owned and operated by the same folks who own the site: private companies, individuals in their homes, government agencies, and so on. They are generally used indoors and historically have been solely for the movement of data, though the recent implementation of voice-over-Internet-protocol technologies has allowed LANs to provide voice service as well. Ethernet (see section 2) is by far the most prevalent LAN technology.

- *Metropolitan area networks* (MANs): MANs connect different buildings and facilities within a city or populated region together. There is a significant technological and historical discontinuity between LANs and MANs: LANs were invented by the computer-using community for data transfer, whereas MANs descended primarily from the telephone network, traditionally organized to move time-synchronous voice traffic. MANs are generally owned by local telephone exchanges (incumbent local exchange companies) or their competitors (competitive local exchange companies). They are organized around a large number of feeders to a small number of telephone central offices, where traffic is aggregated. Most MANs deployed today are based on a hierarchical system of increasingly faster synchronous data technologies. T-1 lines (in the United States) provide 1.5 megabits per second (Mbps) of data over twisted pairs of copper wires and in sheer numbers still dominate over all other connections to the MAN: there are thousands or tens of thousands of these in a large central office. T-1 and T-3 (45 Mbps) are further aggregated into faster connections, usually using fiber optic transmission over synchronous optical network links: OC-3 at 155 Mbps, OC-12 at 622 Mbps, and so on. Traditional MANs support both voice and data transfer.

- *Wide area networks* (WANs): WANs connect cities and countries together. They are descended from the long-distance telephone services developed in the mid-twentieth century and are generally owned and operated by the descendants of long-distance telephone providers or their competitors, where present. Almost all long-distance telecommunications today is carried over fiber-optic cables, commonly at OC-12 (622 Mbps), OC-48 (2.5 Gbps), or OC-192 (10 Gbps) rates; OC-768 (40 Gbps) is in the early stages of deployment in 2004. WAN connections cross the oceans and continents and carry the voice and data commerce of the world. The Internet initially evolved separately from WANs, but its explosive growth in the late 1990s resulted in a complex commingling of the traditional WAN and LAN communities, businesses, and standards bodies.

In the simplest view, there are small networks (PANs and LANs) and big networks (MANs and WANs). Small networks deliver best-effort services over short distances and are cost sensitive. Big networks deliver guaranteed-reliable services over long distances and are quality sensitive and cost competitive. This book focuses on how small networks are converted from wires to wireless links.

All networks, big or small, have certain elements in common. A message (typically a packet in a data network) must be addressed to the destination station. The message must have a format the destination station can understand. The message has to get access to some physical medium to be sent. Errors in transmission must be corrected. These activities can be grouped into a hierarchical arrangement that helps provide structure for designing and operating the networks. *Applications* live on the top of the hierarchy and use the networking services. *Networks* support applications and deal with routing messages to the destination station. *Links* transfer data between one station and another in the network and are responsible for access to the physical medium (a wire or a radio link), packaging the packet in an appropriate format readable by the receiving station and then reading received packets and producing the actual voltages or signals. (Note that an important function of networks, *correcting errors* in messages, can go anywhere in the hierarchy and may be going on simultaneously at some or all of the levels, leading on occasion to regrettable results.)

A very widely used example of this sort of hierarchical arrangement is the *Open Systems Interconnect* (OSI) protocol stack. A simplified view of the OSI stack, arranged assuming a wireless LAN (WLAN) link to one of the stations, is depicted in Figure 3-1. The arrangement is in general correspondence with the requirements noted above, though additional layers have been interposed to provide all the manifold functions needed in complex networks.

The standards work of the Institute of Electrical and Electronics Engineers (IEEE), with which we shall be greatly concerned in this chapter, generally divides the data link layer into an upper *logical link control* layer and a lower *medium access control* (MAC) layer. In this book, we are almost exclusively interested in what goes on in the link layer (in this chapter) and the physical (PHY) layer (here and elsewhere in the book). The logical link control layer is focused mainly on the requirements of the higher layers, and in practice the logical link control layer used for many WLAN technologies is the same as the one used for wired technologies. We do not discuss it further. However, the MAC

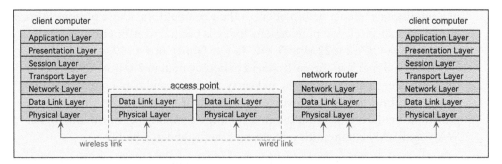

Figure 3-1: The OSI Protocol Stack; the Configuration Shown Is Typical of an 802.11 Infrastructure Network

layer is intimately involved in the efficient use of the wireless medium, and it is of great interest for any user of a wireless link.

2. WLANs from LANs

LANs connect nearby computers together so that they can exchange data. Ethernet, by far the most popular LAN technology today, was invented by Bob Metcalfe of Xerox Palo Alto Research Center in 1973, in connection with the prescient work on personal computers then being done at Palo Alto Research Center. Ethernet is an *asynchronous* technology: there are no fixed time slots assigned to users. There is also no central control of access to the transmitting medium. Instead, Ethernet exploits three basic ideas to minimize the impact of two or more stations trying to use the medium at the same time; two of them depend on the ability of a station to listen to the voltages on the cable while it sends its message. *Carrier sensing* is used by each station to make sure no one is using the medium before beginning a transmission. (Note that in fact the most popular forms of Ethernet use baseband signaling, that is, the voltages on the cable correspond directly to binary bits and are not modulating a high-frequency carrier, so there's no carrier to sense. Why did you expect it to make sense?) *Collision detection*—noticing that the signal on the cable isn't the one the station is sending—allows the station to detect that another station has tried to send a message at the same time it is transmitting. (The combination of these two techniques is often abbreviated *CSMA/CD*, for carrier-sense multiple access with collision detection.) Finally, *random backoff* is used after a collision to ensure that stations won't simply continue to collide: after all stations become silent, each station that has a pending packet to send randomly chooses a time delay before attempting to send it. If another collision results, the stations do the same exercise, but over a larger time window. This approach has the effect of forcing the *offered traffic* to the network to go down when collisions are frequent, so that data transfer continues at a lower rate but the network doesn't get tangled up: ethernet networks fail gracefully when they get busy. Thus, Ethernet was an early example of a MAC layer designed to allow peer stations to share a medium without central coordination.

Wireless stations are mobile, and access to the wireless medium cannot be as readily constrained as a wired connection. Wireless links are much noisier and less reliable in general than wired links. The MAC and PHY layers of a wireless network must therefore deal with a number of issues that are rare or absent from the provision of a wired link:

- *Getting connected*: Wireless stations by default are likely to be mobile. How does a wireless station let other mobile or fixed stations know it is present and wants to join the network? This is the problem of *associating* with the network. A closely related problem is the need of portable mobile stations to save power by shutting down when they have no traffic, without losing their associated status; the network coordinator must remember who is awake and who is asleep, allow new stations to join, and figure out that a station has left.

- *Authentication*: In wired networks, a station plugs into a cable placed by the network administrator. If you've entered the building, you're presumed to be an authorized user. Wireless propagation is not so well controlled (see Chapters 6–8), so it is important for the network to ensure that a station ought to be allowed access and equally important for the station to ensure that it is connecting to the intended network and not an impostor.

- *Medium access control*: The wireless medium is shared by all (local) users. Who gets to transmit what when? A good wireless network protocol must efficiently multiplex the shared medium.

- *Security*: Even if the network refuses to talk to stations it doesn't recognize, they might still be listening in. Wireless networks may provide additional security for the data they carry by *encryption* of the data stream. It is important to note that security may also be provided by other layers of the network protocol stack; the lack of local encryption in the wireless link does not necessarily imply insecurity, and encryption of the wireless data doesn't ensure security from eavesdropping at some other point in the network.

- *Error correction*: The wireless medium is complex and time varying. There is no way to transmit all packets without errors. The link layer may insist on receiving a positive *acknowledgment* (ACK) before a station may consider a transmission successful. Incoming data packets may be *fragmented* into smaller chunks if packet errors are high; for example, in the presence of microwave oven interference (see Chapter 7), which peaks in synchrony with power lines at 60 Hz, fragmentation may be the only way to successfully transmit very long packets. Fragmented packets must be identified as such and reassembled at the receiving station.

- *Coding and interleaving*: Errors are inevitable in the wireless medium. Protection against bit errors can be provided by encoding the data, so that errors are detected by the receiving station. Codes often allow errors of up to a certain size to be corrected by the receiver without further discussion with the transmitting station: this is known as *forward error correction*. Interleaving is the process of redistributing bits into an effectively random order to guard against bursts of errors or interference destroying all the neighboring bits and thus defeating the ability of the codes to correct local errors.

- *Packet construction*: Data packets get *preambles* prepended onto them. The preambles typically contain synchronization sequences that allow the receiving station to capture the timing of the transmitter and, in the case of more sophisticated modulations, may also allow the receiver to determine the carrier phase and frequency. The preambles also contain digital information specific to the wireless medium.

- *Modulation and demodulation*: The resulting packets must then be modulated onto the carrier, transmitted on the wireless medium, and received, amplified, and converted back into bits.

Let us examine how these problems are addressed in some current and upcoming WLAN and WPAN technologies.

3. 802.11 WLANs

In 1985 the U.S. Federal Communications Commission (FCC) issued new regulations that allowed unlicensed communications use of several bands, including the 2.4-GHz band, that had previously been reserved for unintended emissions from industrial equipment. Interest in possible uses of this band grew, and in the late 1980s researchers at NCR in Holland, who had experience in analog telephone modems, initiated work to develop a wireless data link. The initial experimental units used an existing Ethernet MAC chip. As discussed in more detail in section 3.2, there is no easy way to detect a collision in a wireless network; the experimenters worked around this limitation by informing the MAC chip that a collision occurred any time a positive ACK was not received, thus initiating the Ethernet backoff algorithm. In this fashion they were able to make working wireless links with minimal modification of the existing Ethernet MAC. The early workers understood that to make a true volume market for these products, standardization was necessary. When standardization efforts were introduced at the IEEE shortly thereafter, the MAC was elaborated to deal with the many wireless-specific issues that don't exist in Ethernet but remained based on Ethernet, which had been introduced because of its availability and was successful because of its simplicity and robustness. Thus, the standard today continues to reflect both the strengths and limitations of the original Ethernet MAC.

The IEEE 802 working group deals with standards for wired LANs and MANs. IEEE 802.3 is the formal standardization of the Ethernet wired LAN. The IEEE decided to incorporate WLANs as part of the 802 working group and created the 802.11 activity, culminating in the first release in 1997 of the 802.11 standard. In recent years, in addition to various elaborations of the 802.11 standards, which are discussed below, other working groups have formed within 802 to consider related applications of wireless data links. (In practice, WPANs are also included as part of the 802.15 activity.)

The 802.11 standard actually allowed three physical layers: an infrared link, a *frequency-hopping* (FH) radio link, and a *direct-sequence spread-spectrum* (DSSS) radio link. These links supported data rates of 1 to 2 Mbps. The infrared link physical layer has had little commercial significance (in fact, the author has been unable to find any evidence of any commercial product using this protocol ever having been manufactured). Commercial products were deployed with the FH radio, and the FH approach does offer certain advantages in terms of the number of independent collocated networks that can be supported with minimal interference. However, in terms of current commercial importance, the DSSS physical layer is completely dominant, due to its later extension to higher data rates in the 802.11b standard in 1999. Therefore, here we shall concentrate only on this physical layer. The IEEE standards bodies do many wonderful things, but creation of convenient nomenclature is not one of them; therefore, we introduce the terminology *802.11 classic* to refer to the original 1997 802.11 specification and its later releases and in our case predominantly to the DSSS variant. We can then unambiguously use the unmodified moniker "802.11" to refer to all the 802.11 working group standards, including the alphabet soup of elaborations released in later years.

3.1. 802.11 Architecture

The 802.11 standard allows both infrastructure networks, which are connected to a wired network (typically Ethernet) using an *access point,* and independent networks connecting peer computers wirelessly with no wired network present. Most installations are of the infrastructure variety, and we focus on them. From the point of view of the Ethernet network, which is usually connected to both ends of an 802.11 link, the wireless link is just another way of moving an Ethernet packet, formally known as a *frame* (Figure 3-2), from one station to another (Figure 3-3).

However, the possibility (well, likelihood) that some or most wireless stations are mobile stations means that the architecture of a wired Ethernet network with wireless stations is likely to be fundamentally different from a conventional wired network. The stations associated to a particular access point constitute a *basic service set* (BSS).

BYTES:	7	1	6	6	2	<1500	to 64	4
	Preamble	Start Frame	Destination address	Source address	Length or type	Payload	Pad	Error check

Figure 3-2: An Ethernet (802.3) Frame (Packet)

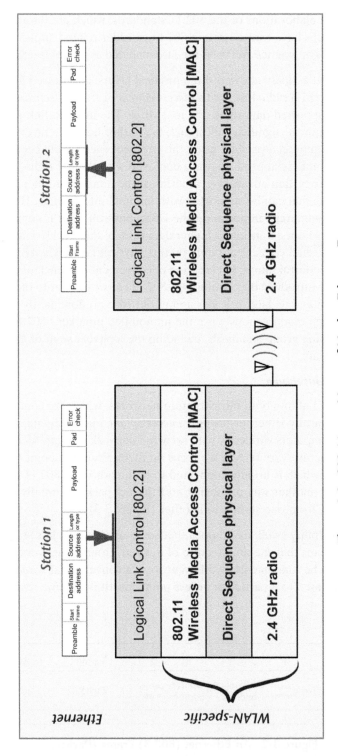

Figure 3-3: 802.11 as a Means of Moving Ethernet Frames

The collection of BSSs connected to a single wired network forms an *extended service set* (ESS) (Figure 3-4). A BSS has a unique identifier, the *BSSID*. An ESS also has an identifier, the *ESSID*, that is unique to that ESS but shared by all the component BSSs (is that enough acronyms in one sentence for you?). Unfortunately, interfaces for many commercial 802.11 implementations use the nomenclature SSID, which in practice usually refers to the ESSID but is obviously somewhat ambiguous and potentially confusing. The ESS architecture provides a framework in which to deal with the problem of mobile stations *roaming* from one BSS to another in the same ESS. It would obviously be nice if a mobile station could be carried from the coverage area of one access point (i.e., one BSS) to that of another without having to go through a laborious process of reauthenticating, or worse, obtaining a new Internet protocol (IP) address; ideally, data transfers could continue seamlessly as the user moved. A distribution system is assigned responsibility for keeping track of which stations are in which BSSs and routing their packets appropriately. However, the original and enhanced 802.11 standards did not specify how such roaming ought to be conducted; only with the release of the 802.11 f *Inter-Access Point Protocol* (IAPP) standard in 2003 did the operation of the distribution system receive a nonproprietary specification. The IAPP protocol uses the nearly universal internet protocol (TCP/IP) combined with RADIUS servers to provide secure communication between access points, so that moving clients can reassociate and expect to receive forwarded packets from the distribution system. In the future, local roaming on 802.11 networks should be (hopefully) transparent to the user, if compliant access points and network configurations become widely established.

3.2. MAC and CSMA/CA

Despite its name, Ethernet was conceived as a cable-based technology. Putting all the stations on a cable and limiting the cable length allowed meant that all stations could always hear one another, and any collision would be detected by all the stations on the cable. However, in a wireless network it is unlikely that all stations will be able to receive all transmissions all the time: this is known as the *hidden station* problem. An example is depicted in Figure 3-5: the access point can communicate with stations A and B, but A and B cannot receive each other's transmissions directly. Because of this fact, collision detection would fail if A and B both attempted to transmit at the same time. Furthermore, as we've already seen, received signals are tiny. Provisions to separate a tiny received signal from a large transmitted signal are possible if the two are at differing frequencies, but this requires bandwidth that was not available in the limited 2.4-GHz ISM band. It is difficult and relatively expensive to reliably receive a tiny transmitted signal while simultaneously transmitting at the same frequency, so collision detection is not practical for a WLAN station. Thus, in a wireless implementation, CSMA/CD cannot be used, because there is no way to confidently associate the absence of a carrier with a free medium or reliably detect all collisions.

To get around this difficulty, 802.11 stations use *carrier-sense multiple access with collision avoidance* (CSMA/CA, Figure 3-6). All stations listen to the current radio

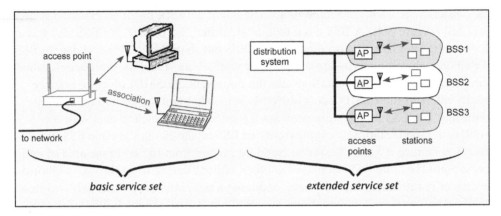

Figure 3-4: Definition of the Basic Service Set (BSS) and Extended Service Set (ESS)

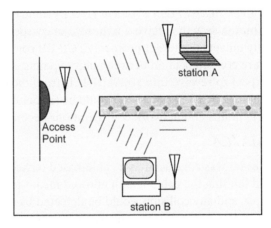

**Figure 3-5: Stations A and B Are Hidden From One Another,
Though Both Are Visible to the Access Point**

channel before transmitting; if a signal is detected, the medium is considered to be busy and the station defers its transmission. A virtual carrier sense mechanism—the *network allocation vector* (NAV)—is provided to further reduce the likelihood of collisions. Each packet header involved in an exchange that lasts longer than a single frame, such as a data frame that expects to be acknowledged by the receiver, will provide a NAV value in the header of the frame. All stations that receive the frame note the value of the NAV and defer for the additional required time even if they can't detect any signal during that time.

Requirements on the timing between frames—interframe spaces—are used to enable flexible access control of the medium using the carrier sense mechanisms. Once a transmission has begun, the transmitting station has captured the medium from all other stations that can hear its transmissions using the physical carrier sense mechanism. A station wishing to send a new packet must first ensure that both the physical carrier sensing mechanism and the NAV indicate the medium to be free for a time equal to the *distributed interframe space (DIFS)*. The distributed interframe space is relatively long, so that stations that are completing exchanges already in progress, which are permitted to transmit after a *short interframe space (SIFS)*, can capture the medium to transmit their response packets before any new station is allowed to contend for the right to transmit. The NAV value of each packet in a sequence is also set to capture the medium for the entire remaining sequence of packets. Once all the necessary parts of an exchange are complete, the medium may then become free for the distributed interframe space, at which point stations waiting to transmit frames randomly select a time slot and start transmission if no carrier has been detected.

Because collisions cannot be detected, the opposite approach is taken: 802.11 depends on positive acknowledgement of a successful transmission through the receipt of an ACK packet by the transmitting station. Failure to receive an ACK, whether due to poor signal strength, collisions with other stations, or interference, is considered to be indicative of a collision; the transmitting station will wait until the medium is again free, choose a random time slot within a larger possible window (the Ethernet backoff mechanism), and attempt to transmit the packet again.

Two additional optional provisions can be taken to improve performance when, for whatever reason, packets are lost frequently. First, the sending station (typically a client rather than an access point) can precede its data transmission with a *request to send* (RTS) packet. This sort of packet informs the access point and any other stations in range that a station would like to send a packet and provides the length of the packet to be sent. The access point responds after the short interframe space with a *clear to send* (CTS) packet, whose NAV reserves the medium for the time required for the remainder of the data transmission and ACK. The advantage of the scheme is that one can hope that all associated stations can hear the access point CTS packet even if they are not able to detect the RTS packet. An exchange like this is depicted schematically in Figure 3-6. The exchange begins with the client's RTS packet, which reserves the NAV for the whole of the envisioned exchange (the length of the data packet, three short interframe spaces, a CTS frame, and an ACK). The access point responds with clearance, reserving the medium for the remainder of the exchange with its NAV setting. All stations then defer until the end of the exchange. After a distributed interframe space has passed without any more transmissions, a station waiting to send data can use one of the contention slots and transmit.

The second backup mechanism for improving the chances of getting a packet through is *fragmentation*. A large packet from the wired medium can be split into smaller fragments, each of which is more likely to be transmitted without a problem. If an error is encountered in one of the fragments, only that fragment needs to be resent. Like the

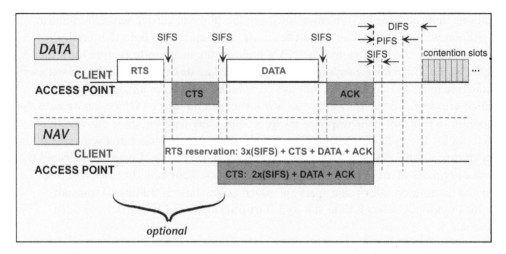

Figure 3-6: Example of Packet Exchange Under CSMA/CA

RTS/CTS exchange, a station transmitting a packet sets the NAV value to reserve the medium for the whole exchange consisting of all the fragments of the packet and an ACK for each fragment.

An important consequence of the MAC is that an 802.11 radio is never transmitting and receiving simultaneously: it is a *half-duplex* system in radio terms. This choice simplifies the design of the radio front end by eliminating the need to distinguish between the powerful transmitted signal and a tiny received signal.

The MAC layer defined by the original 802.11 standard is used almost unchanged for the enhanced versions (802.11b, a, and g) of the standard and thus is worth a moment's reflection. The MAC, although not as simple as the original Ethernet MAC, is nevertheless a fairly basic object. We have not discussed the point-coordination function because it is rarely used. Absent this, the MAC provides no central coordination of contending stations and no guarantees of performance except for the continuity of an ongoing packet exchange.

3.3. 802.11 Classic Direct-Sequence PHY

To maximize the probability that multiple users could share the unlicensed bands without unduly interfering with each other, the FCC placed restrictions on communications systems that could operate there. A fundamental part of the requirement was that the systems not transmit all their energy in one narrow segment of the band but should spread their radiation over a significant part of the band, presumably a much larger segment than actually required by the data bandwidth, that is, users were required to apply *spread spectrum* techniques (a proviso since relaxed; see Appendix 1). In this fashion, it was hoped that interference between collocated systems would be minimized.

One approach to spreading the spectrum is to operate on many channels in some pseudo-random sequence; this is known as *frequency hopping* and is used in the classic FH PHY and the Bluetooth (802.15) PHY.[1] It is obvious that if two neighboring radios are operating on separate channels, interference ought to be minimized except when by chance they happen to select the same frequency. The disadvantage of this approach is that the channels must be quite narrow if there are to be a large number of them and therefore little chance of overlap; the 1-MHz maximum width specified in the regulations limits the data rate that can be transmitted at the modest signal-to-noise ratios (S/Ns) expected on low-cost, unlicensed, wireless links (see Chapter 2 for a more detailed discussion of the relationship between bandwidth and data rate).

A quite distinct and rather more subtle approach is called *direct-sequence spread* DSSS. Direct-sequence methods were developed for military applications and are also used in those cellular telephones that are based on code-division multiple access (CDMA) standards. In DSSS, the relatively slow data bits (or more generally symbols) are multiplied by a much faster pseudo-random sequence of *chips*, and the product of the two is used as the transmitted signal. Recall from Chapter 2 that the bandwidth of a signal is determined by the number of symbols per second transmitted, whether or not those symbols contain useful information. Thus, the bandwidth of a DSSS signal is determined by the chip rate, which in general is much larger than the data rate; a DSSS signal can satisfy the FCC's requirement that the signal be spread. Further, the received signal can be multiplied again by the same sequence to recover the original lower data rate. In spectral terms, multiplying the received wide-bandwidth signal by the chip sequence collapses all its energy into a narrower bandwidth occupied by the actual data while simultaneously randomizing any narrowband interfering signal and spreading its energy out. Thus, the use of direct sequence codes provides *spreading gain*, an improvement in the link budget due to intelligent use of extra bandwidth beyond what is needed by the data.

In 802.11 classic DSSS, each data bit is multiplied by an 11-chip Barker sequence, shown in Figure 3-7. (Here data are shown as +1 or −1 both to assist the reader in thinking in terms of multiplication and due to the direct translation of the result into

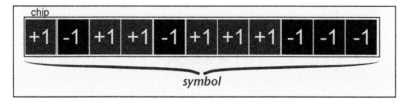

Figure 3-7: Barker Sequence

---

[1] In a remarkable example of unexpected ingenuity, frequency hopping was invented by the Austrian-born actress Hedy Lamarr, aided by composer George Antheil (U.S. patent 2,292,387), though she received little credit during her lifetime for this achievement.

binary phase-shift keying [BPSK] transmitted chips; a formally equivalent treatment in terms of modulo-2 addition of binary bits is also possible.) Thus, a 1 bit becomes the sequence $(+1\ -1\ +1\ +1\ -1\ +1\ +1\ +1\ -1\ -1\ -1)$ and a 0 bit becomes $(-1\ +1\ -1\ -1\ +1\ -1\ -1\ -1\ +1\ +1\ +1)$.

The Barker sequence is chosen for its *autocorrelation* properties: if two Barker sequences offset in time are multiplied together and the result added (a *correlation* of the two sequences), the sum is small except when the offset is 0. Thus, by trying various offsets and checking the resulting sum, one can locate the beginning of an instance of the sequence readily; that is, synchronization to the transmitter is easy. It is also possible to have several DSSS signals share the same frequency band with little interference by using different *orthogonal* spreading codes for each signal: codes whose correlation with each other is small or zero. This technique is key to the operation of CDMA cellular phone standards but is not used in the 802.11 standards.

The basic symbol rate is 1 Mbps; each symbol consists of 11 chips, so the chip rate is 11 Mbps. The *spreading gain*, the ratio of the actual bandwidth to that required by the underlying data, is thus 11:1 or about 10.4 dB, meeting the FCC's original minimum requirement of 10 dB. Because the chip rate is 11 Mbps, one would expect the bandwidth required to be modestly in excess of 11 MHz. The actual bandwidth of the transmitted signal depends on the details of precisely how the signal is filtered and transmitted; rather than specifying such implementation-specific aspects, the standard simply defines a *spectral mask*, which provides a limit on how much power a compliant transmitter is allowed to radiate at a given distance from the nominal center frequency. The spectral mask for 802.11 classic is shown in Figure 3-8. The frequency reference is the nominal frequency for the given channel (about which we'll have more to say in a moment); the amplitude reference is the power density in the region around the nominal frequency. The boxes represent the maximum allowed power density; the smooth line is a cartoon of a typical power spectrum. Observe that although the standard allows the spectrum to be as much as 22 MHz wide, a typical real signal would have a bandwidth measured, for example, 10 dB from the peak of around 16 MHz.

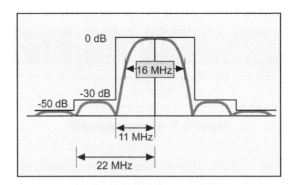

Figure 3-8: 802.11 Spectral Mask

The nominal center frequency for a transmission is chosen from a set of channel frequencies spaced by 5 MHz, shown in Table 3-1. (In the United States, channels 1–11 are available; channels 1–13 are allowed in most European jurisdictions, and channel 14 is for Japan; more details can be found in Appendix 1.) Note that if the bandwidth is 16 MHz, transmissions on adjacent channels will interfere with each other quite noticeably. A separation of five channels (25 MHz) is needed to remove most overlap; thus, there are actually only three *nonoverlapping* channels available in the United States, channels 1, 6, and 11 (Figure 3-9).

Two data rates are supported in 802.11 classic: 1 and 2 Mbps. Both rates use the same chip rate of 11 Mbps and have essentially the same bandwidth. The difference in data rate is achieved by using differing modulations. Packets at the basic rate of 1 Mbps, and the preambles of packets that otherwise use the extended rate of 2 Mbps, use *differential BPSK* (DBPSK), as shown in Figure 3-10. The data portion of extended-rate packets uses differential quaternary phase-shift keying (DQPSK). Recall from Chapter 2 that the bandwidth of a signal is mainly determined by the symbol rate, not by the nature of the individual symbols; because QPSK carries 2 bits per symbol, it is possible to roughly double the data rate without expanding the bandwidth, at a modest cost in required (S/N). In each case, the data bits are scrambled in order before transmission; this procedure avoids the transmission of long sequences of 1s or 0s that might be present in the source data, which would give rise to spectral artifacts.

The 802.11 modulations are differential variants of the BPSK and QPSK schemes we discussed in Chapter 2, that is, the phase of each symbol is defined only with respect to the symbol that preceded it. Thus, if the phase of a BPSK signal is the same for two

Table 3-1: 802.11 Channels

Channel	f (GHz)
1	2.412
2	2.417
3	2.422
4	2.427
5	2.432
6	2.437
7	2.442
8	2.447
9	2.452
10	2.457
11	2.462
12	2.467
13	2.472
14	2.484

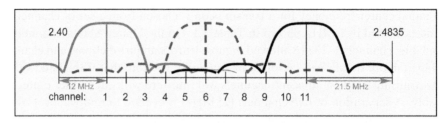

Figure 3-9: Nonoverlapping 802.11 Channels, U.S. ISM Band

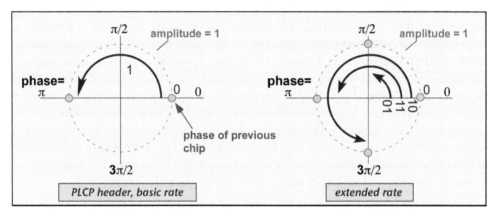

Figure 3-10: 802.11 Classic Modulations: DBPSK (Left) and DQPSK (Right)

consecutive chips, a 0 has been transmitted. (Note that the arrows in Figure 3-10 showing the differences in phase are indicative only of the phase change between the moments at which the signal is sampled, not of the path the signal takes between sample times. This apparently subtle point has significant consequences in terms of the requirements on the radio, as we discuss in Chapter 4.) A moment's reflection will clarify why such a change might be desirable. Imagine that we wish to accurately state the absolute phase of a received QPSK signal by, for example, comparing it with a local oscillator that at the beginning of the packet is locked to the phase of the carrier. A 1500-byte (12,000 bit) packet at 2 Mbps lasts 6000 μsec. During this time, the total phase change of the carrier is 2π (2.4 GHz) $(6000 \times 10^{-6}$ sec$) = 9 \times 10^7$ radians. If we wish the phase of the local oscillator to remain accurate to, for example, $\pi/12$ radians, so that this drift is small compared with the $\pi/2$ radian changes we're trying to resolve between chips, we need the oscillator to maintain phase to about 3 parts per billion. This is implausibly difficult and expensive. (Of course, more clever methods are available to perform such a task at more modest requirements on the hardware; the calculation is done to provide a frame of reference.) On the other hand, to maintain phase lock between successive chips at 11 Mbps, we only need hold this accuracy over 1/11 μsec: 220 cycles, or 1400 radians. The required accuracy in this case is 200 parts per million, which is easily achieved with inexpensive crystal-referenced synthesized sources. The price of this alternative is merely that we lose any information present in the first symbol of a sequence: because we can exploit our synchronization sequence (which doesn't carry any data anyway) for this purpose, it is obvious that differential modulation carries considerable benefits and little penalty.

The maximum transmit power of an 802.11 radio is limited in the standard to comply with the requirements of regulatory bodies. In the United States, power must be less than 1 W, but in practice a typical access point uses a transmit power of about 30 to 100 mW. The standard then requires that a receiver achieve a frame error rate of less than 8% for 1024-byte frames of 2 Mbps QPSK at a signal power of -80 decibels from a milliwatt (dBm). That sounds like a lot of errors, but because each frame has $1024 \times 8 = 8192$ bytes, the bit error rate is a respectable $0.08/8192 = 10^{-5}$. To understand what the transmit and receive power levels imply, let's do a link budget calculation like the one we performed in Chapter 2. Let the transmitted power be 100 mW (20 dBm). Assume the transmitting antenna concentrates its energy in, for example, the horizontal plane, achieving a 6-dB increase in signal power over that of an ideal isotropic antenna (we discuss directive gain in more detail in Chapter 5). Assume the receiving antenna has the same effective area as an ideal isotropic receiver. Our allowed path loss—the link budget—is then $(20 + 6 - (-80)) = 106$ dB. Recalling that an isotropic antenna has an equivalent area of around 12 cm^2, this is equivalent to a free space path of 2000 m or 2 km! However, recall that propagation indoors is unlikely to achieve the same performance as unimpeded ideal propagation in free space. We investigate the distinctions in more detail in Chapter 6; for the present let us simply add 30 dB of path loss for a long indoor path, leaving us with an allowed range of $(2000/10^{1.5}) = 63$ m. The transmit power levels and receiver sensitivities envisioned in the 802.11 standard make a lot of sense for indoor communications over ranges of 10 s to perhaps 100 m, just as one might expect for a technology designed for the LAN environment (see section 1).

The receive sensitivity requirement also tells us something about the radio we need to build. Recall that a QPSK signal needs to have an (S/N) of about 12.5 dB for reasonably error-free reception. At first blush, one would believe this implies that the noise level of the receiver must be less than $(-80 - 12.5) = -92.5$ dBm. However, recall that 11 QPSK chips at 11 Mbps are "averaged" together (correlated with the Barker sequence) to arrive at one 2-bit data symbol. This averaging process provides us with some extra gain—the spreading gain described above—so that we can tolerate an (S/N) of $12.5 - 10.5 = 2$ dB and still get the data right. The noise level we can tolerate in the receiver is thus $-80 - 2 = -82$ dBm. Recalling that the signal is about 16 MHz wide, the unavoidable thermal noise in the receiver is $(-174 \text{ dBm} + 10(\log 16) + 60) = -102$ dBm. The specification has left room for an additional 20 dB of excess noise in the receiver. Receivers with noise figures of 20 dB at 2.4 GHz are very easy to build, and in fact 802.11 commercial products do much better than this. The standard has specified a level of performance that can be reached inexpensively, appropriate to equipment meant for an LAN. The link budget calculation is displayed graphically in Figure 3-11.

3.4. 802.11 Alphabet Soup

Products implementing the 802.11 classic standard were manufactured and sold in the late 1990s by vendors such as Proxim, Breezecom (now part of Alvarion), Lucent (under the WaveLAN brand), Raytheon, Symbol Technologies, and Aironet (now part of Cisco). However, the maximum data rate of 2 Mbps represented a significant obstacle to the intended target usage model of wirelessly extending Ethernet: even the slowest variant of Ethernet has a native bit rate of 10 Mbps and a true throughput of around 8–9 Mbps. Work had started as early as 1993 on improved physical layers, and by 1999 two new higher performance PHY layers were released as enhancements to the original 802.11 classic interfaces. *802.11a* was a fairly radical modification of the physical layer, having essentially no commonalty with the DSSS or FH PHY layers of the classic standard and targeted for operation in a different slice of spectrum around 5 GHz. *802.11b* was a much more conservative attempt to improve performance of the 802.11 PHY layer without changing it very much. The 802.11b standard received the most early commercial attention, including support in the consumer market from Apple Computer's Airport product line and Cisco's Aironet products in the "enterprise" market (LANs for large industrial companies). The elaboration of compatibility efforts begun in association with the release of 802.11 classic led to the formation of the Wireless Ethernet Compatibility Alliance, or WECA, which (recognizing the

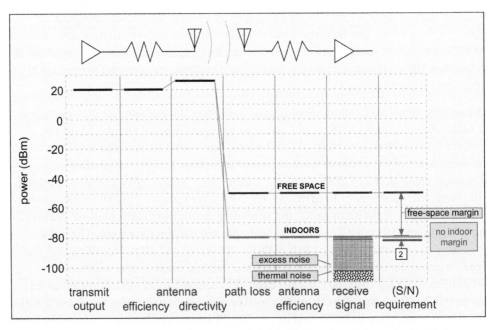

Figure 3-11: Link Budget Estimates for an 802.11–Compliant Radio Link

awkwardness of the IEEE's nomenclature) provided *Wi-Fi* compliance certification to ensure interoperable 802.11b products from multiple vendors. By 2000 it was clear that 802.11b had found a "sweet spot," combining adequate data rates and performance with simplicity and low cost, and was poised for major commercial successes and widespread deployment. Sales volumes grew rapidly, and prices for both client cards and access points fell even more rapidly as numerous vendors entered the field.

The more audacious 802.11a standard, which would provide enhanced data rates of up to 54 Mbps, was seen by most industrial participants as a tool to hold in reserve for future use, but the start-up company Atheros proceeded to release chips implementing the 802.11a standard, with products available from Intel and Proxim in 2002. These products, although modestly successful at introduction, clarified the desirability of combining the high data rates of the 802.11a standard with backward compatibility to the installed base of 802.11b products. The "G" task group, which was charged to take on this awkward task, after some struggles and dissension approved the 802.11g standard in 2003, noticeably after the release of several prestandard commercial products. At the time of this writing (early 2004), 802.11g products have nearly replaced 802.11b products on the shelves of consumer electronics stores: the promise of backward compatibility in conjunction with a much higher data rate at nearly the same price is an effective sales incentive, even though many consumer applications do not currently require the high data rates provided by the standard. Dual-band access points and clients, supporting both 802.11a and 802.11b/g, are also widely available.

A number of task groups are also actively attempting to fill in holes in the original standards work and add new capabilities. Task group "I" is charged with improving the limited authentication and encryption capabilities provided in the original standard, about which we have more to say in section 3.8. Task group "F" is responsible for providing the missing description of the distribution system functions, so that roaming between access points can be supported across any compliant vendor's products; an IAPP specification was approved in June 2003.

Recall that the original Ethernet standard, and the 802.11 MAC derived from it, deliver *best-effort* data services, with no guarantees of how long the delivery will take. Services such as voice or real-time video delivery require not only that data be delivered but that it arrive at the destination within a specified time window. This requirement is commonly placed under the not-very-informative moniker of *quality of service*. Task group "E" is defining quality of service standards for time-sensitive traffic over 802.11 networks, although to some extent that effort has refocused on the 802.15 work we describe in the next section. Task group "H" is defining two important additions to the 802.11a standard: *dynamic frequency selection* and *transmit power control*. These features are required for operation in most European jurisdictions, and the recent FCC decisions on spectrum in the United States will also require such capability for operation in most of the 5-GHz bands.

In the next three subsections we examine the important 802.11b and 802.11a PHY layers and briefly touch on their admixture in 802.11g. We also provide a cursory discussion of WLAN (in)security and some cures for the deficiencies revealed therein. We must regrettably refer the reader to the IEEE web sites described in section 7 for more details on the other task groups within 802.11.

3.5. The Wi-Fi PHY (802.11b)

The 802.11b physical layer uses the same 2.4-GHz band and channelization as the classic PHY. Furthermore, the basic signaling structure of 11 megasamples/second (Msps) of either BPSK or QPSK symbols is unchanged, and so it may be expected that the frequency spectrum of the transmitted signals will be similar to those of the classic PHY. However, the use of these symbols is significantly different.

To maintain compatibility with classic systems, packets with preambles transmitted at the lowest rate of 1 Mbps, using DBPSK modulation, must be supported. However, the new PHY adds the option to use short preambles with 2 Mbps DQPSK modulation to reduce the overhead imposed by the very slow long preamble on short high-rate packets. More importantly, the 802.11b PHY introduces two completely new approaches to encoding the incoming data onto the QPSK symbols: *complementary code keying* (CCK) and *packet binary convolutional coding* (PBCC). Each method may be used to support two new data rates, 5.5 and 11 Mbps. Both methods completely abandon the Barker sequence and conventional direct-sequence spreading.

Let us first examine CCK. CCK is a *block code*: chunks of symbols of fixed size are used as code words, and the subset of *allowed* code words is much smaller than the total *possible* set of code words. Errors are detected and corrected by comparing the received code word with the possible code words: if the received code word is not an allowed word but is close to an allowed word, one may with good confidence assume that the nearby allowed code word is in fact what was transmitted. Block codes are relatively easy to implement.

We look at the high-rate 11 Mbps coding in detail. In the particular code used in the 802.11b CCK-11 Mbps option, transmitted QPSK symbols are grouped into blocks of eight to form code words. Because each symbol carries 2 bits, there are $4^8 = 65,536$ possible code words. Of this large domain, only 256 code words, corresponding to the 8 input bits that define the chosen CCK block are allowed. They are found in the following fashion, shown schematically in Figure 3-12. (Figure 3-12 is headed "even-numbered symbol" because alternate code words have slightly different phase definitions, important for controlling the spectrum of the output but confusing when one is trying to figure out how the scheme works.) The input bits are grouped in pairs (dibits). Each dibit defines one of four possible values of an intermediate phase φ. The intermediate phases are then added up (modulo-2π) as shown in the chart to determine each QPSK symbol c of the transmitted code word. Thus, the last symbol c7 shares the phase of $\varphi1$ (and is used as the phase reference for the remainder of the code word).

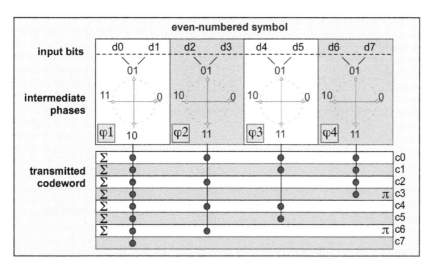

Figure 3-12: CCK-11 Code Word Encoding Scheme (Even-Numbered Symbols)

The phase of c4 is $(\varphi1 + \varphi2 + \varphi3)$. Symbols c3 and c6 have π radians added to the argument (i.e., they are multiplied by $\varphi1$).

This procedure is a bit confusing to describe but easy to implement and ensures that the resulting allowed code words are uniformly distributed among all the possible code words. Recall that we defined the distance between two QPSK symbols by reference to the phase-amplitude plane; for a normalized amplitude of 1, two nearest neighbors are $2/\sqrt{2}$ apart (Figure 2-20, Table 2-1); if we square the distance we find nearest neighbors differ by 2 in units of the *symbol energy*, E_s. We can similarly define the difference between code words by adding up the differences between the individual QPSK symbols, after squaring to make the terms positive definite (> 0). Having done so, we find that CCK-11 code words have their nearest neighbors at a squared distance of 8. This means that no possible single-chip QPSK error could turn one allowed code word into another, so single-chip errors can always be detected and corrected. In RF terms, the bit error rate is reduced for a fixed (S/N), or equivalently the (S/N) can be increased for the same error rate. The change in (S/N) is known as *coding gain*. The coding gain of CCK is about 2 dB. A slight variant of the above scheme is used to deliver 5.5 Mbps.

The standard also provides for a separate and distinct method of achieving the same 5.5- and 11-Mbps rates: PBCC. PBCC is based on a *convolutional code* (Figure 3-13). The coding is performed using a series of shift registers, shown here as z^1 through z^6. At each step, a new input bit is presented to the coder and all bits are shifted to the right one step. The "+" symbols represent modulo-2 additions that produce the output bits y_0 and y_1 from the input and the stored bits. The coder can be regarded as a machine with $2^6 = 64$ possible states. In any given state of the coder, two transitions to the next state are possible (which occurs being determined by the new input bit) and two of the four possible outputs are allowed (again determined by the input). Thus, the sequence

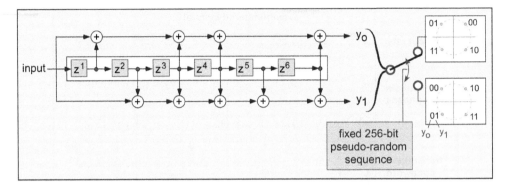

Figure 3-13: PBCC-11 Convolutional Code and Cover Sequence

of allowed outputs is highly constrained relative to the sequence of possible outputs. This code is known as a *rate 1/2* code, because each input bit produces two output bits.

The output bits are mapped onto the next QPSK symbol in two possible ways (rotated 90 degrees with respect to each other) according to a 256-bit fixed sequence. Because there are 11 megachips per second and each chip carries one input data bit coded into two y's, the net data rate is 11 Mbps. The 256-bit sequence helps to remove any periodicity in the output signal and thus helps smooth the spectrum of the output. The relatively long period of the sequence means that any other interfering transmission is unlikely to be synchronized with it and will appear as noise rather than as valid code words, making it easier to reject the interference.

Convolutional codes are usually decoded with the aid of a *Viterbi trellis* decoder. (Trellis algorithms can also be used to decode block codes.) The trellis tracks all the possible state transitions of the code and chooses the trajectory with the lowest total error measure. It would seem at first glance that such a procedure would lead to an exponentially growing mess as the number of possible trajectories doubles with each additional bit received, but by pruning out the worst choices at each stage of the trellis, the complexity of the algorithm is reduced to something manageable. However, implementation is still rather more complex than in the case of a block code like CCK.

The PBCC-11 code has slightly better performance than the corresponding CCK code: about 3.5 additional dB of coding gain (Figure 3-14). However, the computational complexity of decoding the convolutional code is about 3.5 times larger than the CCK code. To the author's knowledge, the PBCC variant has been implemented only in a few products from D-Link and US Robotics and has enjoyed very little commercial success. Commercial politics may have played a significant role in this history, as Texas Instruments enjoyed certain intellectual-property rights in PBCC that may have made other vendors reluctant to adopt it. It is also important to note that in most applications of a WLAN, performance is secondary to price and convenience; the modest advantage

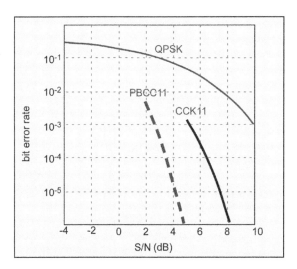

Figure 3-14: Error Rate of Uncoded QPSK, PBCC11, and CCK 11 vs. (S/N) (After Heegard et al., in Bing 2002)

in link budget of PBCC was not enough to produce a commercial advantage in the marketplace.

3.6. 802.11a PHY

The 802.11a PHY is a radical departure from the approaches above. The first major change is the use of the Unlicensed National Information Infrastructure (UNII) band at 5.15–5.825 GHz instead of the 2.4-GHz ISM band. At the time the PHY was promulgated, 5-GHz radios implemented in standard silicon processing were not widely available, so this choice implied relatively high costs for hardware. The motivation for the change was the realization that a successful commercial implementation of 802.11 classic and/or 802.11b devices, combined with the other proposed and existing occupants of the 2.4-GHz band (Bluetooth devices, cordless telephones, and microwave ovens, among others), would eventually result in serious interference problems in this band. The UNII band was (and is) relatively unoccupied and provides a considerable expansion in available bandwidth: in the United States, 300 MHz at the time versus 80 MHz available at ISM.

Rules for operating in the UNII band have changed significantly in the United States since the promulgation of the 802.11a standard: in November 2003, the FCC added 255 MHz to the available spectrum and imposed additional restrictions on operation in the existing 5.250- to 5.350-GHz band. The original and modified U.S. assignments are depicted in Figure 3-15. The allowed bands at the time of the standard consisted of a lower, middle, and upper band, each with 100 MHz of bandwidth. The lower band was dedicated to indoor use only and limited to 40-mW output power. The middle

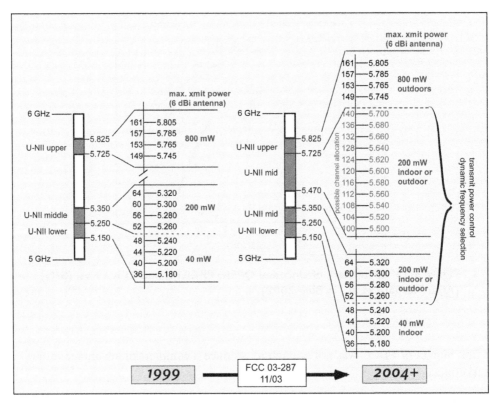

Figure 3-15: UNII Band in the United States, Then and Now

band allowed dual use, and the upper band was targeted to outdoor uses with a much higher allowed output of 800 mW.

The FCC's changes in late 2003 added an additional band, almost as big as what had been available, for dual use at 200 mW. Use of this band, and retroactively of the old UNII mid-band, is only allowed if devices implement transmit power control and dynamic frequency selection. The former reduces the power of each transmitting device to the minimum needed to achieve a reliable link, reducing overall interference from a community of users. The latter causes devices to change their operating frequency to an unoccupied channel when possible. These changes in band definitions and usage requirements improve consistency between U.S. and European (ETSI) requirements in the 5-GHz band and will presumably increase the market for compliant devices. The 802.11h standard, approved in 2003 to allow compliant devices that meet European standards, adds several capabilities relevant to power control and channel management. Clients inform access points about their power and channel capabilities; quiet periods are added when no station transmits to enable monitoring of the channel for interference, and access points can use the point coordination function interframe space to capture the medium to coordinate switching channels when interference is detected.

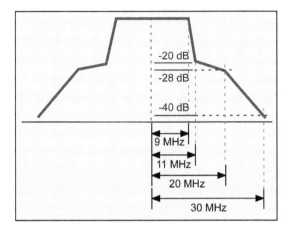

Figure 3-16: 802.11a Transmit Spectral Mask

Devices that are compliant to 802.11h should also meet the new FCC standards for UNII operation in the United States.

The spectral mask limiting the bandwidth of the transmitted signal is shown in Figure 3-16. Note that although the mask is somewhat more complex and apparently narrower than the old 802.11 mask (Figure 3-8), in practice both signals end up being around 16-MHz wide at 10 dB down from the maximum intensity at the center of the transmitted spectrum. This design choice was made to allow the same analog-to-digital conversion hardware to be used for the baseband in either standard, which is beneficial in those cases where a single baseband/MAC design is to serve both 802.11a and 802.11b radio chips. This important fact allowed the 802.11a PHY to be kidnapped and transported whole into the ISM band in the 802.11g standard, as we discuss in section 3.7.

Nonoverlapping channels in the old band definition, and a possible channel arrangement in the new bands, are shown in Figure 3-17. The assumed channel shape is based on the spectral mask in Figure 3-16. We can see that the old definition had room for as many as 12 nonoverlapping channels, though operation at the high band at full power in the lowest and highest channels would result in out-of-band emissions exceeding FCC allowances, so that in practice fewer channels might be used. In the new band assignment, there is room for up to 19 nonoverlapping channels in the lower and middle bands alone. Recall that the ISM band allows only three nonoverlapping 802.11 channels (Figure 3-9).

The availability of more than three channels is a significant benefit for implementing networks that are intended to provide complete coverage to a large contiguous area. With four nonoverlapping channels, one can construct a network of individual access points according to a *frequency plan* that ensures that access points on the same channel are separated by at least four cell radii; with seven channels to work with, a minimum

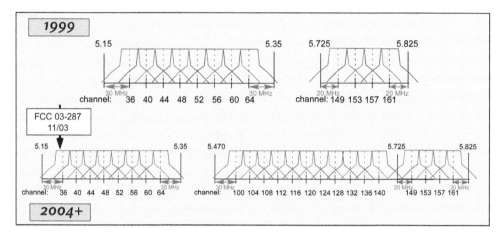

Figure 3-17: Possible Nonoverlapping Channel Assignments in Old and New UNII Bands

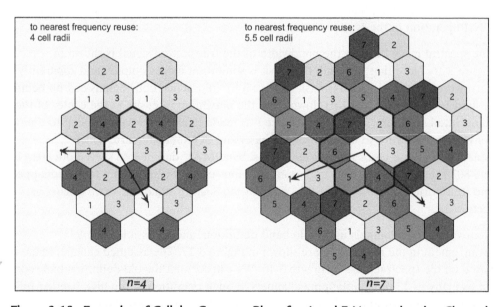

Figure 3-18: Examples of Cellular Coverage Plans for 4 and 7 Nonoverlapping Channels

spacing of 5.5 radii is achieved (Figure 3-18). Thus, interference between adjacent cells is minimized, and each cell is able to provide full capacity to its occupants. With the large number of independent channels now available, individual access points can be allocated more than one channel each to increase capacity in heavily used cells while still maintaining minimal interference.

Coding in 802.11a is somewhat similar to the PBCC option of the 802.11b PHY: after scrambling, incoming bits are processed by rate 1/2 convolutional encoder as shown in

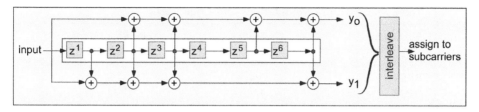

Figure 3-19: 802.11a Convolutional Encoder

Figure 3-19. Like the PBCC encoder, the 802.11a code has a six-stage shift register, though the sequence of taps that define the output bits (formally, the *generating polynomials* of the code) have been changed. Codes with rates of 3/4 and 2/3 (fewer output bits per input bit and thus higher throughput of useful data) are created by *puncturing* the output of the rate 1/2 code, that is, some of the output bits are simply ignored. Such a procedure is roughly equivalent to intentionally introducing some bit errors in the data stream, though it is easier to correct for because the location of the missing bits is known; if done cleverly, it allows for higher throughput with little performance loss.

The use of the resulting output bits is significantly different from PBCC, because instead of single-carrier modulation, the 802.11a standard uses orthogonal frequency-division multiplexing (OFDM), a technique introduced in Chapter 2, section 3.2. A data symbol is composed of 48 subcarriers, which carry data, and 4 additional *pilot* subcarriers, which transmit a known pseudo-random sequence to assist the receiver in maintaining synchronization with the transmitter. The subcarriers are assigned to 64 frequency slots, separated by 312.5 KHz; the slots at the band edges and the slot in the center of the band (i.e., at the carrier frequency) are not used. If all the subcarriers were occupied, the signal would be approximately 64×312.5 KHz = 20 MHz wide (63 spacings between subcarriers + half of the width of a subcarrier on each side); the actual (ideal, undistorted) signal is about 16.6 MHz wide, thus fitting nicely within the spectral mask in Figure 3-16.

Recall from Chapter 2 that to use OFDM, the time over which a symbol is integrated must be an integer number of cycles for all the subcarriers in order that orthogonality is maintained. The native symbol time of the 802.11a OFDM symbol is thus one cycle of the lowest frequency subcarrier, or 3.2 μsec. A cyclic prefix is appended to the symbol, as we described in Chapter 2, to allow for a guard interval that eliminates the effects of multipath. The cyclic prefix is 0.8 μsec long (about 20% of the total symbol), so that the total symbol length is 4 μsec. Therefore, the OFDM symbol rate is 250 Ksps. Each symbol contains 48 active subcarriers, so this is equivalent to sending 12 million single-carrier symbols per second.

To allow for varying conditions, many different combinations of code rate and subcarrier modulation are allowed. These are shown in Table 3-2. Recall from Chapter 2

Table 3-2: 802.11a Modulations and Code Rates

Modulation	Code Rate	Data Rate (Mbps)
BPSK	1/2	6
BPSK	3/4	9
QPSK	1/2	12
QPSK	3/4	18
16QAM	1/2	24
16QAM	3/4	36
64QAM	2/3	48
64QAM	3/4	54

that BPSK transports one bit per (subcarrier) symbol, QPSK two bits, and so on up to 64 quadrature-amplitude-modulation (QAM), which carriers 6 bits per symbol. The product of the bits per symbol of the modulation and the rate of the convolutional code gives the number of bits each subcarrier contributes to each OFDM symbol; multiplication by 48 provides the number of bits transported per OFDM symbol.

Note that to avoid bursts of bit error from disruption of a few neighboring subcarriers, the output bits from the code are *interleaved*—that is, distributed in a pseudo-random fashion over the various subcarriers.

Several views of a simulated OFDM frame are shown in Figure 3-20. The frame consists of a preamble and simulated (random) data. The spectrum of the output signal is about 16-MHz wide, in good agreement with the simple estimate given above. Recall that a signal in time can be described by its in-phase and quadrature components, I and Q. The I and Q amplitudes for the frame are shown at the lower left. The frame begins with a set of simplified synchronization characters, which use only 12 subcarriers, and are readily visible at the left of the image. The remainder of the preamble and data frame appear to vary wildly in amplitude. This apparently random variation in the signal is also shown in the close-up of I and Q amplitudes over the time corresponding to a single OFDM symbol shown at the top right of the figure.

At the bottom right we show the *combined cumulative distribution function* for the signal power, averaged over the packet length, for a packet at 54 and 6 Mbps. The combined cumulative distribution function is a graphic display of the frequency with which the instantaneous power in the signal exceeds the average by a given value. The 54-Mbps packet instantaneous power is greater than 9 dB above the average around 0.1% of the time. The 6-Mbps packet displays an 8-dB enhancement in power with this frequency of occurrence. Although that may not seem like much, recall that the spectral mask (Figure 3-16) requires that the transmitted signal 20 MHz away from the carrier must be reduced by more than 28 dB. As we learn in Chapter 4, a high-power signal may become distorted, resulting in spectral components far from the carrier. If such components were comparable in size with the main signal and occurred only 1% of the time, their intensity would be 20 dB below the main signal. Infrequent power peaks contribute significantly to the output spectrum when distortion is present. This problem of a high ratio of peak power to average power represents one of the important disadvantages of using OFDM modulations, because it forces transmitters to reduce their average output power and requires higher precision and linearity from receivers than is the case for simpler modulations.

Recall that the OFDM symbol is constructed using an inverse fast Fourier transform (FFT), with the reverse operation being performed on the received signal to recover the subcarriers. In the case of 802.11a, a 64-entry FFT involving some hundreds of arithmetic operations must be performed at fairly high precision every 4 μsec, followed by a trellis decode of the resulting 48 data symbols. The rapid progress of digital integrated circuit scaling has permitted such a relatively complex computational system to be implemented inexpensively in standard silicon complementary metal-oxide semiconductor (CMOS) circuitry, a remarkable achievement.

3.7. 802.11g PHY

The great disadvantage of the 802.11a PHY is its incompatibility with the older classic and 802.11b installed base. To overcome this obstacle while preserving the attractive high peak data rates of the 802.11a standard, task group "G" provided two enhancements to the Wi-Fi PHY. The first variant is essentially the transfer of the 802.11a OFDM physical layer *in toto* to the ISM band. This trick is possible because the bandwidth of the 802.11a symbols (Figure 3-20) is about 16 MHz, just as the classic symbols were.

The problem presented by this radical reinvention is that older systems have no ability to receive and interpret the complex OFDM symbols and simply see them as noise. Because the CSMA/CA MAC layer is heavily dependent on all stations being able to hear (at least) the access point, invisible preambles are a serious obstacle: receiving stations cannot set their NAV to reserve the medium (Figure 3-6) if they can't read the packet preambles. In order for "g" stations to coexist with older "b" stations, several options are possible. First, in the presence of mixed traffic, "g" stations can use the RTS/CTS approach to reserve the medium using 802.11b packets and then use the reserved time to send faster 802.11g packets. This approach is simple but adds overhead to the "g" packet exchanges. Alternately, a station can send a CTS packet with itself as the destination: this approach, known as a CTS-to-self, is obviously more appropriate for an access point, which expects that every associated station can hear it, than a client.

A "g" station can also use a mixed frame format, in which the preamble is sent in a conventional 802.11b form and then the appended data are transferred as OFDM. In this case there is additional overhead because of the low data rate of the older preamble

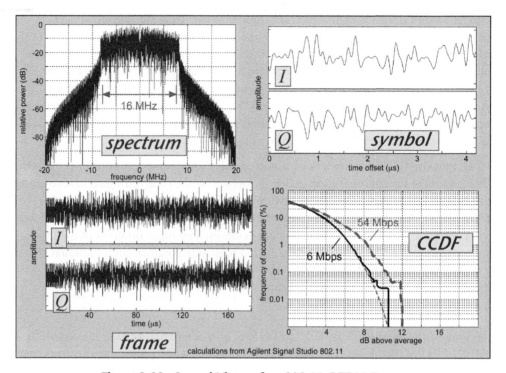

Figure 3-20: Several Views of an 802.11 OFDM Frame

but no additional packet exchange. The final chip of the preamble is used as the phase reference for the first OFDM symbol.

Finally, "g" also defines a PHY using an extension of the PBCC coding system; in the highest-rate variant, 8PSK is substituted for QPSK modulation. It is perhaps too early in the history of 802.11g to draw confident conclusions about its evolution, but in view of the limited success of PBCC in 802.11b deployment, and successful if limited deployment of OFDM-based 802.11a chipsets by major manufacturers, it seems plausible that the OFDM variant will dominate 802.11g devices.

Trade-offs are necessarily encountered in the enhancement of the PHY. Higher data rates use higher modulation states of the subcarriers and thus require better (S/N) than the old QPSK chips. This implies that higher rate communications will have a shorter range than the lower rates. The use of OFDM signals gives rise to higher peak-to-average power ratios, allowing less transmission power from the same amplifiers to avoid distortion. The benefit of coexistence is achieved at the cost of reduced actual data rates in the presence of legacy stations due to the overhead of any of the coexistence protocols discussed above. These problems are fairly minor bumps in the road, however: it seems likely that 802.11g products will dominate the WLAN marketplace until crowding in the ISM band forces wider adoption of 802.11a/dual-band devices.

3.8. 802.11 (In)Security

Most participants in wireless networking, or indeed networking of any kind, are at least vaguely aware that 802.11 presents security problems. One can hardly have a discussion of 802.11 without dealing with encryption and security; however, as whole books have already been written on this subject, the treatment here will be quite cursory.

We must first put the problem in its wider context. Most enterprise LANs have minimal internal authentication and no security, because it is presumed that physical security of the Ethernet ports substitutes for systemic provisions of this nature (though those hardy souls charged with administering networks at universities may find such presumptions laughable at best). Access to the network from the outside, however, is typically limited by a firewall that controls the type and destination of packets allowed to enter and exit the local network. Remote users who have a need to use resources from an enterprise network may use a *virtual private network* (VPN), which is simply a logical link formed between two clients, one at the remote site and the other at the enterprise network, with all traffic between them being encrypted so as to defeat all but the most determined eavesdropper. Traffic on the Internet is similarly completely open to interception at every intermediate router; however, reasonably secure communications over the Internet can be managed with the aid of the Secure Sockets Layer (SSL), another encrypted link between the user of a web site and the site server.

It was apparent to the developers of wireless data devices that the wireless medium is more subject to interception and eavesdropping than the wired medium, though they were also aware of the many vulnerabilities in existing wired networks. They therefore attempted to produce a security mechanism that would emulate the modest privacy level obtained with physically controlled Ethernet wired ports in a fashion that was simple to implement and would not incur so much overhead as to reduce the WLAN data rate by an objectionable amount.

The resulting *Wired Equivalent Privacy* (WEP) security system was made optional in the classic standard but was widely implemented and extended. The system is a *symmetric key* system, in which the same key is used to encrypt and decrypt the data, in contrast to asymmetric systems in which a public key is used to encrypt data and a separate private key is used to decrypt it. Public key systems are much more difficult to penetrate but very computation-intensive and are used for the secure exchange of symmetric keys but not for packet-by-packet encryption. The basic idea of the encryption approach is to use the key to create a pseudo-random binary string (the *cipher stream*) of the same length as the message to be encrypted. The data bits (the *plaintext*) can then be added modulo-2 (which is the same as a bit-by-bit exclusive-or [XOR]) to the cipher stream to create the encrypted data or *cipher text*, which when added again to the same cipher stream will recover the original data (plaintext). The general scheme is shown in Figure 3-21.

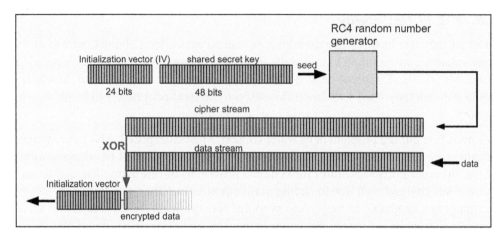

Figure 3-21: Simplified Schematic of WEP Encryption

A simple way to achieve such a result is to use the key as the seed value for a random number generator algorithm. The problem with this naive implementation is that every instance of the (pseudo)random sequence would then be identical; it wouldn't be very hard for an eavesdropper to reconstruct the cipher stream once a large number of packets, many containing known plaintext such as the contents of standard Internet packets, had been intercepted. Therefore, to avoid using the same cipher stream for every packet while also avoiding the complexity of exchanging a new key between the sender and receiver each time a packet needs to be transmitted, the WEP algorithm uses an *initialization vector* (IV): 24 additional bits that are prepended to the 48-bit secret key to form the seed for the random number generator. In this fashion, each new packet gets a new seed for the random number generator and thus a unique cipher stream.

In addition to security from interception, a secure communications system should provide some means of *authentication*: each side should have some confidence that they are in fact communicating with the machine (and ultimately the person) that is intended. The 802.11 standard appropriates WEP to support authentication by allowing an access point to challenge a potential client with a plaintext packet, which the client then returns encrypted with the shared WEP key. If the decrypted packet matches the original plaintext, the client is presumed to have knowledge of the shared key and thus be privileged to make use of the resources of the access point.

As has been so vividly shown with the public release of the remarkable means by which British and American researchers cracked the German Enigma code during World War II, attacks on cryptographic systems are rarely frontal but instead exploit weaknesses in the system and implementation. WEP has a number of weaknesses and limitations, some of which were quickly apparent and others more subtle and only revealed after significant work by outside researchers.

The first weakness is in the definition of the IV. The IV is necessarily sent in the clear because it is needed to decrypt the message content. As we noted above, to get decent security out of a random-number-generator scheme, it is necessary to avoid reusing the seed value, because this will generate a reused cipher stream. However, the IV space is only 24 bits. Furthermore, there is no requirement in the specification about how the IV is to be treated between successive packets; any IV value may be used and must be accepted by the receiving station. Many implementations simply incremented the IV value by 1 for each packet sent. In such a network, the IV is bound to be reused after $2^{24} = 16.8$ million packets. A heavily loaded network would deliver that much traffic in a couple of hours. Over the course of a week or two of eavesdropping, a huge number of packets with identical IVs and thus identical cipher streams could be collected and analyzed.

Once any cipher stream is obtained, it can be reused even in the absence of knowledge of the shared key, because any IV must be accepted. Any packets sent with the same IV can be decrypted (at least to the length of the known cipher stream); further, a packet of the same length with the intruder's data can be encrypted, prepended with the same IV, and injected into the system.

The authentication system is another weakness. If an eavesdropper can hear the challenge and the response, the XOR of the challenge (plaintext) and corresponding response (cipher text) reveals the cipher stream. In this fashion one can accumulate cipher streams for packet injection or other attacks.

Obviously, it is even better to obtain the shared key and thus encrypt and decrypt freely than to extract a few cipher streams. Attacks to find the shared key can exploit the fact that the key is defined as a bit stream, but human users can't remember bit streams very well and tend to prefer words or streams of recognizable characters. Many 802.11 management systems allow the entry of a text password, which is processed (hashed) into a 48-bit shared key. Early hashing algorithms did not use the whole key space—that is, not all 48-bit numbers were possible results—so that the effective size of the keys to be searched was only 21 bits, which is about 2 million possibilities. (Some early cards even converted the ASCII values of letters directly into a key, resulting in a greatly reduced key space that is also easily guessed.) A very simple brute force recitation of all possible passwords is then possible, given that one can automate the recognition of a successful decryption. Because certain types of packets (standard handshakes for IPs) are present in almost any data stream, the latter is straightforward.

Because IVs can be reused, if there isn't enough traffic on a network to crack it, an attacker can always introduce more packets once he or she has acquired a few cipher streams, thus generating additional traffic in response.

Certain attacks are fixed or rendered much less effective by simply increasing the length of the shared key and the IV. Many vendors have implemented WEP variants with longer keys, though these are not specified in the standard and thus interoperability is questionable. For example, the larger space for keys, combined with improved hashing algorithms, renders the brute force password guessing attack ineffective on 128-bit WEP key systems.

The widely publicized "cracking" of the WEP algorithm by Fluhrer, Mantin, and Shamir (FMS) was based on a demonstration that the particular random number generator, RC4, used in WEP is not completely random. RC4 is a proprietary algorithm that is not publicly released or described in detail in the standard. It has been reverse engineered and shown to be based on fairly simple swaps of small segments of memory. Fluhrer and colleagues showed that certain IV values "leak" some information about the key in their cipher streams. To benefit from this knowledge, an attacker needs to know a few bytes of the plaintext, but because of the fixed nature of formats for many common packets, this is not difficult. The number of weak IVs needed to extract the shared key is modest, and the frequency of weak IVs increases for longer key streams, so that for a network selecting its IVs at random, the time required to crack, for example, a 104-bit key is only roughly twice that needed for a 40-bit key. Long keys provide essentially no added security against the FMS attack. Shortly after Fluhrer et al. became available, Stubblefield, Ioannidis, and Rubin implemented the algorithm and showed that keys could be extracted after interception of on the order of 1,000,000 packets.

Finally, we should note that there is no provision whatsoever in the standard to allow a client to authenticate the access point to which it is associating, so that an attacker with a shared key can set up a spoof access point and intercept traffic intended for a legitimate network.

All the above attacks depend on another weakness in the standard: no provision was made for how the shared secret keys ought to be exchanged. In practice, this meant that in most systems the secret keys are manually entered by the user. Needless to say, users are not eager to change their keys every week, to say nothing of every day, and manual reentry of shared keys more than once an hour is impractical. Further, manual key exchange over a large community of users is an administrative nightmare. As a consequence, it is all too likely that WEP shared secret keys will remain fixed for weeks or months at a time, making the aforesaid attacks relatively easy to carry out.

Some of the problems noted above can be solved completely within the existing standards. Weak IVs can be avoided proactively, and some vendors have already implemented filters that ensure that weak IVs are never used by their stations. If all stations on a network avoid weak IVs, the FMS attack cannot be carried out. Weak hashing algorithms for short keys can be avoided by entering keys directly as hexadecimal numbers. In home networks with few users and relatively light traffic, manual key exchange on a weekly or even monthly basis will most likely raise the effort involved in a successful attack above any gains that can be realized.

The simple measures cited above can hardly be regarded as adequate for sensitive industrial or governmental data. In 2003, products conforming to a preliminary standard promulgated by the WECA industrial consortium, known as Wi-Fi Protected Access (WPA), became available. WPA is a partially backward-compatible enhancement of WEP. WPA uses the IEEE 802.1x standard as a framework for authentication of users and access points; within 802.1x various authentication algorithms with varying complexity and security can be used.

WPA uses a variant of the *Temporal Key Integrity Protocol* created by Cisco to improve security of the packet encryption process. An initial 128-bit encryption key, provided in a presumably secure fashion using the 802.1x authentication process, is XOR'd with the sending station's MAC address to provide a unique known key for each client station. This unique intermediate key is mixed with a 48-bit sequence number to create a per-packet key, which is then handed over to the WEP encryption engine as if it were composed of a 24-bit IV and 104-bit WEP key. The sequence number is required to increment on each packet, and any out-of-sequence packets are dropped, preventing IV-reuse attacks. The 48-bit sequence space means that a sequence number will not be reused on the order of 1000 years at today's data rates, so there are no repeated cipher streams to intercept. A sophisticated integrity checking mechanism is also included to guard against an attacker injecting slight variations of valid transmitted packets.

WPA addresses all the currently known attacks on WEP, though total security also depends on proper selection and implementation of algorithms within the 802.1x authentication process. It is certainly secure enough for home networks and for most enterprise/industrial implementations. At the time of this writing (mid-2004), the 802.11i task group has approved an enhanced standard based on the Advanced Encryption Standard rather than the WEP RC4 algorithm, which will provide an adequate level of security for most uses.

Enterprises that have serious concerns about sensitive data can also implement end to end security through the use of VPNs and SSL web security. The advantage of this security approach is that protection against eavesdropping at any stage of the communications process, not just the wireless link, is obtained. However, VPNs are complex to set up and maintain and may not support roaming of the wireless device from one access point to another, particularly if the IP address of the mobile device undergoes a change due to the roaming process.

4. HiperLAN and HiperLAN 2

Some of the researchers who participated in defining the 802.11a standard were also active in similar European efforts: the resulting standards are known as HiperLAN. The HiperLAN 2 physical layer is very similar to that used in 802.11a: it is an OFDM system operating in the 5-GHz band, using an almost identical set of modulations and code rates to support nearly the same set of data rates. The physical layer is also compliant with ETSI requirements for dynamic frequency selection and power control.

However, the MAC layer is quite different. Instead of being based on the Ethernet standard, it is based on a totally different traffic approach, *asynchronous transfer mode* (ATM). ATM networking was developed by telephone service providers in the 1980s and 1990s in an attempt to provide a network that would support efficient transport of video, data, and voice, with control over the quality of service and traffic capacity assigned to various users. Native ATM is based on fixed 53-byte packets and virtual connections rather than the variable packets and global addressing used in IP networking over Ethernet. The HiperLAN MAC was constructed to provide a smooth interface to an ATM data network, by using fixed time slots assigned to stations by an access point acting as the central controller of the network.

ATM was originally envisioned as extending all the way from the WAN to the desktop but in practice has seen very little commercial implementation beyond the data networks of large telecommunications service providers. Similarly, no commercial products based on HiperLAN have achieved significant distribution, and the author is not aware of any products in current distribution based on HiperLAN 2. It seems likely, particularly in view of recent FCC decisions bringing U.S. regulations into compliance with European regulations, that 802.11a products will achieve sufficient economies of scale to prevent the wide distribution of HiperLAN-based products in most applications. Although HiperLAN provides superior quality-of-service controls to 802.11, many major vendors for the key quality of service-sensitive service, video delivery, have transferred their attention to the ultrawideband (UWB) standardization activities in the 802.15.3a, discussed in the next section.

5. From LANs to PANs

WLANs are an attempt to provide a wireless extension of a computer network service. WPANs are intended for a somewhat different purpose and differ from their WLAN cousins as a consequence. The basic purpose of a WPAN is to replace a cable, not necessarily to integrate into a network attached to that cable. WPANs seek to replace serial and parallel printer cables, universal serial bus connections, and simple analog cables for speakers, microphones, and headphones with a single wireless digital data connection. Personal areas vary from person to person and culture to culture but are usually much smaller than a building, so PANs do not need to have the same range expected of WLANs. The cables being replaced are generally very inexpensive, so even more than WLANs, WPAN products must be inexpensive to build and use. Many WPAN products are likely to be included in portable battery-powered devices and should use power sparingly.

In this section we discuss three differing WPAN standards activities, all (at least today) taking place under the aegis of the 802.15 working group of IEEE. As is often the case, with the passing of time the definition of a category is stretched: what we've defined as a

PAN for replacing low-rate cables is certainly evolving to replace higher rate cables and may be becoming a path to the universal home data/entertainment network that many companies have already unsuccessfully sought to popularize.

5.1. Bluetooth: Skip the Danegeld, Keep the Dane

The first WPAN effort to achieve major visibility was the Bluetooth Special Interest Group, initiated in 1998 by Ericsson with support from Intel, IBM, Nokia, and Toshiba. The Bluetooth trademark is owned by Ericsson and licensed to users. Although the group clearly displayed more marketing savvy than all the IEEE working groups put together in its choice of a memorable moniker—the name refers to Harald "Bluetooth" Blätand, king of Denmark from about AD 940 to 985—the separation of this activity from the IEEE helped produce a standard that was not compatible in any way with 802.11 networks. Given the likelihood of collocated devices using the two standards, this is unfortunate. The Bluetooth standard is now also endorsed by the IEEE as 802.15.1.

To make life simple for users, the Bluetooth PAN was intended to operate with essentially no user configuration. Thus, Bluetooth networks are based on *ad hoc* discovery of neighboring devices and formation of networks. Simple modulation, low transmit power, low data rate, and modest sensitivity requirements all contribute to a standard that can be inexpensively implemented with minimal power requirements. Frequency hopping was used to allow multiple independent collocated networks. The network is based on synchronized time slots allocated by the master device, so that resources could be provided on a regular deterministic basis for voice and other time-sensitive traffic.

The architecture of Bluetooth networks is based on the *piconet* (Figure 3-22). In terminology also perhaps harkening back to the time of King Blätand, a piconet consists of one device that acts as the *master* and a number of other devices that are *slaves*. Each piconet is identified by the FH pattern it uses. The hopping clock is set by the master device, and slaves must remain synchronized to the master to communicate. Provisions are made to permit devices both to discover nearby piconets and spontaneously form their own.

Only seven slave devices can be active at any given time. Some of the active devices can be placed in a power-saving *sniff* state, in which they only listen for their packets occasionally, or *hold* states, in which they do not actively participate in the piconet for some period of time before checking back in. Additional members of the piconet may be *parked*: in the parked state, slaves maintain synchronization with the master clock by listening to beacons periodically but do not otherwise participate until instructed to become active.

Piconets can physically overlap with one another with minimal interference, as they use distinct hopping schemes. A slave in one piconet could be a master in another. By having some devices participate in multiple piconets, they can be linked together to

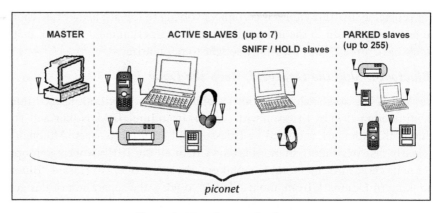

Figure 3-22: Bluetooth Piconet

form *scatternets*. Devices within a piconet can exchange information on their capabilities, so that the user may use printers, cameras, and other peripherals without the need for manual configuration. Devices use *inquiry* and *paging* modes to discover their neighbors and form new piconets.

The Bluetooth/802.15.1 PHY layer operates in the 2.4-GHz ISM band, which is available at least partially throughout most of the world. The band in the United States is divided into 79 1-MHz channels (Figure 3-23), with modest guard bands at each end to minimize out-of-band emissions.

During normal operation, all members of the piconet hop from one frequency to another 1600 times per second. The dwell time of 625 μsec provides for 625 bit times between hops. When a device is in inquiry or paging mode, it hops twice as fast to reduce the time needed to find other devices.

The signal is modulated using *Gaussian minimum-shift keying*, GMSK (also known as Gaussian frequency-shift keying). GMSK can be regarded as either a frequency modulation or phase modulation technique. In each bit time, frequency is either kept constant or changed so as to shift the phase of the signal by π by the end of the period T (Figure 3-24). Because the amplitude remains constant during this process, the peak-to-average ratio of a GMSK signal is essentially 1, so such signals place modest

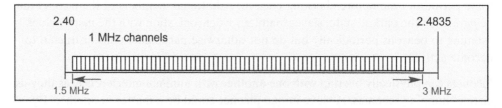

Figure 3-23: Bluetooth Channelization

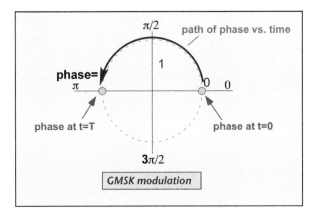

Figure 3-24: Gaussian Minimum-Shift Keying

requirements on transmitter and receiver linearity. GMSK is similar to BPSK, but BPSK systems have no such constraint on the trajectory taken between phase points and thus may have large amplitude variations during the transition from one symbol to another. Each GMSK symbol carries only a single bit, and thus 1 Msps = 1 Mbps of data transfer capacity.

Bluetooth radios come in three classes: class 1 can transmit up to 20 dBm (100 mW), class 2 is limited to 4 dBm (2.5 mW), and class 3 devices, the most common, operate at 1 mW (0 dBm). Bluetooth receivers are required to achieve a bit error rate of 0.1% for a received signal at −70 dBm: that's a raw bit error rate of about 60% for 600-bit packets. These specifications are rather less demanding than the corresponding 802.11 requirements: recall that an 802.11 receiver is required to achieve a sensitivity of −80 dBm at 2 Mbps and typical transmit power is 30–100 mW. A link between two compliant class 3 devices has 30 dB less path loss available than a link between two 802.11 devices. The receiver noise and transmit power requirements are correspondingly undemanding: Bluetooth devices must be inexpensive to build.

An example of a link budget calculation is shown in Figure 3-25. We've assumed a transmit power of 1 dBm and modest antenna directivity (1 dB relative to an isotropic antenna [dBi]) and efficiency (80%) as the antennas are constrained to be quite small to fit into portable/handheld devices. At 3 m the free-space path loss is about 51 dB; we assumed that if a person is in the way of the direct path, an additional 10 dB of obstructed loss is encountered. If we assume a modest 4 dB (S/N) is required by the GMSK modulation, we find that we have room for 39 dB of excess noise from the receiver while still providing 10 dB of link margin in the presence of an obstruction. This is a huge noise figure; practical Bluetooth devices can do much better. The Bluetooth specifications are appropriate for inexpensive devices meant to communicate at moderate rates over ranges of a few meters.

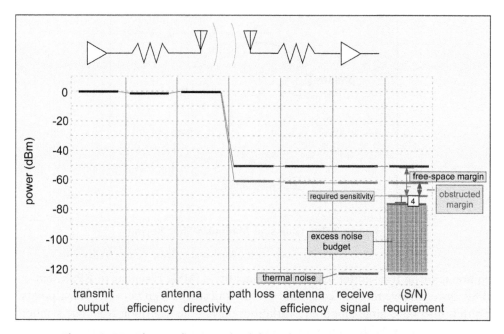

Figure 3-25: Bluetooth Example Link Budget: 10 m, 0 dBm TX Power

A few words about the security of a Bluetooth link seem appropriate. Bluetooth links have two fundamental security advantages over 802.11 irrespective of any encryption technology. The first is a physical limit: a Bluetooth class 3 transmitter at 0 dBm is harder to intercept from afar than an 802.11 access point at 20-dBm transmit power. The second is related to the usage of the devices. Most 802.11 devices are connected to a local network of some kind, which provides many resources of interest to an attacker: servers containing possibly valuable data and a likely connection to the global Internet. Many Bluetooth devices may only provide access to a single data stream or a few neighbor devices. The attacker who intercepts data sent to a printer from a few meters away might just as well walk over to the printer and look at what's coming out. Finally, any given Bluetooth link is likely to be used less intensively and provides a lower peak data rate than an 802.11 link, so an attacker has a harder time gathering data to analyze.

That being said, Bluetooth provides security services that are in some ways superior to those in 802.11. Authentication and encryption are both provided and use distinct keys. The cipher stream generation uses the E0 sequence generator, which is at the present regarded as harder to attack than RC4. However, just like in 802.11, weaknesses are present due to the usage models. Keys for devices with a user interface are based on manually entered identifier (PIN) codes. Users won't remember long complex codes and don't do a very good job of choosing the codes they use at random, so links are likely to be vulnerable to brute force key-guessing attacks. Devices without a user interface have fixed PIN codes, obviously more vulnerable to attack. Devices that wish to communicate securely must exchange keys in a pairing operation; if an eavesdropper overhears this operation, they are in an improved position to attack the data.

Although there hasn't been as much publicity about cracking Bluetooth, as was the case with 802.11, it is prudent to use higher layer security (VPN or SSL, etc.) if large amounts of sensitive data are to be transferred over a Bluetooth link.

5.2. Enhanced PANs: 802.15.3

The 802.15.3 task group was created to enhance the capabilities of 802.15.1 (Bluetooth) while preserving the architecture, voice support, low cost of implementation, and range of the original. Data rates as high as 55 Mbps were targeted to allow applications such as streaming audio and video, printing high-resolution images, and transmitting presentations to a digital projector. Because the work was now under the aegis of the IEEE, the group sought to create a PHY with some compatibility with existing 802.11 devices; thus a symbol rate of 11 Msps was chosen, and the channels were to be compatible with the 5-MHz 802.11 channels. However, the group sought to use a slightly reduced bandwidth of 15 MHz, so that more nonoverlapping channels could be provided in the same area than 802.11 classic or b/g allow.

In July 2003, the 802.15.3 task group approved a new PHY layer standard that provides higher data rates while otherwise essentially using the Bluetooth MAC and protocol stack. The general approach is to use more sophisticated coding than was available a decade ago to allow more data to be transferred in less bandwidth. As with 802.11a and g, higher modulations are used to send more data per symbol; 32QAM uses the 16QAM constellation with the addition of partial rows and columns.

A novel coding approach, *trellis-coded modulation* (TCM), is used to achieve good performance at modest (S/N). In the conventional coding schemes we examined so far, each set of bits determines a modulation symbol (for example, a dibit determines a QPSK point) and the received voltage is first converted back to a digital value (an estimate of what the dibit was) and then used to figure out what the transmitted code word probably was. Coding is a separate and distinct operation from modulation. TCM mixes coding and modulation operations by dividing the constellation into subsets, within each of which the distance between signal points is large. The choice of subset is then made using a convolutional code on some of the input bits, whereas the choice of

Table 3-3: 802.15.3 Data Rates

Modulation	Coding	Code Rate	Data Rate (Mbps)	Sensitivity (dBm)
QPSK	TCM	1/2	11	−82
DQPSK	None	–	22	−75
16QAM	TCM	3/4	33	−74
32QAM	TCM	4/5	44	−71
64QAM	TCM	2/3	55	−68

points within the subset is made using the remainder of the input bits uncoded. The subset points are widely separated and noise resistant, so uncoded selection works fine. The subsets are more difficult to distinguish from one another, but this choice is protected by the code, making it more robust to errors. The net result is typically around 3 dB of improved noise margin in the same bandwidth versus a conventional modulation.

The benefits of this approach are revealed in the spectral emission mask (Figure 3-26). The 802.15.3 signal is able to deliver the same data rate as 802.11g but fits completely within a 15-MHz window. The practical consequence is that one can fit five nonoverlapping channels into the U.S. ISM band, providing more flexibility in channel assignment and reducing interference. Some exposure to increased intersymbol interference and sensitivity to multipath may result from the narrower bandwidth, but because 802.15.3 is a PAN technology not intended to be used at long ranges, little performance limitation should result.

5.3. UWB PANs: A Progress Report

Although 55 Mbps may seem like a lot of bits, it is a marginal capacity for wireless transport of high-definition television signals (a technology that may finally reach mainstream deployment in this decade in conjunction with large-screen displays). Task group 802.15.3a was charged with achieving even higher data rates, at minimum 100 Mbps with a target of 400 Mbps, to support versatile short-range multimedia file transfer for homes and business applications.

To achieve such ambitious goals, the task group turned to a new approach, based again on recent FCC actions. In 2002 the FCC allowed a novel use of spectrum, *ultrawideband transmission*. UWB radios are allowed to transmit right on top of spectrum licensed for other uses (Figure 3-27). However, UWB radios operate under several restrictions

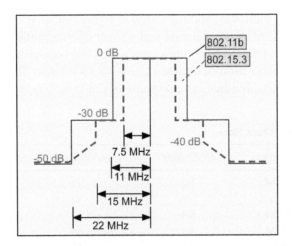

Figure 3-26: Comparison of 802.11b and 802.15.3 Emission Spectral Mask

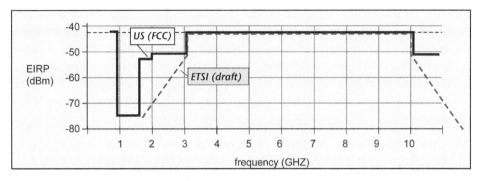

Figure 3-27: U.S. FCC and Proposed ETSI Limits for Ultrawideband Emitters (EIRP Is Equivalent Isotropic Radiated Power; see Chapter 5 for a Discussion of this Concept)

designed to minimize interference with legacy users of the spectrum. First, the absolute power emitted at any band is restricted to less than −40 dBm, which is comparable with the emissions allowed by nonintentional emitters like computers or personal digital assistants. The restrictions are more stringent in the 1- to 3-GHz region to protect cellular telephony and global position satellite navigation. Finally, like the original requirements on the ISM band, UWB users are required to spread their signals over at least 500 MHz within the allowed band.

The wide bandwidth available for a UWB signal means that the peak data rate can be very high indeed. However, the limited total power implies that the range of transmission is not very large. As with any transmission method, UWB range can be extended at the cost of reduced data rate by using more complex coding methods to allow a tiny signal to be extracted from the noise.

The resulting trade-off between range and rate is shown in Figure 3-28. Here a full band transmission is one that exploits the whole 7-GHz bandwidth available under the FCC

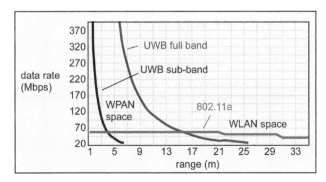

Figure 3-28: UWB Range vs. Data Rate, Shown for Both Full Band (7 GHz) and Subband (500 MHz) Approaches (After Gandolfo, Wireless Systems Design Conference, 2003)

specs, whereas a subband transmission uses only the minimum 500-MHz bandwidth. It is clear that a subband approach can deliver 100 Mbps at a few meters, appropriate to a PAN application. As the demanded range increases, even the full UWB band cannot provide enough coding gain to substitute for transmit power, and the performance of the UWB transmitter falls below that of a conventional 802.11a system. UWB systems do not replace conventional WLAN systems but have ideal characteristics for WPAN applications.

At the time of this writing, two distinct proposals are under consideration by the 802.15.3a task group as a WPAN standard. We first examine the direct-sequence CDMA proposal, based on 802.15–03 334/r3 due to Welborn et al., with some updates from /0137…r0. This proposal divides the available bandwidth into two chunks, a low band and a high band, avoiding the 5-GHz UNII region to minimize interference with existing and projected unlicensed products (Figure 3-29). The radio can operate using either the low band only, the high band only, or both bands, achieving data rates of 450, 900, or 1350 Mbps, respectively. Alternatively, the dual-band operating mode can be used to support *full-duplex* communications, in which the radio can transmit in one band and simultaneously receive in the other.

The transmissions use Hermitian pulses, the nomenclature referring to the pulse shape and consequent spectrum (Figure 3-30). The high band pulse is shown; a longer pulse is used for the low band.

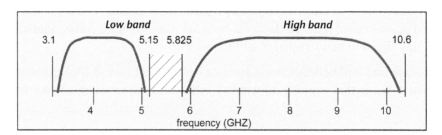

Figure 3-29: Pulse UWB Band Structure

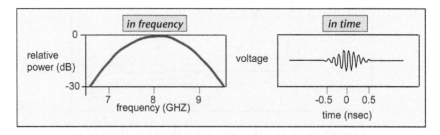

Figure 3-30: High-Band Modified Hermitian Pulse

The modulation techniques used are either BPSK or a variant of the biphase keying described in Chapter 2: *M-ary biorthogonal keying* (MBOK). In this scheme, the allowed symbols are specific sequences of positive and negative pulses; the sequences can be arranged in a hierarchy of mutually orthogonal symbols, in the sense that if they are multiplied together timewise and the sum taken (i.e., if they are correlated), the sum is 0 for distinct sequences. An example for $M = 4$ is shown in Figure 3-31. Each possible symbol is composed of four pulses, shown in cartoon form on the left. In the shorthand summary on the right, an inverted pulse is denoted "−" and a noninverted pulse is "+." Clearly, the first two symbols (00 and 01) are inversions of each other, as are the last two (10 and 11). Either symbol in the first group is orthogonal to either symbol in the second, for example, $00.10 = -1 + 1 - 1 + 1 = 0$, where we have assumed the product of a positive pulse and an inverted pulse to be -1.

The proposed radio architecture uses the MBOK symbols as chips in a CDMA system. Each data bit is multiplied by a 24- or 32-chip code. This scheme is similar to the Barker code used to multiply data bits in the 802.11 classic PHY, but unlike 802.11, in this case multiple codes are provided to enable simultaneous noninterfering use of the bands. Four independent 24-chip codes are provided, allowing four independent users of each band or eight total collocated piconets. A *rake receiver* can be used to separately add the contributions of sequences of chips arising from paths with different delays; a *decision feedback equalizer* is assigned to deal with interference in the data symbols resulting from multipath. Both convolution and block code options are provided; the codes may be *concatenated* for improved performance at the cost of computational complexity.

Link performance at a transmit-receive distance of 4 m and data rate of 200 Mbps is summarized in Table 3-4. Note that the transmit power, −10 dBm, is 10 dB less than the lowest Bluetooth output and about 25 dB less than a typical 802.11 transmitter. The required bit energy over noise, E_b/N_o, is comparable with what we have encountered in conventional BPSK or QPSK modulation. Allowing for a couple of decibels of "implementation loss" (antenna limitations, cable losses, etc.) and a noise figure of 6.6 dB, the link should still achieve a respectable −75 dBm sensitivity, allowing some margin at a few meters range.

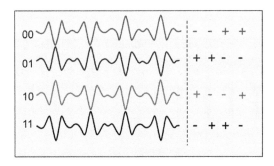

Figure 3-31: 4-Ary Biorthogonal Symbols

Table 3-4: Link Performance, DS-CDMA UWB at 200 Mbps

Parameter	Value
Data rate	200 Mbps
TX power	−10 dBm
Path loss	56 dB (4 m)
RX power	−66 dBm
Noise/bit	−91 dBm
RX noise figure	6.6 dB
Total noise	−84.4 dBm
Required E_b/N_o	6.8 dB
Implementation loss	2.5 dB
Link margin	8.7 dB
RX sensitivity	−75 dBm

The second current proposal is based on multiband OFDM, as described in IEEE 802.15–03/267r2, r6 Batra et al., with some updates from 802.15–04/0122r4 of March 2004. In this case, the available UWB bandwidth is partitioned into 528-MHz bands (recall that 500 MHz is the smallest bandwidth allowed by the FCC for an UWB application), as shown in Figure 3-32. The bands are assembled into five groups. Group 1 devices are expected to be introduced first, with more advanced group 2–5 devices providing improved performance. All devices are required to support group 1 bands; the higher bands are optional.

The implementation requires a 128-point inverse FFT to create subcarriers spanning slightly more than 500 MHz of bandwidth. To reduce the computational demands, QPSK modulation on each subcarrier is used, instead of the higher QAM modulations used in 802.11 a/g. Furthermore, for data rates less than 80 Mbps, the subcarrier amplitudes are chosen to be conjugate symmetric around the carrier. That is, the nth subcarrier at positive frequency (relative to the carrier) is the complex conjugate of the nth subcarrier at negative frequency. A signal composed of complex conjugate frequencies produces a pure real voltage (this is just a fancy way of reminding us that $e^{i\omega t} + e^{-i\omega t} = 2\cos(\omega t)$). This means that the transmitter and receiver don't need

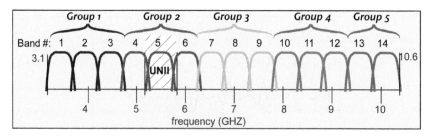

Figure 3-32: Band Definitions for OFDM UWB Proposal

separate branches to keep track of the in-phase and quadrature (I and Q) components of the signal and can be simplified.

Successive OFDM symbols are sent on different bands within a band group. Within each band group 1–4 there are four hopping patterns, each six symbols long, allowing for four collocated piconets. (Group 5 supports only two hopping patterns.) Thus, a total of 18 collocated piconets is possible.

The implementation parameters for the 200 Mbps are summarized in Table 3-5. A convolutional code with rate 5/8 (8 output bits for 5 input bits) is used. Of the 128 subcarriers, 22 are used as pilot tones and guard tones and 6 are not used at all. The resulting symbol is about 1/4 of a microsecond long. The cyclic prefix is no longer included, because this prefix adds a periodicity to the signal, giving rise to spectral lines which reduce the amount of power that can be transmitted. Instead, an equivalent amount of zero padding is added in frequency space (*zero-padded OFDM*). Each data symbol is spread by being sent twice, in separate frequency subbands. A brief guard interval is provided between symbols to allow the radio to change subbands. The use of separate subbands allows multiple collocated piconets.

In Table 3-6 we compare link performance for the two proposals for a data rate of 200 Mbps at 4 m. OFDM has a slight advantage in noise tolerance, giving it modestly improved sensitivity. However, as we noted in connection with the 802.11b PHY, small distinctions in radio performance are likely to be less important than cost and ease of use in determining commercial viability.

Recent modifications (March 2004) to the direct-sequence proposal have changed the MBOK signaling and proposed a common signaling mechanism between the multiband OFDM and direct sequence approaches. The selection process appears to

Table 3-5: Wideband OFDM Parameters, 200 Mbps Rate

Parameter	Value
Data rate	200 Mbps
Modulation constellation	OFDM/QPSK
FFT size	128 tones
Coding rate	R = 5/8
Spreading rate	2
Pilot/guard tones	22
Data tones	100
Information length	242.4 nsec
Padded prefix	60.6 nsec
Guard interval	9.5 nsec
Symbol length	312.5 ns
Channel bit rate	640 Mbps

Table 3-6: Link Performance, 200 Mbps

Parameter	Wideband OFDM	DS-CDMA
Information data rate	200 Mbps	200 Mbps
Average TX power	−10 dBm	−10 dBm
Total path loss	56 dB (4 m)	56 dB (4 m)
Average RX power	−66 dBm	−66 dBm
Noise power per bit	−91.0 dBm	−91 dBm
Complementary metal-oxide semiconductor RX noise figure	6.6 dB	6.6 dB
Total noise power	−84.4 dBm	−84.4 dBm
Required E_b/N_o	4.7 dB	6.8 dB
Implementation loss	2.5 dB	2.5 dB
Link margin	10.7 dB	8.7 dB
RX sensitivity level	−77.2 dBm	−75 dBm

be favoring the multiband OFDM proposal, but it has not yet been approved. At the time of this writing, it is not known whether either of these proposals will be approved by the IEEE as standards or whether the resulting standard will enjoy any success in the marketplace. However, even if never deployed, these proposals provide an indication of the likely approaches for future high-rate short-range radio technologies.

6. Capsule Summary: Chapter 3

WLANs are constructed to extend LANs and share some of the properties of their wired forebears. However, the wireless transition places requirements on any WLAN protocol for connection management, medium allocation, error detection and correction, and additional link security. IEEE 802.11b WLANs provided a good combination of range and data rate at low cost and easy integration with popular Ethernet-based wired LANs; the successful WECA consortium also provided assurance of interoperability between vendors. These factors made 802.11b Wi-Fi networks popular despite limited support for time-sensitive traffic and porous link security. Enhancements of the 802.11b standard to higher data rates (802.11g) and different bands (802.11a) provide an evolution path to support more users and applications; WPA and eventually 802.11i enable a more robust encryption and authentication environment for protecting sensitive information.

Contemporaneous with the evolution of WLANs, WPANs have been developed to replace local cables. For historical and technical reasons, a distinct PAN architecture has evolved, using master/slave self-organizing piconets. WPAN radios are targeted for short-range communications and require modest transmit power and limited receiver sensitivity. Support for time-sensitive traffic is provided in the design. WPANs are evolving toward higher data rates and in a possible UWB implementation may provide

the basis for the long-envisioned convergence of the various home entertainment/
information media into a single short-range wireless network.

7. Further Reading

802.11 Networks

802.11 Networks: The Definitive Guide, Matthew Gast (cited in chapter 1): *The best
overall introduction. Goes far beyond what we have provided here in examining
framing and network management.*

Wireless Local Area Networks, Benny Bing (ed.), Wiley-Interscience, 2002: *Because this
is a collection of chapters by different authors from different organizations, the quality of
the chapters varies considerably, but the book covers a lot of ground. Chapter 2
(Heegard et al.) provides a very nice discussion of the performance of 802.11b codes.*

How Secure is Your Wireless Network?, Lee Barken, Prentice–Hall, 2004: *Provides a
somewhat more detailed discussion of the flaws in WEP and a practical introduction to
the use of WPA, EAP, and VPN setup.*

WEP Attacks

"Weaknesses in the Key Scheduling Algorithm of RC4," S. Fluhrer, I. Mantin, and
A. Shamir, Selected Areas in Cryptography conference, Toronto, Canada, 16–17
August 2001

"Using the Fluhrer, Mantin and Shamir Attack to Break a WEP," A. Stubblefield,
J. Ioannidis, and A. Rubin, Network and Distributed Security Symposium
Conference, 2002

Bluetooth

Bluetooth Revealed, Brent Miller and Chatschik Bisdikian, Prentice–Hall, 2001: *Nice
description of how the Bluetooth protocols work to provide discovery, association, and
network traffic management.*

Trellis-Coded Modulations

Digital Modulations and Coding, Stephen Wilson (cited in chapter 2): *See section 6.6 for
a more detailed examination of how TCM works.*

Standards

http://www.ieee802.org/ is the home page for all the 802 standards activities. At the time
of this writing, 802 standards are available for free download as Adobe Acrobat PDF
files 6 months after they are approved. Many working documents, such as those cited
in the UWB description in section 5.3, are also available on the respective task group
web pages. Draft standards that are not yet available for free download can be
purchased from the IEEE store (but beware of awkward digital-rights-management
limitations on these files, at least as of 4Q 2003).

http://www.etsi.org/SERVICES_PRODUCTS/FREESTANDARD/HOME.HTM provides access to ETSI standards as PDF downloads. The HiperLAN standards are described in a number of ETSI documents; three useful ones to start with are as follows:

ETSI TS 101 475 v 1.3.1 (2001): "Broadband Radio Access Networks...HiperLAN 2 PHYSICAL LAYER"

TR 101 031 v 1.1.1 (1997) "Radio Equipment and Systems...(HiperLAN)..." and

ETSI EN 301 893 v 1.2.3 (2003) "Broadband Radio Access Networks...5 GHz High Performance RLAN"

http://www.bluetooth.com/ is the home page for the Bluetooth Special Interest Group (SIG), but at the time of this writing it appears that standardization activity is not included at this web site.

Radio Transmitters and Receivers

1. Overview of Radios

1.1. The Radio Problem

It's easy to build a radio. When I was a kid, one could purchase "crystal radio" kits, consisting of a wirewound inductor antenna, a diode, and an earpiece; by tuning a plug in the inductor, one could pick nearby AM radio stations quite audibly (at least with young pre-rock-amplifier ears). To build a good radio is harder, and to build a good cheap radio that works at microwave frequencies is harder still. That such radios are now built in the tens of millions for wireless local area networks (WLANs) (and the hundreds of millions for cellular telephony) is a testament to the skills and persistence of radio engineers and their manufacturing colleagues, the advancement of semiconductor technology, and the large financial rewards that are at least potentially available from the sale and operation of these radios.

The key requirements imposed on a good WLAN radio receiver are as follows:

- *Sensitivity*: A good radio must successfully receive and interpret very small signals. Recall that the thermal noise in a 1-MHz bandwidth is about—114 decibels from a milliwatt (dBm) or about $4\,\text{fW}$ ($4 \times 10^{-15}\,\text{W}$). Depending on the application, the requirements on the radio may be less stringent than that, but in general extremely tiny wanted signals must be captured with enough signal-to-noise ratio (S/N) to allow accurate demodulation.

- *Selectivity*: Not only must a radio receive a tiny signal, it must receive that signal in the presence of vastly more powerful *interferers*. An access point on channel 1 should be able to receive the signal from a client across the room and down the hall at $-90\,\text{dBm}$, even though a client of a different access point on channel 6 is right below the access point, hitting it with a signal at $-40\,\text{dBm}$ (that's a hundred thousand times more power).

- *Dynamic range*: The same access point should be able to receive its client radio's transmissions as the user moves their laptop from their office to the conference room directly under the antenna—that is, the receiver must adapt to incoming wanted signal power over a range of 60 or 70 dB. The same receiver that can accurately detect a few femtowatts needs to deal with tens of microwatts without undue distortion of the signal.

- *Switch states*: All current WLAN and wireless personal area network technologies are *half-duplex*—the same channel is used for transmitting and receiving, and the transmitter and receiver alternate their use of the antennas. A receiver needs to be turned off when the transmitter is on, both to minimize any chance of a damaging overload from leakage of the transmitted signal and to minimize power consumption; it then needs to return to full sensitivity when the transmission is done, quickly enough so that no data are missed. Most WLAN radios also allow for the use of either of two *diversity* antennas (discussed further in Chapter 6): the receiver must decide which antenna to use quickly enough to be ready for the transmitted data.

A WLAN radio transmitter has a different set of requirements:

- *Accuracy*: The transmitter must accurately modulate the carrier frequency with the desired baseband signal and maintain the carrier at the desired frequency.

- *Efficiency*: The transmitter must deliver this undistorted signal at the desired absolute output power without wasting too much DC power. The final amplifier of the transmitter is often the single largest consumer of DC power in a radio.

- *Spurious radiation*: Distortion of the transmitted signal can lead to radiation at frequencies outside the authorized bands, which potentially can interfere with licensed users and is frowned upon by most regulatory authorities. (We discuss in more detail how this *spurious* output arises in section 2.2.) Production of clean *spur*-free signals is often a trade-off between the amount of radio frequency (RF) power to be transmitted and the amount of DC power available for the purpose.

- *Switch states*: Like the receiver, the transmitter should turn off when not in use to save power and avoid creating a large interfering signal and turn back on again quickly, so as not to waste a chance to send data when there are data to send.

These requirements are actually relatively undemanding compared with other applications of microwave radios. For example, the use of a half-duplex architecture results from the limited bandwidth historically available in the unlicensed bands. Licensed cellular telephony generally uses *paired bands* for transmission and reception; for example, in the United States, a phone operating in the "cellular" bands will transmit at some channel in the range 825–849 MHz and receive at a corresponding channel 44 MHz higher in the 869–894 band. Older digital phones use separate time slots for transmission and reception and are therefore also half-duplex, but old analog phones as well as modern General Packet Radio Service and code-division multiple access phones are *full-duplex*: they transmit and receive at the same time. As WLAN evolves toward simultaneous operation in both the 2.4-GHz ISM and 5-GHz Unlicensed National Information Infrastructure (UNII) bands, simultaneous transmission and reception may become more common in WLAN. However, constructing a filter to separate 2.48 and 5.2 GHz (a 96% change) is much easier than playing the same trick between, for example, 830 and 874 MHz (a 5% change). In addition, for 802.11 radios, the medium access control (MAC) interprets all failed transmissions as collisions and tries again: the system fails gracefully in the presence of radio deficiencies and so can tolerate performance that might be unacceptable for other applications.

1.2. Radio Architectures

All modern digital radios have one or more antennas at one end and a set of *analog-to-digital converters* (ADCs) and *digital-to-analog converters* (DACs) at the other. The signal at the antenna is (for most WLAN and wireless personal are network applications) at a frequency in the GHz range, whereas the ADCs and DACs operate with signals up to a few tens of megahertz. Among their other services, radios must provide *frequency conversion* between the antenna and the digital circuitry.

Ever since its invention by Edwin Armstrong in 1917, the *superheterodyne* architecture (superhet for short) has dominated the design of practical radios. The superhet receiver uses two frequency conversion steps: first, the received frequency is converted to an *intermediate frequency* (IF), and second, after some amplification and filtering, the IF is again converted to baseband. For a WLAN application, the RF frequency of 2.4 GHz might be converted to a common IF of 374 MHz and then down to DC (Figure 4-1). Note that the final down-conversion is performed twice, to produce an I and Q output (recall I and Q are the in-phase and quadrature components of the signal). This is necessary because the down-conversion operation can't tell the difference between frequencies above and below the carrier, so they end up on top of each other when the carrier is converted to zero frequency. By doing two conversions using different phases of the carrier, we can preserve information about the sidebands. We provide a detailed example of the use of this sort of trick in our discussion of the image-reject mixer in section 2.3.

A WLAN superheterodyne receiver might look like the block diagram in Figure 4-2. Here for simplicity we show only one final down-converter. Mixers are used to convert

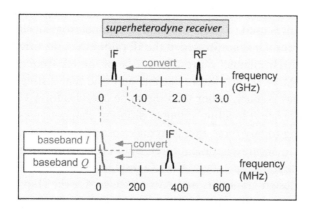

Figure 4-1: WLAN Superheterodyne Frequency Plan

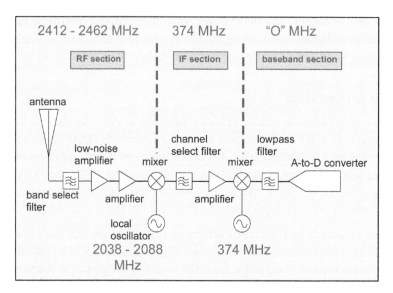

Figure 4-2: Block Diagram of a Superheterodyne Receiver

the RF signal to IF and the IF signal to a baseband ("0" frequency) output. The band select filter typically accepts the whole RF band of interest (e.g., the whole 2.4- to 2.483-GHz ISM band), and the following amplifiers should work over this band. The first *local oscillator* (LO) must be tunable. The IF is the difference between the wanted RF frequency and the LO frequency; to receive, for example, channel 1 at 2412 MHz, we set the LO at (2412—374) = 2038 MHz. However, once this conversion is complete, the IF is always the same, and DC is DC: all other radio parts can be fixed in frequency. (In modern radios, the baseband filters may have adaptive bandwidths to improve performance in varying conditions.)

The use of an IF has a number of benefits. In most cases, we'd like to listen to signals on a particular channel (e.g., channel 1 at 2.412 GHz) and not neighboring channels. An electrical filter circuit is used to select the wanted channel from all channels. The task of filtering is much easier if it is performed at the IF rather than the original RF frequency. For example, a 20-MHz channel represents 0.83% of the RF frequency of 2.4 GHz but 5% of the 374-MHz IF; it is both plausible and true that distinguishing two frequencies 5% apart is much easier than the same exercise for a 1% difference. Furthermore, a radio needs to allow *tuning* to different channels: a user would hardly be thrilled about having to buy three radios to make use of the three channels in the ISM band. It is difficult to accurately tune a filter, but in a superhet architecture we don't need to: we can tune the frequency of the LO instead. The fact that the IF is fixed also makes it easier to design amplifiers and other circuitry for the IF part of the radio. Finally, all other things being equal, it is easier to provide a given amount of gain at a low frequency than at a high frequency, so it costs less to do most of one's amplification at the IF than at the original RF frequency. The choice of an IF is a trade-off between selectivity, filtering, requirements on the LO, and component cost; we discuss this issue, *frequency planning*, in section 3.

The performance of a superhet receiver is dominated by certain key elements. The *low-noise amplifier* (LNA) is the main source of excess noise and typically determines the sensitivity of the radio. The cumulative distortion of all the components (amplifiers and mixers) before the channel filter plays a large role in determining how selective the receiver is: a distorted interfering signal may acquire some power at the wanted frequency and sneak through the channel filter. Finally, enough gain adjustment must be provided to allow the output signal to lie within the operating limits of the ADC, which are typically much narrower than the range of input RF powers that will be encountered.

A superhet transmitter is the opposite of a receiver: the I and Q baseband signals are mixed to an IF, combined, and then after amplification and filtering converted to the desired RF (Figure 4-3; again we show only one branch of the baseband for clarity). In a WLAN radio, the IF is generally the same for both transmit and receive to allow the use of a single channel filter in both directions and to share LOs between transmit and receive functions.

As with the superhet receiver, in the transmitter all components before the final mixer operate at fixed frequencies (baseband or IF). The final mixer and amplifiers must operate over the whole RF bandwidth desired. The transmitter performance is often dominated by the power amplifier. Distortion in the power amplifier causes radiation of undesired spurious signals, both into neighboring channels and into other bands. To

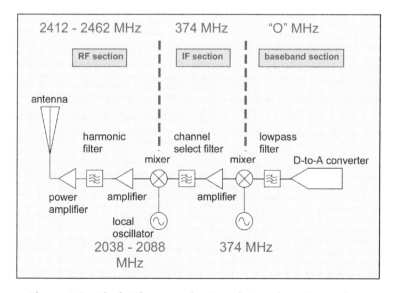

Figure 4-3: Block Diagram of a Superheterodyne Transmitter

minimize distortion, the output power can be decreased at fixed DC power, but then the efficiency of the power amplifier is reduced. Because the power amplifier dominates transmitter performance, the superhet architecture offers fewer benefits in transmission than in reception.

Even though IF filtering is easier than RF filtering, IF filters still constitute an expensive part of a superhet radio; IF amplifiers are also more expensive than amplifiers at baseband. Why not skip the IF stage? Two increasingly common architectural alternatives to superhet radios, *direct conversion* and *near-zero IF* (NZIF), attempt exactly this approach. A direct-conversion radio is what it purports to be: RF signals are converted to baseband in a single step (Figure 4-4). The channel filter becomes an inexpensive low-pass filter from zero frequency to a few megahertz; furthermore, at these low frequencies, active filters using amplifiers can be implemented, allowing versatile digital adjustment of bandwidth and gain.

Direct-conversion radios encounter serious challenges as well. Because there is no IF stage, the gain that the IF provided must be placed somewhere else, typically at baseband because it is cheaper to do so. DC offset voltages, which may appear for various reasons, including distortion of the LO signal, may be amplified so much at baseband that they drive the ADC voltage to its rail, swamping the wanted signal. Similarly, low-frequency noise from many causes, including electrical noise in metal-oxide-semiconductor field-effect transistors (MOSFETs) and *microphonic* noise (the result of mechanical vibrations in the radio), can be amplified to an objectionable level by the large baseband gain. Because the LO is at the same frequency as the RF signal in a direct conversion system, it cannot be filtered out and may radiate unintentionally during the receive operation, interfering with other users. The radiated signal may bounce back off external objects in a time-varying fashion and interfere with the wanted signal.

One solution to some of these problems is not to convert to zero but to a very low frequency, just big enough to allow the whole of the received signal to fit. For a WLAN signal, we would choose an IF of about 8 MHz: an *NZIF* receiver. The IF is low enough that no additional conversion is needed; the ADC can accept the IF signal directly. Because now there is no signal at DC, we can filter out any DC offsets without affecting the receiver. Low-frequency noise at less than a few megahertz (encompassing much

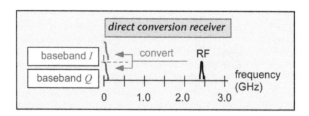

Figure 4-4: WLAN Direct Conversion Frequency Plan

electrical noise and all microphonic noise) can also be filtered out. However, because the IF in this case is small, the *image* of the wanted frequency is very close to the wanted frequency and can't be filtered. Some other means of *image rejection* must be provided. We discuss these alternative architectures in more detail in section 3.

1.3. A "Typical" WLAN Radio

A radio consists of a transmitter, receiver, and whatever ancillary functions are needed to support their operation. A complete WLAN radio might look something like the block diagram in Figure 4-5. The radio is composed of a superhet transmitter and receiver. In general, the IFs for transmit and receive would be the same so that the LOs and filters could be shared between the two functions, though here we depicted them as separate components for clarity. Both transmit and receive convert in-phase and quadrature (I and Q) signals separately, because as noted above this step is necessary to provide separate access to sidebands above and below the carrier frequency. The heart of the radio is typically implemented in one or few chips. Older systems used silicon integrated circuits with bipolar transistors or heterostructure transistors with a silicon-germanium (SiGe) base layer. In recent years, improvements in integrated circuit technology have allowed most radio chips to migrate to less-expensive MOSFET implementations.

The power amplifier is often a separate chip, using either a SiGe process or a compound semiconductor such as GaAs. Because WLANs are half-duplex, the antenna must be switched between the transmitter and receiver with a *transmit/receive* (T/R) switch. Most WLAN radios provide the option (at least for the receiver) of using one of two

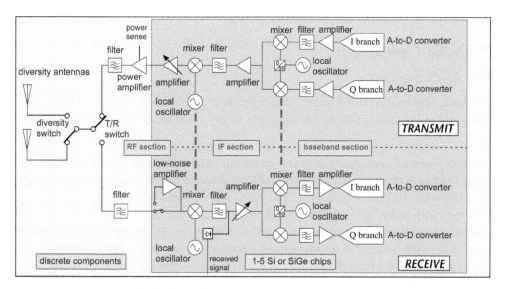

Figure 4-5: Simplified Block Diagram of an Archetypal Superhet WLAN Radio

diversity antennas, requiring an additional switch. Switches are currently implemented using one or more separate chips, though we discuss an example of an integrated complementary metal-oxide semiconductor (CMOS) implementation in section 2.6. Finally, an important capability of a WLAN radio is the provision of a *received signal strength indication* to the MAC layer to help determine when the channel is clear for transmission.

Many variants of the simple block diagram of Figure 4-5 are possible. Today, multiband radios are available, which may provide simultaneous operation at 2.4 and 5.2 GHz with a single set of antennas and a single baseband/MAC chip. Some WLAN systems use antenna arrays much more complex than a simple diversity pair and may use modified radio architectures. As we noted in section 1.2 above, many radios use direct-conversion or NZIF architectures. However, all these radios depend on the same set of functional blocks:

- *ADCs and DACs* to convert between the analog and digital worlds
- *Amplifiers* to increase signal power
- *Mixers* to convert between frequencies
- *Oscillators* to provide the means for conversion and define the frequency of the transmitted or received signals
- *Filters* to select desired frequencies from a multitude of interferers and spurs
- *Switches* to select the required input at the right time

In the next section we examine the key performance parameters that determine the suitability of each type of component for its application and how each block affects the operation of the overall radio system.

2. Radio Components

2.1. ADCs and DACs

This book is primarily about analog components and issues, but because ADCs and DACs are only half-digital devices we ought to say a few words about them.

Any time-dependent signal, such as a varying voltage or current, that is band limited (i.e., composed of frequencies within a restricted bandwidth) can be completely reconstructed from its value at a tiny number of discrete points, that is, from *samples* of the waveform. According to another theorem of the inevitable Dr. Nyquist, the frequency with which the samples must be acquired is twice the bandwidth of the signal. An example of a sampled signal is depicted in Figure 4-6. The rather complex analog signal is the sum of a signal at normalized frequency 1, another component at $f = 2$, a third at $f = 5$, and a final component at $f = 7$. We show the sampling operation taking place on each component separately, which is correct because sampling is a linear

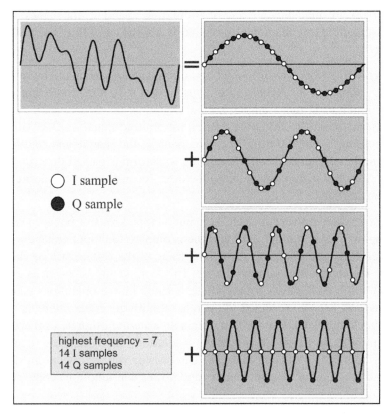

Figure 4-6: Nyquist Sampling of a Band-Limited Signal

operation (a sample of $(A + B)$ = the sum of a sample of A and a sample of B). Because the highest frequency is 7, 14 samples are needed to define the signal. It is obvious that this is more than we need to determine the amplitude of the $f = 1$ and $f = 2$ components but just enough to find the $f = 7$ component. Note that the theorem assumes that the sampled values are complex; in terms of real signals, this means that the signal must be sampled both in-phase and at quadrature or equivalently that the signal must be mixed to produce an I and Q output, each of which is sampled separately at 2(bandwidth) to provide the whole.

In principle, a modulated carrier (i.e., the received RF signal) could be directly sampled at a sampling rate equal to the bandwidth of the modulation, because the information is contained in the modulation rather than the carrier. Such a scheme, known as a *subsampling mixer*, simultaneously achieves frequency conversion and ADC. However, this approach is not commonly used; most often, the RF signal is down-converted to baseband using analog components, and the resulting I and Q signals are sampled at (at least) twice the information bandwidth. Note that the bandwidth required to send a data rate R is approximately $R/2$. (You can see why this should be so by imagining a string of alternating 1s and 0s, the highest frequency digital data stream: the fundamental is obviously a sinusoid of frequency equal to half the data rate.)

The same reasoning works in reverse: a band-limited analog signal can be reconstructed by providing samples at the Nyquist rate. Thus, similar requirements are imposed on the DAC: a minimum of 11 megasamples/second (Msps) in the case of 802.11 classic or 802.11b must be output to define the desired transmitted signal. A DAC attempts to produce a constant voltage starting at the moment of the sample and continuing until the next sample output; the resulting stairstep waveform is passed through a filter to smooth the transitions from one value to the next, reconstructing the desired band-limited analog signal.

The simplest way to construct an ADC is to put a tapped resistor between two voltages (e.g., the supply voltage and ground) and then compare the input voltage with the voltage on each tap using a comparator. The value of the voltage will be shown as the tap at which the comparator outputs change. This architecture is known as a *flash ADC* (Figure 4-7). Flash ADCs are extremely fast, because all the operations take place in parallel, and have been constructed to operate at multi-gigahertz sampling rates. However, they don't scale very well: to obtain an accurate estimate of the signal voltage—equivalently, to have many bits of resolution—many taps are required, with an amplifier and comparator for each tap. The resistor voltage divider must be very accurate, and the number of components (and power consumption) doubles with each bit of resolution.

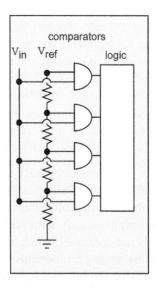

Figure 4-7: Flash ADC

Two alternative architectures allow the use of low-resolution ADC modules to achieve high-resolution overall conversion and provide better scaling and more sensible implementation for WLAN applications. The first is the *pipelined ADC* (Figure 4-8). A low-resolution ADC (here a 4-bit example is shown) digitizes the input signal to provide a coarse estimate—that is, the most significant bits—of the input sampled voltage. The result, as well as being output, is sent to a DAC, which provides modest resolution but high precision for the few states it does produce. The DAC output is subtracted from the input voltage and after amplification (not shown) is applied to a second-stage ADC, again low resolution, to get a fine estimate of the residual voltage— the least significant bits.

Thus, one can combine two 4-bit ADCs to get a total of 8 bits of resolution. In practice, additional stages are used, and the actual output resolution is somewhat less than the sum of the resolution of all the ADCs to allow correction for overlap errors. Because several steps are involved in converting each sample, pipelined ADCs have some latency between the receipt of a voltage and the transmission of a digital value. By loading several samples into the pipeline at once and converting in parallel, the overall delay associated with the ADC operation can be minimized. Pipelined ADCs can permit sample rates of tens of megasamples per second at reasonable size and power consumption.

Another common architecture, the *sigma-delta ADC*, uses a feedback loop to provide high resolution from a modest-resolution conversion stage (Figure 4-9). At each sampling time, a voltage is provided to the input of the ADC. The voltage is subtracted from the current estimate of the sampled value, as provided by a DAC. The resulting difference is then integrated and provided to an ADC, which may be low resolution because its job is only to correct the estimate. After a time, hopefully short compared with the sample time, the estimate converges to the sampled input, and the correction voltage goes to 0.

The sigma-delta design trades speed for resolution. One-bit converters are often used, and the sampling rate is much higher than the desired output rate. Digital filters on the

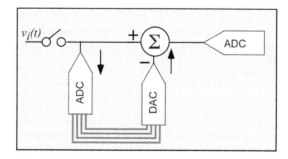

Figure 4-8: Pipelined ADC

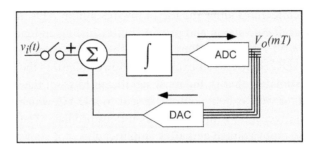

Figure 4-9: Sigma-Delta Analog-to-Digital Converter

output remove the resulting noise to produce a reasonably accurate estimate of the sampled voltage.

Each ADC architecture has a corresponding DAC architecture. For example, the analog of a flash ADC is a *current-steering DAC*, in which binary-weighted current sources are directed by switches to the output based on the binary value of the current output string. Demands on DACs are generally somewhat less stringent than ADCs, because there are no interferers, wild variations in input signal, or other unpredictable conditions to deal with. It should be noted, however, that in higher power access points, the broadband noise of the DACs can become a significant contributor to out-of-band power, particularly in direct conversion architectures where the noise cannot be readily filtered. More discussion of noise and architecture can be found in section 3 of this chapter.

What speed and resolution are required for a WLAN application? In the case of an 802.11 classic signal, at least 11 Msps would be required to capture the 11 Mchips/ second signal. Many systems take more than the minimum number of samples (*oversampling*) to obtain an improved estimate of the received signal. An 802.11a or g (or HiperLAN 2) signal contains an orthogonal frequency-division multiplexing (OFDM) symbol every 4 μsec, and each symbol must be sampled (at least) 64 times to produce the 64 possible subcarriers: $(64/4 \times 10^{-6}) = 16$ Msps.

Recall from Chapter 2 that a binary phase-shift keying (BPSK) signal (802.11 classic) requires about 9.5 dB of signal to noise for accurate demodulation: this is equivalent to only 2–3 bits of ADC resolution (recalling that the ADC samples a voltage not a power, so 9.5 dB = a factor of 3 in voltage). Higher rate quaternary phase-shift keying (QPSK) signals need about 12 dB, or around 4 bits. It is more subtle to decide how much resolution is required for an OFDM signal. Recall that a 64 quadrature-amplitude-modulation (QAM) signal (a subcarrier in the high-rate modes) needs about 26 dB (S/N) for accurate demodulation: that's about 5 bits of resolution. However, (S/N) is measured relative to the average signal; the peak (corner) points of a 64QAM constellation are about 4 dB above the average power, so we're up to 6 bits. Then we add a bunch of these signals together to form the OFDM symbol. One would naively expect that on the order of 6 extra bits would be needed to describe the 52 active carriers ($2^6 = 64$, after all), but simulations show that about 7–9 total bits are sufficient. (Details are provided in Côme et al. in section 6.) If we design the ADCs to provide exactly the necessary resolution to demodulate the RF signal, we'd better have exactly the right signal amplitude coming in, requiring perfect gain adjustment. In practice it seems prudent to provide a few bits of resolution to allow for imperfect adjustment of the analog signal gain. Thus, we'd expect to need around 6 bits for 802.11 classic/b and perhaps 8–10 bits for 802.11a or g.

A summary of ADC and DAC parameters for various reported 802.11 chip sets is provided in Table 4-1. A comparison with recent commercial stand-alone ADCs may be made by reference to Table 4-2. We see that our expectations for resolution are roughly met: 802.11b requires 6 bits or so of ADC resolution, whereas 802.11a or g chips provide 8 bits of resolution, both with oversampling used when the sampling rate is reported.

The performance of integrated ADCs is modest compared with the state of the art for stand-alone devices (Table 4-2), and by comparison of power consumption one can see why: for example, the high-performance AD6645 provides 14 bits of resolution at 105 Msps, adequate to enable a very simplified radio architecture with reduced gain control, but its power consumption is greater than that of the whole Broadcom radio chip. Functional integration demands trade-offs between the performance, power consumption, and cost of each of the components that make up the radio. Nevertheless, performance achievable in integrated implementations is adequate for WLAN requirements.

2.2. Amplifiers

The key properties of an amplifier are gain, bandwidth, noise, distortion, and power. Gain is the amplifier's *raison d'être*: it is the ratio of the size of the output to that of the input. Amplifiers must respond rapidly to be part of the RF chain; bandwidth is less significant in the rest of the radio. Noise is what an amplifier should not add to the signal but does. The output signal ought to be just a multiplied version of the input signal, but it isn't; distortion in the time domain shows up as spurious outputs in the frequency domain. Distortion usually represents the main limitation on amplifier output power. Let's consider each of these aspects of amplifier performance in turn.

Gain is the ratio of the output signal to the input signal. In RF applications, power gain is usually reported, because power is easier to measure than voltage: $G = $ (power out)/(power in). In radio boards and systems, one can usually assume that the input and output of the amplifier are matched to a $50\,\Omega$ environment; however, amplifiers integrated into a chip may use differing impedance levels. At low IF frequencies or at baseband, high impedances may be used, and power gain becomes less relevant than voltage gain. Gain is harder to get at RF than at IF or baseband, so most designs use only as much high-frequency gain as necessary to achieve the desired noise performance.

Table 4-1: Survey of 802.11 ADC/DAC Performance

Vendor	Intersil/Virata	Broadcom	Broadcom	Marvell	Thomson
Protocols supported	802.11b	802.11b, g	802.11a	802.11b	802.11a
DAC resolution/speed	6 b 22 Msps	8 b	8 b	9 b 88 Msps	8 b 160 Msps
ADC resolution/speed	6 b 22 Msps	8 b	8 b	6 b 44 Msps	8 b 80 Msps
DC power: transmit	720 mW	144 mW	380 mW	1250 mW	920 mW
DC power: receive	260 mW	200 mW	150 mW	350 mW	200 mW
Reference	Data sheets	Trachewsky et al. HotChips 2003	Trachewsky et al. op. cit.	Chien et al. ISSCC 2003 paper 20.5	Schwanenberger et al. ISSCC 2003 paper 20.1

Table 4-2: Survey of Commercial Stand-Alone ADCs

Part	Resolution (bits)	Rate (Msps)	(S/N) (dB)	Quantization (S/N) (dB)	Power (mW)
AD6645	14	105	73.5	90	1500
SPT7722	8	250	46	54	425
MAX1430	15	100	77	96	2000
ADC10080	10	80	60	66	75
ADC10S040	10	40	60	66	205

After "High Speed ADC's...," David Morrison, Electronic Design 6/23/03, p. 41.

Two general classes of devices are available for providing gain at gigahertz frequencies: field-effect transistors (FETs) and bipolar junction transistors (BJTs) (Figure 4-10). FETs are lateral devices: electrons move sideways in a *channel* under a *gate* from a *source* to a *drain*, with the number of electrons being determined by the potential on the gate. The output current is the product of the number of electrons in the channel and their velocity. The velocity the electrons can reach is mainly determined by the type of semiconductor used for the device. (The velocities of holes are so much lower than those of electrons that p-type devices are rarely if ever used for high-frequency amplifiers.) The number of electrons is roughly linear in the gate-source voltage.

Bipolar transistors are vertical devices: electrons are injected from an n-type *emitter* into a p-type *base*, in which they drift until they are collected by the n-type *collector* or (more rarely) recombine with a hole in the base region. The number of electrons in the base is set by the likelihood that an electron will be able to thermally hop over the emitter-base potential barrier and is thus exponentially dependent on the barrier height. Therefore, the current in a bipolar transistor is exponential in the input voltage; for small variations in input voltage, the corresponding change in output current grows linearly in the current:

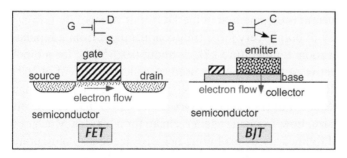

Figure 4-10: Field-Effect and Bipolar-Junction Transistors

$$\frac{d}{dv_{in}} I_{out} \approx \frac{d}{dv} e^{\frac{-qv_{in}}{kT}} = \frac{-q}{kT} e^{\frac{-qv_{in}}{kT}} = \frac{-q}{kT} I_{out} \qquad [4.1]$$

where q is the electron charge, k Boltzmann's constant, and T the absolute temperature.

To compare them, it is useful to think of the semiconductor device as a *transconductance* amplifier: it accepts a small RF voltage and produces a large output current. The ratio of the change in current to the change in voltage that produces it is the transconductance, $g_m = dI_{out}/dv_{in}$. To make a transductor into a voltage amplifier, we put the current through a load resistor to convert it into a voltage. The power dissipated is the product of current and voltage. The gain of the amplifier at frequencies low enough that parasitic capacitances can be ignored is

$$G = \frac{P_{out}}{P_{in}} = \frac{(g_m v_{in})^2 R_L}{v_{in}^2 / R_{in}} = g_m^2 R_L R_{in} \qquad [4.2]$$

To get high gain, one needs high transconductance. The transconductance of a FET device is approximately the product of the gate capacitance (the change in charge due to a change in voltage) and the velocity of the charge and is thus roughly determined by the width of the channel and the type of semiconductor. To achieve higher electron velocities, one can progress from silicon ($V_{el} \approx 8 \times 10^6$ cm/sec) to GaAs ($V_{el} \approx 1.5 \times 10^7$ cm/sec). To go farther, one can construct *high-electron-mobility transistors* (HEMTs) in which the electrons are located at a heterojunction between two semiconductors of differing bandgap, removing them from dopant atoms and consequent scattering. However, the amount of improvement is limited by the fact that semiconductors with the highest electron velocities, such as InGaP, have different crystal lattice sizes from common substrates such as GaAs. The lattice mismatch can be tolerated if the layers of InGaP are sufficiently thin: the resulting *pseudomorphic* HEMT, or pHEMT, can achieve effective electron velocities of around 2×10^7 cm/sec. Heterostructure devices also benefit from much improved electron mobility (the ratio of velocity to electric field at low velocities), which can be many times higher than mobility in silicon. In practice, FET devices can obtain transconductance of tens (silicon) to hundreds (pHEMT) of milliSiemens for a channel width of a few hundred microns.

In a bipolar device, transconductance increases with collector current (equation [4.1]), and collector current is exponential in the DC input voltage. By turning up the current density, it is easy to obtain very large transconductances from a bipolar transistor: recalling that at room temperature kT/q is about 0.026 V, g_m for a bipolar is about $40(I_{out})$, irrespective of the technology used. A collector current of 100 mA (which is quite reasonable for a transistor with a periphery of a few hundred microns) will provide 4000 mS or 4 S of transconductance, much more than can be obtained from a FET of similar size. Low-frequency gain is cheap for a bipolar. Much of this gain is then often converted to linearity by exploiting negative feedback configurations. On the other hand, parasitic capacitances in BJTs are large relative to FET capacitances and vary strongly with applied voltage. This makes design of broadband amplifiers somewhat more difficult with BJTs than with FETs. Bandwidth is roughly determined by the ratio of some measure of the gain (for example, g_m) to some measure of the parasitics, typically a capacitance. Bipolar devices can have more transconductance but also more capacitance than FETs; the net result is that BJTs display similar bandwidth to FETs of similar dimensions. One important advantage of BJTs is that, being vertical devices, the critical length in the direction in which the electrons flow is determined by the growth of a layer or diffusion of dopants into a layer, both of which are relatively easy to control compared with the width of a submicron feature such as a FET gate. Therefore, BJTs have historically been easier and cheaper to produce for high-frequency applications and are often the first technology applied to a microwave problem, with silicon-based FET technology catching up as lithographic dimensions decrease.

In the receiver, the job of the amplifier chain is to deliver an output voltage within the dynamic range of the ADC. Because the input signal power can vary over such a large range, gain adjustability must be provided. Some adjustment in either the magnitude of the digital output signal or the gain or both must also be provided for the transmitter, though in this case the output range is modest in comparison with that faced by the receiver: 10–20 dB variation is typical.

In Table 4-3 we show typical gain for some single-stage commercial amplifiers operating at 2 GHz, using various technologies and compared with a representative silicon CMOS result. Note that GaAs FET devices are fabricated with the metal gate directly on the surface of the semiconductor and are thus often referred to as MEtal-Semiconductor FETs or MESFETs. Power gain of around 15 dB for a single stage is a reasonable expectation at the sort of frequencies used in WLAN radios.

To put this number in context, remember that a typical input signal of -80 dBm needs to be amplified to a convenient value for digital conversion: on the order of 1 V at 1 KΩ, or 1 mW (0 dBm). Thus, we need around 80 dB total net gain, partitioned between the RF sections, IF if present, and baseband. (Actually, more is needed to overcome losses in conversion and filtering.) We can see that several gain stages are likely to be needed, though at this point it isn't obvious how to allocate the gain between the parts of the radio.

Table 4-3: Gain at 2 GHz, Commercial Devices, Various Technologies

Technology	Gain (2 GHz)	Reference
GaAs pHEMT	16 dB	Agilent 9/03
GaAs MESFET	14 dB	WJ Comm 9/03
SiGe HBT	17 dB	Sirenza 9/03
Si CMOS, 0.8 μm	15 dB	Kim MTT '98

To elucidate one of the key considerations in determining the amount of RF gain needed, let us proceed to consider noise and its minimization. Recall from Chapter 2, section 4.2, that a 50 Ω input load (a matched antenna) delivers −174 dBm/Hz of thermal noise at room temperature into the radio. The radio has its own resistors, as well as other noise sources, and thus adds some excess noise over and above amplifying the noise at the input. The amount of noise added by an amplifier is generally quantified by defining a *noise factor*, the ratio of the output (S/N) divided by the input (S/N) (Figure 4-11). The noise factor is 1 if the amplifier contributes no excess noise and grows larger when excess noise is present. Note that the noise factor of a single amplifier is independent of the amplifier gain.

It is common to define the *noise figure* as the logarithm of the noise factor, reported in decibels. Because log (1) = 0, the noise figure *NF* is 0 for an ideal noiseless amplifier and grows larger as more excess noise is added.

Amplifier gain G rears its head when we try calculating the noise factor of two amplifiers in sequence—*cascaded* amplifiers—as shown in Figure 4-12. The excess noise contributed by the first amplifier is amplified by both G_1 and G_2, whereas the excess noise of the second amplifier is magnified only by the second-stage gain G_2. In consequence, the noise factor of the cascaded combination is the sum of the noise factor of the first stage, F_1, and the excess noise of the second stage divided by the gain of the first stage. *If the first stage gain is large, overall system noise factor is dominated by the noise in the first stage.*

$$F = \frac{S/N_{in}}{S/N_{out}} = \frac{\left(\frac{P_{in}}{N_o}\right)}{\left(\frac{GP_{in}}{G[N_o+N_{ex}]}\right)} = \left(\frac{GN_o+GN_{ex}}{GN_o}\right) = 1 + \frac{N_{ex}}{N_o}$$

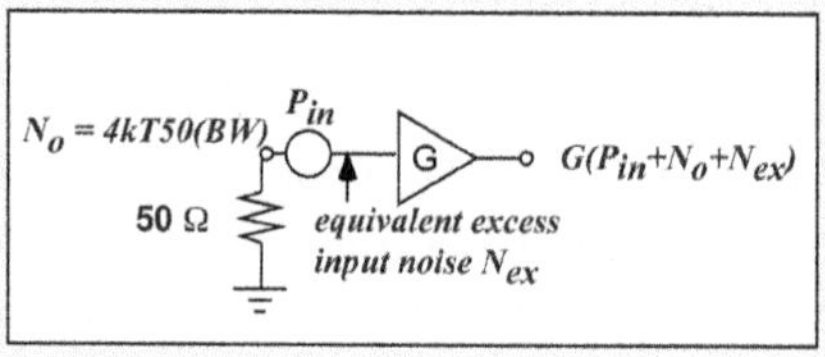

Figure 4-11: Noise Factor Definition

$$F = \frac{S/N_{in}}{S/N_{out}} = \frac{\left(\frac{P_{in}}{N_o}\right)}{\left(\frac{G_1G_2P_{in}}{G_1G_2[N_o+N_{ex1}]+G_2N_{ex2}}\right)}$$

$$= 1 + \frac{N_{ex1}}{N_o} + \frac{N_{ex2}}{G_1N_o} = F_1 + \frac{F_2-1}{G_1}$$

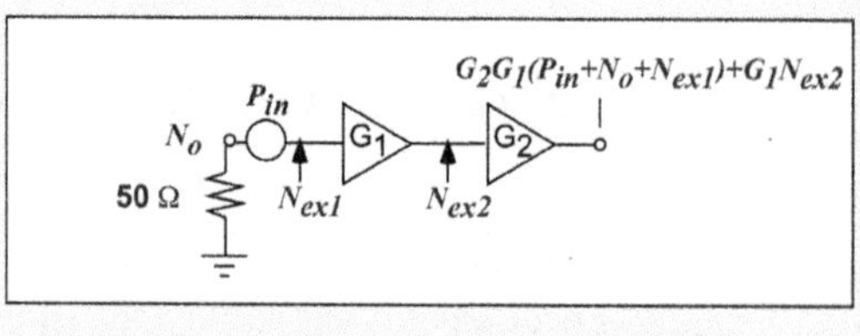

Figure 4-12: Noise Factor of Two Amplifiers in Cascade

For this reason, we always need to allocate enough gain in the first RF amplifier stage or two to minimize the noise contribution of the rest of the chain. A typical first-stage gain of 15 dB (a factor of 30) is usually enough to ensure that the first stage—the *low-noise amplifier (LNA)*, so named because its (hopefully small) noise figure determines the noise in the whole receiver—does indeed fulfill its dominant role.

The noise figure of the chain, once obtained, provides a quick insight into the noise performance of the radio: the effective noise floor is just the thermal noise (in dBm) plus the noise figure (in dB). For example, a radio with a 1-MHz input bandwidth and a 3-dB noise figure has a noise floor of $(-174 + 60 + 3) = -111$ dBm.

In Table 4-4 we include noise figure in our amplifier comparison. pHEMTs are the champions of the low-noise world and are widely used in applications like satellite receivers where noise figure is a key influence in overall performance. The other device technologies all readily achieve noise figures of a few decibels, adequate for typical WLAN/wireless personal network applications. (Note that these are not the lowest noise figures that can be achieved in each technology but are those obtained in practical devices optimized for overall performance.)

Noise sets the lower bound on the signals that can be acquired by a receiver. The upper bound, and more importantly the upper bound on interfering signals that can be present without *blocking* the reception of the tiny wanted signal, is set by distortion. Distortion redistributes power from the intended bandwidth to other *spurious* frequencies. In the receiver, distortion spreads the power from interferers on other channels into the wanted channel; in the transmitter, distortion splatters transmitted power onto other folks' channels and bands, potentially earning the wrath of neighbors and regulators. How does distortion do its dirty work, how is it measured, and how does the radio designer account for its effects?

Let us consider an amplifier with a simple *memoryless* distortion, depicted in Figure 4-13. (Memoryless refers to the fact that the output depends only on the amplitude of the input and not on time or frequency; real amplifiers remember what is done to them—though they often forgive—but incorporating phase into the treatment adds a lot of complexity for modest rewards and is beyond our simple needs.)

Table 4-4: Gain and Noise Figure at 2 GHz; Devices From Table 4-3

Technology	Gain (2 GHz)	NF (2 GHz)	Reference
GaAs pHEMT	16 dB	1.2 dB	Agilent 9/03
GaAs MESFET	14 dB	3 dB	WJ Comm 9/03
SiGe HBT	17 dB	3 dB	Sirenza 9/03
Si CMOS, 0.8 μm	15 dB	3 dB	Kim MTT '98

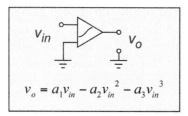

Figure 4-13: Amplifier With Simple Memoryless Polynomial Transfer Characteristic

The output of the amplifier is taken to be accurately expressed as a polynomial function of the input voltage. The coefficients are taken here to be real numbers for simplicity. Further, we assume in most cases that the input voltage is small enough so that the nonlinear (quadratic and cubic) contributions are much smaller than the ideal linear amplification.

What happens when we plug a pure sine wave input into this distorting amplifier? First we'll do the math and then provide some pictures to show what it means. The output voltage is obtained by plugging the sinusoidal input into the polynomial:

$$v_0 = a_1 v_i \sin(\omega_1 t) - a_2 v_i^2 \sin^2(\omega_1 t) - a_3 v_i^3 \sin^3(\omega_1 t) \tag{4.3}$$

To rephrase this equation in terms that have more direct spectral relevance, we exploit a couple of trigonometric identities:

$$\sin^2 x = \frac{1}{2} - \frac{1}{2}\cos(2x) \tag{4.4}$$

and

$$\sin^3 x = \frac{3}{4}\sin x - \frac{1}{4}\sin(3x) \tag{4.5}$$

After a bit of algebra we can partition the output in a fashion that clearly shows how each type of distortion creates specific contributions to the spectrum of the output (Figure 4-14). The quadratic term, or *second-order distortion*, creates a constant (DC)

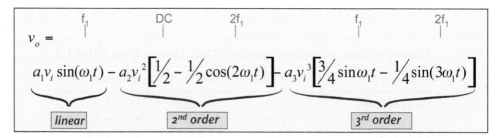

Figure 4-14: Output Voltage of a Distorting Amplifier Displayed as Linear, Second-, and Third-Order Contributions

output voltage and a new component at twice the input frequency. The cubic term, a *third-order distortion*, produces no DC component but adds both an undesired nonlinear term at the input (*fundamental*) frequency and a new term at three times the fundamental. Naturally, the math proceeds in an essentially identical fashion for a cosine input instead of a sine.

Some insight into how these terms arise can be had by reference to Figures 4-15 and 4-16, showing (grossly exaggerated) images of the time-domain distortion. Figure 4-15 shows what would happen with a pure second-order amplifier. The amplifier only produces positive output voltages, so an input signal with an average voltage of 0 obviously must produce an output with a positive average: a DC offset has been introduced by the distortion. The quadratic characteristic takes the negative-going peaks of a sine and inverts them; thus, there are two maxima per (input) cycle instead of 1: the frequency of the input has been doubled.

A pure cubic distortion is shown in Figure 4-16. The transfer characteristic is symmetric about 0, so no DC offset results. Clearly, the output generally resembles the input

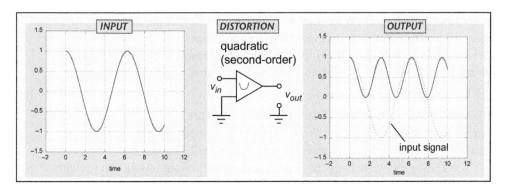

Figure 4-15: Pure Second-Order Distortion of a Sinusoidal Input Signal

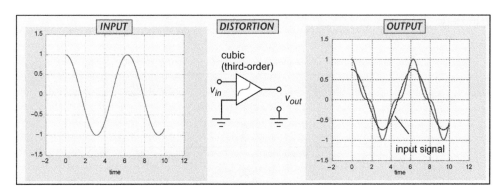

Figure 4-16: Pure Third-Order Distortion of a Sinusoidal Input Signal

signal: a component is present at the input frequency. However, the output is not the same shape as the input. In fact, we can see that the distortion alternates sign six times in one cycle of the fundamental (at about $t = 0.5$, 1.6, 2.6, 3.7, 4.7, and 5.8, where the distorted signal and the input signal cross), that is, there is a component at three times the frequency of the fundamental.

The resulting spectra are depicted in Figure 4-17. The spectrum contains a DC offset due to the presence of second-order distortion, a change in the amplitude of the fundamental (relative to an ideal linearly amplified input) due to the third-order distortion, and second and third *harmonics* of the input frequency.

So far it's not entirely obvious why a WLAN radio designer is very concerned. For example, an 802.11b radio operates in the ISM band. Say the fundamental is at channel 1, 2.412 GHz. The second and third harmonics, at, respectively, 4.824 and 7.236 GHz, are so far from the band of interest that they are easily filtered out; the DC component can be removed by using a capacitor between stages. Only the tiny distortion of the main signal appears to be consequential. (Note that this statement is a bit too glib if a direct-conversion architecture is being used: the DC term in the output may be important if we have converted directly to zero frequency. We discuss DC offsets in section 3 of this chapter. We could also be in for some trouble in an ultrawideband radio, where the distortion products at twice the input frequency are still in our operating band. (Avoiding this situation is a good reason to partition bands into subbands less than an octave [factor of 2 in frequency] wide: see for example the subbands defined in Figure 3.29.)

However, as we discussed in Chapter 2, real signals are modulated and therefore contain a range of frequencies. What happens if an input with more than one frequency is presented to an amplifier with distortion? The smallest integer that isn't 1 is 2: let's see what happens if we insert sinusoidal signals at two (neighboring) frequencies into our distorting amplifier. The result, after some uninteresting algebra again exploiting the trigonometric identities of equations [4.4] and [4.5] and the additional product identity [4.6], is summarized in Figure 4-18. It's a mess. To make any sense of it we need to split up the groups of terms and examine them separately.

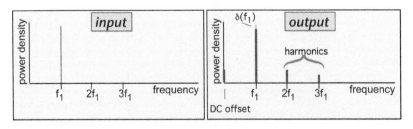

Figure 4-17: Spectrum of a Distorted Signal

linear

$$v_o = a_1 v_{i1}\sin(\omega_1 t) + a_1 v_{i2}\sin(\omega_2 t)$$

2nd order

$$-a_2\left[v_{i1}^2\left(\tfrac{1}{2}-\tfrac{1}{2}\cos(2\omega_1 t)\right)+v_{i2}^2\left(\tfrac{1}{2}-\tfrac{1}{2}\cos(2\omega_2 t)\right)+2v_{i1}v_{i2}\left(\frac{1}{2}\cos(\omega_1 t-\omega_2 t)-\frac{1}{2}\cos(\omega_1 t+\omega_2 t)\right)\right]$$

3rd order

$$-a_3\left[\begin{array}{l} v_{i1}^3\left(\tfrac{3}{4}\sin\omega_1 t-\tfrac{1}{4}\sin(3\omega_1 t)\right)+v_{i2}^3\left(\tfrac{3}{4}\sin\omega_2 t-\tfrac{1}{4}\sin(3\omega_2 t)\right) \\ +3v_{i1}^2 v_{i2}\left(\tfrac{1}{2}-\tfrac{1}{2}\cos(2\omega_1 t)\right)\sin(\omega_2 t)+3v_{i1}v_{i2}^2\sin(\omega_1 t)\left(\tfrac{1}{2}-\tfrac{1}{2}\cos(2\omega_2 t)\right)\end{array}\right]$$

Figure 4-18: The Result of Distorted Amplification of a Two-Tone Input Signal

$$\sin(x)\sin(y) = \frac{1}{2}\cos(x-y) - \frac{1}{2}\cos(x+y) \qquad [4.6]$$

The second-order terms alone are shown in Figure 4-19. They consist of three components. First, there's a DC term and a double-frequency term from the first input tone, just like the single-frequency case. Input tone 2 similarly contributes its own single-tone distortion terms. The new and interesting part is the third block, composed of a term at the sum of the two input frequencies and another term at their difference frequency. Note that all the terms have an amplitude proportional to the square of the input, assuming the two tones to be of equal size.

The resulting spectrum is depicted in Figure 4-20. So far we're still not much worse off than we were. Recall that the bandwidth of a typical 802.11 signal is around 16 MHz, so

$$-a_2\Bigg[\underbrace{v_{i1}^2\left(\tfrac{1}{2}-\tfrac{1}{2}\cos(2\omega_1 t)\right)}_{\text{single-tone, } f_1}+\underbrace{v_{i2}^2\left(\tfrac{1}{2}-\tfrac{1}{2}\cos(2\omega_2 t)\right)}_{\text{single-tone, } f_2}+\underbrace{2v_{i1}v_{i2}\left(\frac{1}{2}\cos(\omega_1 t-\omega_2 t)-\frac{1}{2}\cos(\omega_1 t+\omega_2 t)\right)}_{\text{sum and difference frequencies}}\Bigg]$$

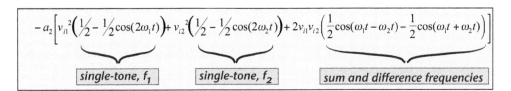

Figure 4-19: Second-Order Two-Tone Distortion Terms

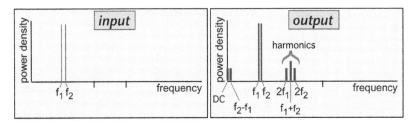

Figure 4-20: Spectrum of a Two-Tone Signal After Second-Order Distortion

the biggest difference in frequency that can result is about 16 MHz: still close to DC from the point of view of an RF signal at 2.4 GHz. The sum frequency is still essentially twice the input and so easily filtered.

This happy circumstance comes to an end when we allow for third-order two-tone distortion (Figure 4-21). Once again we see that each tone contributes the same individual distortion terms, at the fundamental and the third harmonic, that were observed in isolation. All the output terms are proportional to the cube of the input if the two tones are equal in magnitude.

However, the two-tone terms now contain both additional distortion of the fundamental frequencies and new tones at all the possible third-order combinations $(2f_1 + f_2)$, $(2f_2 + f_1)$, $(2f_1 - f_2)$, and $(2f_2 - f_1)$. This is significant because if the two input tones are closely spaced, the differences $(2f_1 - f_2)$ and $(2f_2 - f_1)$ are *close to the input tones and cannot be filtered*. This result is depicted graphically in Figure 4-22, where we show the whole two-tone output spectrum. Third-order distortion increases the apparent bandwidth of signals. Interferers grow wings in frequency that may overlap neighboring wanted signals. Transmitted signals extend into neighboring channels and bands.

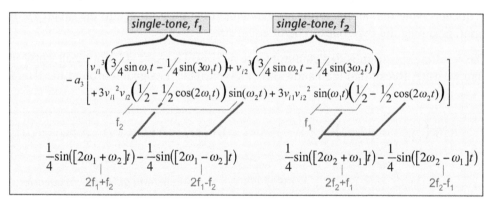

Figure 4-21: Third-Order Two-Tone Distortion Terms

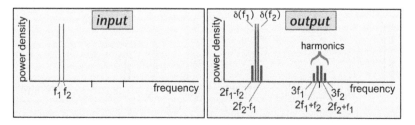

Figure 4-22: Spectrum of a Two-Tone Signal After Third-Order Distortion

We should note that the behavior cited here is characteristic of all even- and odd-order distortion products. Even-order (fourth, sixth, eighth, etc.) distortion terms always create DC or DC-like components and components at all even harmonics of the fundamental frequency; odd-order (fifth, seventh, ninth, etc.) distortion terms always create components at the fundamental frequency, odd harmonics, and, in the case of multiple tone, inputs at the possible difference terms near the fundamental. Some authors have pursued extensive investigations of the results of higher order polynomial distortion (which, when complex values are allowed, are known as Volterra series). However, in real devices it seems more sensible to treat high-order distortion as being due to an effectively abrupt encounter with the limitations of the device output: *clipping* of the input signal, which we discuss momentarily.

Given that it is not normally possible to specify the behavior of an amplifier in terms of a known polynomial, how shall we measure distortion? One method is to simply measure the power in one of the distortion products—an *intermodulation* (IM) product—at a particular input power and report this as the extent of the relevant distortion. Although this method is common and valuable, it requires that we measure the IM product at each power of interest. Recall that all second-order distortion products scale as the square of the input power and all third-order products scale as the cube. If the distortion products are truly of known order, it ought to be possible to measure the IM power at any input power and infer what the IM would have been at a different input power. Because the order of the distortion associated with a given IM product may be inferred from its frequency (hoping that only the lowest order term contributes significantly), this prescription only requires that we adopt some convention about the input power to be used. The most common convention is to specify the input power at which the linearly amplified ideal output and the distortion product would be of equal magnitude: the *intercept* of lines drawn through the linear output and the IM power. This procedure is illustrated in Figure 4-23 for second- and third-order distortion.

It is important to note that intercepts are obtained by extrapolation, not direct measurement. Any device intended for use as a linear amplifier generally has such small distortion terms that long before we could turn the power up enough to make them equal to the undistorted fundamental, some other limitation of the device—typically its maximum output power—would be encountered, invalidating the idea of simple second- or third-order scaling of the distortion product.

The concept of the intercept point is only valid and useful in the region where the IM products scale appropriately: referring to Figure 4-23, where the IM power is in fact linear in decibels with input power, with slope of either 2 or 3 as appropriate. The definition of an intercept is often abused in literature and practice by measuring an IM product at a given power level and extrapolating from that single point to an intercept, without verifying the scaling of the distortion to be correct. When higher order distortion becomes significant, such extrapolations can be misleading: a fifth-order term may fortuitously cancel the third-order distortion over some limited range, resulting in an increase in the naively defined intercept point that misrepresents performance of the device except in the narrow power range in which it was characterized.

As a final caution, one must be alert to what intercept is being referred to, because many definitions are possible. First, referring to Figure 4-23, either the input or output intercept can be cited, the two being different by (approximately) the linear gain of the device. Common usage uses the output intercept for output-power–oriented

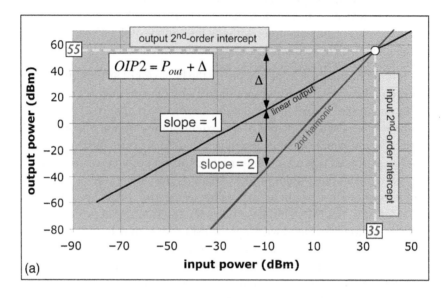

(a)

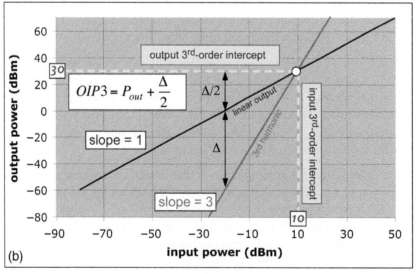

(b)

Figure 4-23: (a) Definition of the Second-Order Intercept (b) Definition of the Third-Order Intercept

applications such as a power amplifier and input intercept for an LNA. These are often abbreviated as *OIP2* and *IIP2* for second-order and *OIP3* and *IIP3* for third-order distortion, respectively. Second, there are several possible intercepts depending on which spurious product is selected for examination. One can define an intercept based on the harmonics that result when a single tone is placed at the input or based on the two-tone mixing products when two tones are present. Although there is no conceptual distinction between these methods and both provide sensible estimates of the nonlinearity of the amplifier in question, they do not produce the same numbers: IP2 from a second harmonic is 6 dB higher than IP2 obtained from a two-tone IM product. Similarly, IP3 from a third harmonic is 9 dB higher than IP3 from a $(2f_x \pm f_y)$ output power. (These distinctions may be derived from a thoughtful perusal of Figures 4.14 and 4.18; they are just the result of the differing coefficients multiplying the respective distortion terms.) Although third-order distortion is usually reported based on the $(2f_x \pm f_y)$, because of its practical importance and ease of measurement, conventions for reporting second-order distortion are less consistent. Finally, it is possible to define a two-tone intercept as occurring either at the point at which a distortion term is equal in power to the output power in one linear tone or equal to the total linear output power, 3 dB larger in the case of equal tones. The most common convention seems to be to define the intercept when one distortion tone is equal in power to one linear output, but this may not be universal. The distinction here is either (3/2 dB) for second-order or (3/3 = 1 dB) for third-order distortion and therefore is of modest practical importance.

Knowing the intercept for a given amplifier, we may obtain values of the IM products expected for any input power where scaling applies. This operation is depicted graphically in Figure 4-24. The operation is conveniently described in terms of the *backoff* from the intercept point: in Figure 4-24(a) the input power is backed off 40 dB from the intercept, and so the second-order IM product is decreased by twice as much—2(40) = 80 dB—relative to the power at the intercept. Similarly, in Figure 4-24(b) 20 dB of backoff in linear power produces 60 dB reduction in third-order spurious output power. Spurious output powers are frequently measured by reference to the linear output power, the carrier in the case of a single-tone signal: thus, the spur is measured as decibels from the carrier, or *dBc*, because it is after all the size of the distortion with respect to the desired linear output that matters. Thus, at the indicated power in Figure 4-24(a) and (b), both the second- and third-order spurs are at −40 dBc.

Now that the reader is either thoroughly conversant with the terminology and concepts used in describing distortion or thoroughly confused, we (finally) present some typical values of third-order intercept, for the same devices we examined previously, in Table 4-5. Different technologies produce wildly different third-order intercepts for otherwise similar values of gain and noise figure. The reader should not conclude from this very simplified comparison that every CMOS amplifier suffers a 38-dB inferiority in OIP3 to every GaAs MESFET device, but it is valid to point out that compound-semiconductor devices in general are capable of higher output power and better linearity for otherwise similar performance because of their higher electron mobility and generally better

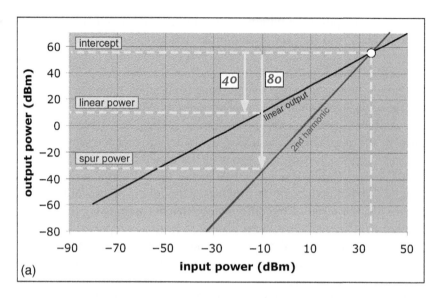

(a)

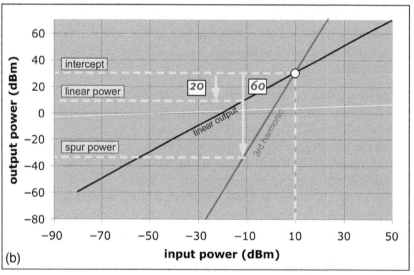

(b)

Figure 4-24: (a) Calculating Spurious Output Power Due to Second-Order Distortion Given OIP2 (b) Calculating Spurious Output Power Due to Third-Order Distortion Given OIP3

resistance to electrical breakdown compared with silicon devices. These benefits must be measured against the expensive starting materials and relatively primitive and high-cost process technology characteristic of even mature GaAs fabrication. In most WLAN radios, the tremendous cost and integration benefits of silicon CMOS result in compound-semiconductor devices being relegated to the power amplifier and RF switches, if they are present at all.

Table 4-5: Comparison of Third-Order Distortion for Devices/Technologies From Table 4-4

Technology	*Gain (2 GHz)*	*NF (2 GHz)*	*OIP3 (2 GHz)*	*Reference*
GaAs pHEMT	16 dB	1.2 dB	28 dBm	Agilent 9/03
GaAs MESFET	14 dB	3 dB	40 dBm	WJ Comm 9/03
SiGe HBT	17 dB	3 dB	14 dBm	Sirenza 9/03
Si CMOS, 0.8 μm	15 dB	3 dB	2 dBm	Kim MTT '98

The noise figure of a device represents some measure of the smallest signal it can usefully amplify. Similarly, the intercept represents a measure of the largest signal that can be linearly amplified. The two values may be combined to estimate the device's dynamic range, the difference between the smallest and largest useful signals. The exact definition of dynamic range can vary depending on what is considered useful. A fairly general but rather demanding approach is to define the *spur-free dynamic range* as the range of input power in which the amplified signal is both greater than the amplifier noise and no distortion is detectable (i.e., the spurious output is less than the amplifier noise floor). For a third-order spur to be undetectable, we must be backed off by one-third of the difference between the intercept and the noise floor (Figure 4-24(b)): we find the awkward formula

$$SFDR = \frac{2}{3}\{OIP3 - [- 174\ dBm + BW + NF]\} \qquad [4.7]$$

where all quantities are measured in dB or dBm as appropriate. The spur-free dynamic range (SFDR) depends on the bandwidth of interest and cannot be defined independent of the application.

We have so far concentrated all our efforts in describing distortion of signals composed of at most two input tones. Real digital signals are the result of an effectively random sequence of bits defining a time series of single-carrier modulation symbols (802.11 classic or b, 802.15.1, or 802.15.3) or the subcarriers of an OFDM symbol (802.11a,g). The spectrum of such a signal is generally rather broadly distributed, containing power in a wide range of frequencies, as depicted for example in Figure 3-20. What happens to such a more-complex signal in the presence of nonlinear distortion?

The effect of third-order distortion on a realistic digital signal spectrum is qualitatively depicted in Figure 4-25. Replicas of the ideal spectrum—*wings*—appear on the sides of the spectrum, extending roughly as far as the original spectrum width in each direction. (Just rewrite the expression: $2f_2 - f_1 = f_2 + (f_2 - f_1)$.) Images of the spectrum at the third harmonic and sum tones are also present but are generally easily filtered. The distorted spectrum is wider in frequency than the original spectrum and may extend into neighboring channels or bands.

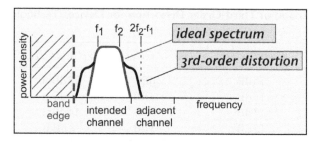

Figure 4-25: Effect of Third-Order Distortion on the Spectrum of a Typical Digital Signal

Power that extends into a neighboring channel is known, sensibly enough, as adjacent-channel power, and the ratio between the power in the intended channel and the adjacent one is known as the *adjacent-channel power ratio* (ACPR). Such spreading is important, even if the ACPR is quite small, because of the near-far problem: a nearby powerful radiator on one channel may, if its spectrum is broadened, block a distant wanted radio on a neighboring channel. Standards generally contain provisions to limit such unwanted interference in the form of the spectral masks defining the allowed shape of the emitted signal. For example, a comparison of the 802.11a spectral mask to the spectrum of a distorted signal (Figure 4-26) may clarify the origin of the heretofore mysterious mask shape: it is designed with the expectation of some third-order (and higher) distortion in the signal, providing a compromise between minimizing adjacent-channel interference and cost of the resulting spec-compliant transmitter.

It is not a trivial matter to evaluate quantitatively how a given degree of distortion will affect a complex digital signal. Fortunately, a rough but useful estimate of the resulting

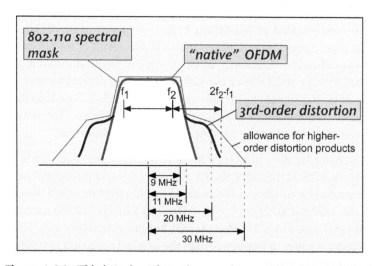

Figure 4-26: Third-Order Distortion vs. the 802.11a Spectral Mask

ACPR can be obtained directly from the two-tone intercept: Pedro and de Carvalho showed that for a fairly general digital-like signal composed of a number of equally spaced randomly phased tones, the ACPR, measured as the ratio of the power in one of the third-order "wings" of the spectrum to the power in the linearly amplified original spectrum, is roughly the same as the ratio of one of the two-tone difference spurs to one of the corresponding linear tones, to within a decibel or two. It is this partially accidental agreement that accounts for the popularity and utility of the simple two-tone measurement of IP3 (Figure 4-7).

A more subtle consequence of third-order distortion, in this case within the receiver, is of some consequence in 802.11 WLAN applications, where the three equally spaced nonoverlapping channels (1, 6, and 11) available in the United States allow for the IM product of two interferers to lie within a wanted channel. An example is shown in Figure 4-28: two nearby transmitters on channels 6 and 11 transmit simultaneously with a wanted signal on channel 1.

The receiver contains a first filter that removes signals outside the ISM band (Figure 4-29), but within this band all three signals are amplified by the LNA and mixed in the first mixer. It is only after mixing that a filter is used to remove channels 6 and 11 from the total. (Note that we have symbolically indicated a radio in a "high-gain" state by switching the LNA into the chain; a real radio would likely have variable-gain amplifiers or multiple stages of amplification with switching.)

Any third-order distortion of the (channel 6 + channel 11) signal present in the LNA or mixer will result in an IM component at $(2.437 - (2.462 - 2.437)) = 2.412\,\text{GHz}$: right on top of the wanted channel 1 (Figure 4-30). We can estimate the consequent requirements placed on the radio: we want the IM product to be less than or

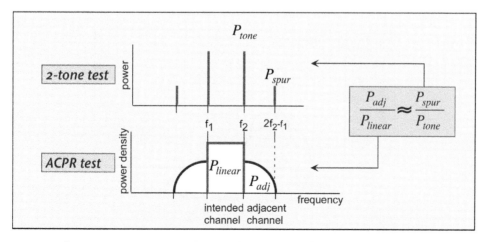

Figure 4-27: Correspondence Between ACPR of a Distorted Digital Signal and Spurious Power From a Two-Tone Test

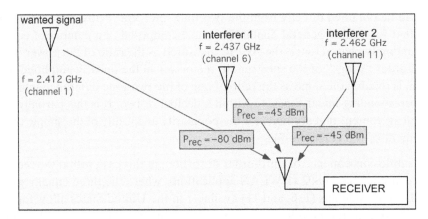

Figure 4-28: Simultaneous Transmitters on 802.11 U.S. ISM Channels

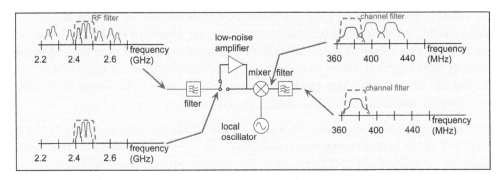

Figure 4-29: Progressive Filtering of Band and Channel Within a Superheterodyne Receiver

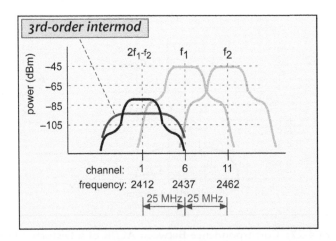

Figure 4-30: IM Product of Channels 6 and 11 at Channel 1

comparable with the noise, which for a QPSK signal is around 10 dB less than the signal for reliable reception. An IM product at −95 dBm seems unobjectionable. This is 50 dBc from the interfering tones at − 45 dBm, so the input intercept required to produce this result is $(−45 + (50/2)) = −20$ dBm. This estimate is in reasonable agreement with actual chipset IIP3, which varies from around −18 to − 8 dBm (see Table 4-10 later in this chapter).

Obviously, the same problem could arise for interferers on channels 1 and 6 and a wanted channel 11. Because only three nonoverlapping ISM channels are available in the United States, this unfortunate occurrence must be regarded as reasonably probable and taken into account in radio design. An 802.11a WLAN deployment has a great deal more flexibility in frequency planning because of the expanded number of available channels; although equally spaced interferers remain possible and should be accounted for in design, they are less likely to be encountered in the field.

The astute reader, in perusing Figure 4-21, may have noted with some concern that both the single-tone and two-tone terms of a third-order–distorted signal produce components at the fundamental. For the rest of us, the calculation is reproduced with this fact highlighted in Figure 4-31: the change in the fundamental power due to distortion is three times larger in amplitude than that at the spur frequency. Third-order distortion has a much larger absolute effect at the fundamental frequency than at the spurs, even though the effect is less noticeable on, for example, a spectrum analyzer because of the presence of the large linearly amplified fundamental. Pedro and de Carvalho showed that for a representative multitone signal, the corresponding result is a discrepancy of about 12 dB between the large effect on the linear signal and the smaller adjacent-channel power. In other words, the presence of third-order distortion in the signal chain results in—dare we say it?—nonlinear distortion of the input signal

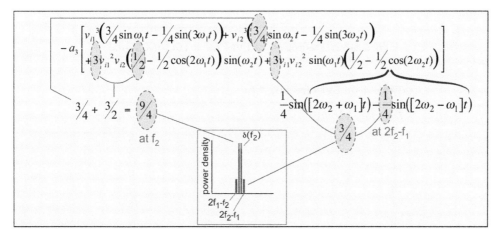

Figure 4-31: Repetition of the Calculation of a Third-Order Two-Tone Distortion; the Distortion Product at f_2 Is 3 Times Larger in Magnitude Than That at $(2f_2 − f_1)$

and, consequently, a difference between the signal that was to be transmitted or received and is. How do we measure this difference and when do we care that it exists?

Distortion of the received signal is quantified as *error vector magnitude* (EVM). EVM measures the difference between the transmitted or received constellation point on the phase-amplitude plane and the intended symbol, normalized by the amplitude corresponding to the average transmitted power (Figure 4-32). The measure is averaged over a pseudo-random sequence of input data to obtain a statistically valid estimate of the accuracy of the system.

Requirements on EVM for the 802.11a signal versus the data rate are summarized in Table 4-6. For higher data rates, EVM requirements are more demanding; this should not be surprising once we recall that the higher data rates use QAM constellations with more points and thus less space between the points (see Table 3-2). How much third-order distortion is allowed by these specifications? For the 54 megabits/second (Mbps) case, we need distortion less than -25 dB. Remember that third-order distortion of the fundamental is actually about 9 dB larger than the adjacent-channel spurious output (Figure 4-21). To get third-order distortion in the same channel (*cochannel* distortion) to be down by 25 dB, we require the adjacent-channel distortion

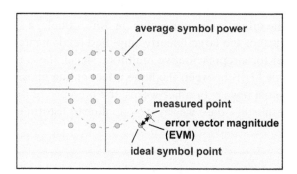

Figure 4-32: Definition of EVM for a 16QAM Signal

Table 4-6: EVM for 802.11a

Data Rate (Mbps)	<EVM> (dB)
6	−5
9	−8
12	−10
18	−13
24	−16
36	−19
48	−22
54	−25

to be at least 34 dB less than the signal. To have a two-tone distortion 34 dB below the signal (Figure 4-23(b), Δ), we need the intercept to be at least 17 dB above the signal power (i.e., $\Delta/2$).

On the receive side, it's not hard to provide a sufficiently large intercept: the maximum input power we expect to see is around -30 to -40 dBm, and the input third-order intercept IIP3 is typically around -10 dBm even in the high-gain state (which we would not be using to receive such a large signal). On transmit it's another story. A typical amplifier might achieve an output intercept point OIP3 that's 10 dB higher than its nominal output power. To keep EVM under control, we'd need to reduce the output power—back off—by another 8 dB. Every dB of backoff means less output power for the same DC power. Achieving EVM specifications is often the limitation on output power for high–data rate OFDM signals.

So far we've dealt only with low-order polynomial distortions. As output power is increased, every amplifier reaches its limit: the input power at which output power can no longer increase. The maximum output power is known as the saturated output power, P_{sat}. Any signal larger than the limit is *clipped*: the output power becomes substantially independent of the input power (Figure 4-33). Folks often use the power at which the gain has decreased by 1 dB, P1dB, as a somewhat less disastrously distorted limit on the output. In an ideal amplifier that is perfectly linear until it clips, the input 1-dB-compressed power would be 1 dB greater than the input power for saturation; however, real amplifiers aren't quite so abrupt, and P1dB is a decent substitute for P_{sat} for many purposes.

Clipping can be analyzed by adding higher order odd distortion terms (fifth, seventh, etc.) to the polynomial transfer function (Figure 4-13). However, for many purposes, Cripps showed that it is simpler and more profitable to use an ideal polynomial transfer function, as shown in Figure 4-33: the amplifier is either linear or fixed in output power (see section 6). In either case the treatment is complex and unenlightening for our

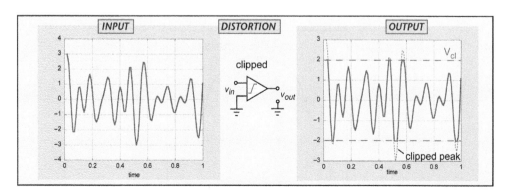

Figure 4-33: Clipping Distortion

purposes; the net result, in addition to the obvious distortion of the desired signal, is a spectrum with wings that extend farther and fall off more slowly than those due to third-order distortion.

Dealing with clipping appears at first sight to be quite simple: turn the power down until there isn't any. However, recall that many signals have a *peak-to-average* ratio P/A > 1. (This quantity is also frequently referred to as the *crest factor*.) The largest value that the signal can take is much larger than the average power for a random input stream. To avoid clipping, it's not enough to keep the power at the saturated power; we need to back off by (roughly) the peak-to-average ratio to keep from clipping. How much is that?

The peak-to-average ratio depends not just on the modulation scheme but also on the path taken between constellation points. Some modulations, such as Gaussian minimum-shift keying, are chosen specifically because the path between constellation points is along a circle of constant power (Figure 4-34), though at the sacrifice of some bandwidth. QPSK is a bandwidth-efficient modulation, but the path between points passes close to the origin (depending on the previous symbols), so the average power is less than the constellation point power.

The actual peak-to-average ratio for a particular modulation scheme depends on the exact approach to making the transitions between symbols and is generally obtained numerically. An example for a QPSK system is depicted in Figure 4-35: the calculated (P/A) = 2.7:1 or about 4.3 dB. An alternative approach, *offset QPSK*, is sometimes used in which the I-branch transition and the Q-branch transition are offset by 45 degrees; in this case, the trajectories never pass through the origin and the (P/A) is somewhat reduced.

For a complex signal like OFDM, the peak-to-average ratio is a statistical quantity: what is of interest is how often a given peak power is encountered in a random data stream, because if a certain level is sufficiently improbable, it is no longer important. The likelihood of a given power level can be succinctly displayed using a cumulative distribution function, the combined likelihood that the power in a sequence is less than a given power level. We encountered some of these before in Figure 3-20; they are reproduced below as Figure 4-36 for convenience.

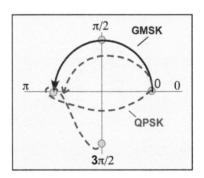

Figure 4-34: Trajectories in Phase-Amplitude Space for Two Modulations

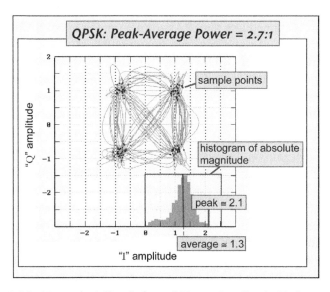

Figure 4-35: Numerical Simulation of Phase-Amplitude Trajectory of QPSK Signal for a Pseudo-Random Input (Simulation by Walter Strifler, Image Courtesy of WJ Communications)

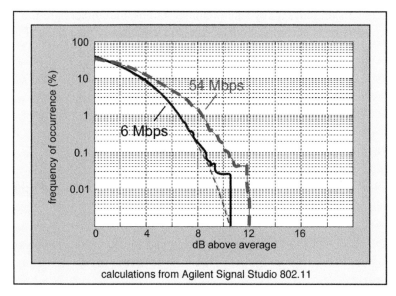

Figure 4-36: Combined Cumulative Distribution Functions for 802.11 OFDM Signals at Low and High Data Rate

If we were to choose a frequency of 0.1% (10^{-3}) as our threshold, we see that a 6 Mbps signal has a (P/A) $\approx$ 8.3 dB; the 54 Mbps signal (P/A) is around 10 dB. To transmit an OFDM signal that is clipped less than 0.1% of the time, it appears we would have to back off about 10 dB—that is, we can only transmit 10% of the power that our output amplifier can deliver as useful OFDM power. Fortunately, things aren't quite this bad. In practice, around 5–8 dB of backoff is sufficient to achieve compliant output spectra. For high data rates (>32 Mbps), it is usually the EVM specification that limits output power rather than the spurious emission due to distortion or clipping.

Let us pause for a moment to briefly summarize this rather long subsection. Amplifiers must provide enough gain on the receive side to deliver a signal the ADC can read and enough power on the transmit side to reach the receiver. The noise in the receiver is dominated by the LNA noise if the gain of the LNA is reasonably large. The noise floor sets the lowest signals that can be received; the upper limit is set by distortion. Typically, we are most concerned with third-order distortion, because the resulting spurious outputs are in and near the original frequency and cannot be filtered. Third-order distortion can be avoided by using more linear amplifiers (which usually costs either current or dollars or both) or by reducing the signal amplitude. Complex signals also have a significant (P/A), and average power must be backed off enough to avoid undue disturbances of the signal resulting from clipping distortion during the relatively rare excursions to high power.

2.3. Mixers and Frequency Conversion

Frequency conversion plays a key role in analog radios, and mixers are the means by which this conversion is accomplished. Several are used in a typical radio (Figure 4-37).

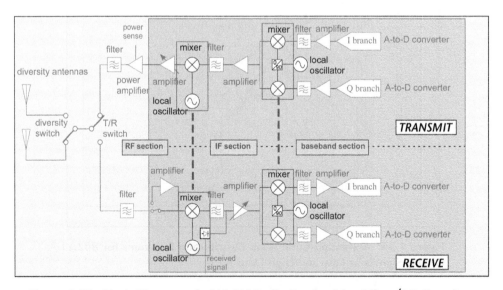

Figure 4-37: Block Diagram of a WLAN Radio Emphasizing Mixer/LO Functions

A mixer combines an incoming transmitted or received signal with a nearly pure sinusoid from a LO to create new frequencies. On the transmit side, these new frequencies are typically much higher than the incoming signal: the mixer is used for *up-conversion*. On the receive side, the output frequency is reduced: *down-conversion*. The efficiency of conversion is normally described as the *conversion loss*: the ratio of the output power at the converted frequency to the input power at the original frequency. (This can be a conversion gain in an active mixer such as the Gilbert cell, discussed below.) Mixers are always nonlinear devices, but good mixers are highly linear. This puzzling apparent contradiction will hopefully be resolved shortly as we examine mixer operation.

How does a mixer work? Recall from Chapter 2 (Figure 2-14 to be exact) that when we modulate a signal, new frequencies appear in its spectrum. A mixer is a modulator in disguise: it modulates a sinusoidal input to produce new frequencies, one of which is typically selected by filtering the output (Figure 4-38).

How are we to produce this modulated signal? Recall that the simplest modulation scheme is on–off keying (see Figure 2-12); similarly, the simplest mixer is just a switch that turns the incoming signal on or off. Such a simple mixer is depicted schematically in Figure 4-39; its use as an up-converter and down-converter is shown in Figure 4-40. The input signal is effectively multiplied by a square wave whose value varies between 1 and 0. Because the square wave and the input are at differing frequencies, their relative phase varies with time, producing a time-varying output with components at the sum and difference frequencies. It is this phase variation that is key to the operation of the mixer as a converter.

In addition to its manifest simplicity, such a mixer has significant advantages. Our ideal switch is either on or off and nothing else. When it is off, there is no output, no matter what the input. When it is on, the output is equal to the input. Thus, even though the

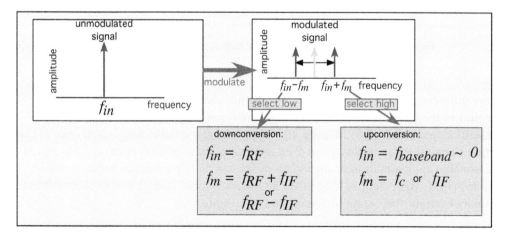

Figure 4-38: Modulation as Frequency Conversion

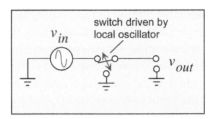

Figure 4-39: Simple Switch Mixer

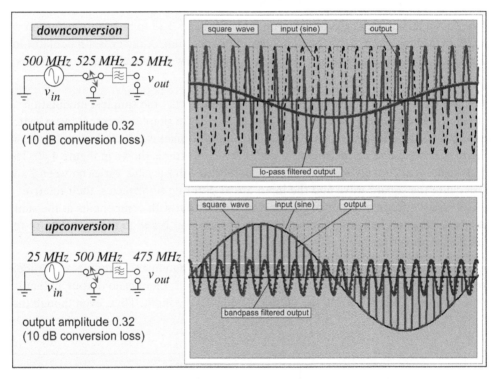

Figure 4-40: Input, Control, and Output Signals for Switch Mixer

operation of the device is highly nonlinear in the sense that the two possible states have very different output–input relationships, within each state the output is a perfect replica of the input (admittedly with a gain of 0 in the "off" state). Once the sharp transitions are removed by filtering, the output signal is a faithful replica of (part of) the input signal: there is no distortion. The mixer is a linear nonlinear device. Further, a reasonable approximation of the switch mixer can be implemented with any active device (FET, BJT, etc.), as long as the LO input is large enough to switch the device rapidly from its "off" state to a saturated "on" state.

The price for this simplicity is apparent in Figure 4-40. The switch is off half the time, so half the input signal power is wasted: it is immediately apparent that the conversion loss must be at least 3 dB. The actual situation is worse because many harmonics are present in the output signal in addition to the desired frequency; when the output is filtered to obtain the wanted signal only, 10 dB of conversion loss has resulted.

Another limitation of the simple switched mixer is that it does not reject *images* of the input or suppress images on the output. The down-converter will also convert a 550-MHz input to 25 MHz; the up-converter will also produce a 525-MHz output signal. A simple mixer of this sort must be combined with filtering to remove undesired image products.

An improved mixer design can be obtained using two switches, and *baluns* to convert the input signal from a *single-ended* voltage referenced to ground to a differential or *balanced*[1] signal symmetrically placed around the ground potential (Figure 4-41).

This mixer effectively multiplies the input signal by a square wave with values $(+1, -1)$ instead of $(1, 0)$ (Figure 4-42). Thus, the input is always connected to the output, albeit with alternating polarity, and the conversion loss is reduced. Because of the balanced design, no power is wasted in a DC offset, so the realized conversion loss is only about 4 dB: a 6-dB improvement. As long as the switches are ideal, the mixer is also linear in the sense described above. The trade-off is that complexity has increased considerably. An additional switching element has been added, and (if the input signal isn't already differential) two baluns are required to convert the single-ended signal to differential and back again. In Figure 4-41, the baluns are shown as transformers. Broadband baluns of this type for frequencies up to a few gigahertz can be constructed by wrapping a few turns of trifilar (three-filament) wire around a donut-shaped ferrite core a few millimeters in diameter, at a cost of some tens of cents; such devices are obviously huge relative to a chip and must be placed separately on the circuit board. Narrowband baluns can also be constructed using lines on the circuit board as transmission lines, with one branch delayed to produce a 180-degree phase shift, again a

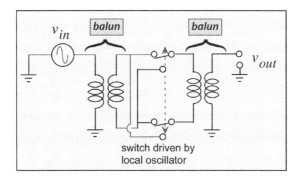

Figure 4-41: Balanced Switch Mixer

[1] Unfortunately it is also common practice in the microwave world to use the term *balanced* to describe a quite distinct trick in which two signals are offset from each other by 90 rather than 180 degrees, done to produce an improved input and output match.

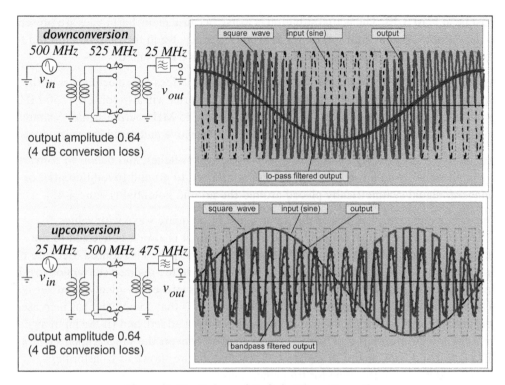

Figure 4-42: Balanced Switch Mixer Operation

large structure requiring space on a board. Active baluns can be fabricated using an amplifier but add current consumption and distortion. Baluns are an inconvenient but often inevitable fact of life for radio designers. The balanced switch mixer is also subject to the same limitations with regard to image rejection noted for the simple (unbalanced) switch mixer.

Both mixers we describe here are known (despite the use of apparently active switch elements) as *passive mixers*. The nomenclature emphasizes the fact that the output is at best equal to the input; no amplification accompanies the mixing operation. We have already seen that balanced mixers have lower conversion loss. This fact is doubly important because the noise figure of a good passive mixer is essentially equal to the conversion loss (plus a slight correction for excess noise in the devices), so a balanced mixer requires less amplification before the mixer to get good overall noise performance.

To achieve good linearity between the input and the converted output, it is necessary that the transitions between switch states occur as rapidly as possible. This is because real switching devices (FETs, BJTs, or diodes) are highly nonlinear during the transition. For example, a FET device typically has high distortion near the pinch-off voltage where the current goes to zero; if the mixer spends a lot of time turning off, considerable distortion in the filtered output signal will result. A linear mixer requires that the switching devices be fast compared with the signal frequency, and the best performance is usually obtained at a LO voltage significantly larger than the minimum necessary to switch states, so that the time spent in the transition is only a small fraction of the LO cycle.

Another important characteristic of mixers is *isolation*, referring not to the existential loneliness of an unwanted image frequency but to the extent to which the output voltage is isolated from the input voltages. Only the converted products should appear at the output: the LO and input signals should not. The LO is usually the biggest problem because it is much higher in power than the other inputs. High LO powers may be used, as noted above, to ensure fast switching of the active devices and thus good linearity. Leakage of the LO into the transmitted signal during up-conversion results in an undesired emitted frequency, or for direct upconversion a peak in the center of the transmitted spectrum that may confuse the demodulation process; allowed leakage is generally specified in the standards. LO leakage into the IF stages on the down-conversion side may be further converted by IF nonlinearities into a DC offset. Balanced mixers generally provide better isolation than unbalanced mixers; for example, a small DC offset in the input signal in Figure 4-40 would be converted by the switch into an output signal at the LO frequency, whereas the same offset in Figure 4-41 would be rejected by the balun and have no effect. Isolation is normally limited by more subtle effects such as parasitic capacitive or inductive coupling within the device; balanced signals are also beneficial in this case, because they contain no net voltage to ground and are thus less likely to couple over long distances.

Unwanted images are not the only undesirable outputs of a mixer. An ideal switch mixer effectively multiplies the input signal by a square wave, which contains not only the fundamental frequency but all odd harmonics (Figure 4-43). The amplitude of higher harmonics decreases rather slowly (as $1/n$ for the nth harmonic). Each of these harmonics multiplies the input signal to produce spurious outputs (*spurs*) at, for example, $3f_{LO} + f_{IF}$. Any nonlinearities present in the device will produce mixing of these spurs with whatever other frequencies are lying around to produce essentially

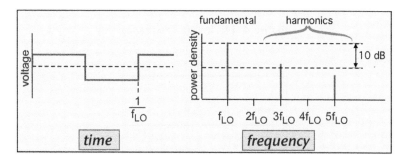

Figure 4-43: Harmonics of Square Wave Signal

every imaginable integer combination of frequencies: $3f_{LO} - 2f_{IF}$, $2f_{LO} - 3f_{IF}$, and so on. Spurs can sneak through filters and accidentally lie on or near wanted signals and cause all sorts of mischief. Balanced mixers are usually better than unbalanced mixers, but all are subject to spurious outputs. The intensity of all the possible low-numbered spurs of a mixer is often measured or simulated and summarized in a *spur table* for a given set of input conditions.

We've mentioned a couple of times that images of the wanted frequency are a problem. A simple mixer with a LO frequency of 2.5 GHz will convert a signal at 2.43 GHz to a 700-MHz IF, but a signal at 2.57 GHz will also be converted to the same IF (Figure 4-44) and interfere with the wanted signal. What measures can be taken to avoid this problem?

One obvious approach is to filter out the offending signal before it reaches the mixer (Figure 4-45). Remember the near–far problem: an interfering signal could be much larger than the wanted signal, so an image filter has to do a very good job of filtering the image. Obviously, the larger the IF frequency, the greater the separation between image and wanted signal and the easier it is to filter the image. On the other hand, use of a very high IF sacrifices some of the benefits of converting in the first place: gain is harder to come by and channel filtering is rendered more difficult. The choice of the IF (*frequency planning*) is an important aspect of radio design. In the limit of an NZIF, radio filtering is impossible, because the image is immediately adjacent to the wanted signal.

A different and much more subtle approach is to construct a mixer that converts only the wanted signal and rejects the image frequency: an *image-reject mixer* (IRM). A conceptually identical trick is to produce only the wanted sideband on up-conversion:

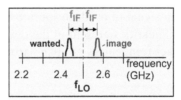

Figure 4-44: Image Frequency

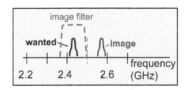

Figure 4-45: Image Filter

a *single-sideband* mixer. IRMs are somewhat more complex than conventional mixers but are key enablers for NZIF architecture and are often useful in other radio designs. To explain how they work, it is helpful to expand upon the idea of complex exponential signals introduced in Chapter 2.

We can write a sinusoidal input voltage as the sum of two exponentials:

$$\cos(\omega t) = \frac{e^{i\omega t} + e^{-i\omega t}}{2}; \sin(\omega t) = \frac{e^{i\omega t} - e^{-i\omega t}}{2i} \qquad [4.8]$$

We can construct a graphic depiction of these operations if we admit the concept of a negative frequency (Figure 4-46). Exponentials of positive frequency are depicted as arrows on the frequency axis to the right of zero and are to be imagined as spinning counterclockwise around the axis at the frequency ω. Negative-frequency terms lie on the negative axis to the left of $f = 0$ and spin clockwise in the complex plane, perpendicular to the frequency axis. A phase shift α displaces positive frequencies and negative frequencies in opposite directions. (We apologize for the rotation of the conventional complex plane to place the real axis vertically, but this is convenient as real signals then point up or down.) The justification of equation [4.7] can be easily constructed: for a cosine, the positive-frequency and negative-frequency arrows both start vertical at time $t = 0$ and counterrotate so that their sum has no imaginary part. A similar construction takes place for the sine, which, however, is rotated onto the imaginary axis.

Because mixing is a multiplicative process, it is necessary to go a bit further and figure out what happens to our picture when we multiply two signals together. Mathematically, this is very simple (which is the reason to use exponential signals)—the product of two exponentials is a third exponential of the sum of the arguments. In the case of harmonic signals, the frequencies add

$$e^{i\omega_1 t} \cdot e^{i\omega_2 t} = e^{i(\omega_1 + \omega_2)t} \qquad [4.9]$$

In our spatial picture, the product of two arrows is another arrow whose angle (phase) is the sum of the constituent phases and whose frequency is the sum of the constituent

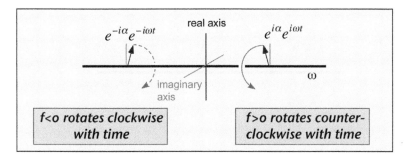

Figure 4-46: Exponentials in Complex–Frequency Space

frequencies. Assume for the present we are interested in a down-conversion application and ignore the terms with the same sign of frequency, which when summed will produce high-frequency components to be filtered out. The product of a positive-frequency term from, for example, a signal at $(\omega + \delta)$ and a negative-frequency term from a LO at frequency $(-\omega)$ will be at positive frequency (Figure 4-47, top set of dashed lines). A negative-frequency signal term will combine with a positive-frequency LO of lesser absolute frequency to produce a negative-frequency term (bottom set of dashed lines). Here the negative-frequency signal is shown as phase shifted by 180 degrees, and the product term inherits this phase shift.

With these preliminaries in mind, let us now consider the problem at hand: we wish to down-convert a wanted signal at some frequency $f = f_{LO} + f_{IF}$ but reject the image at $f_{LO} - f_{IF}$. First, let's mix the signal (regarded as a pair of cosines) with a cosine LO—an *in-phase* conversion. The result is shown in Figure 4-48. At positive frequency we find both the positive-frequency component of the wanted signal and the negative-frequency component of the image (dashed lines); the negative-frequency terms arise from the negative-frequency wanted signal and the positive-frequency image.

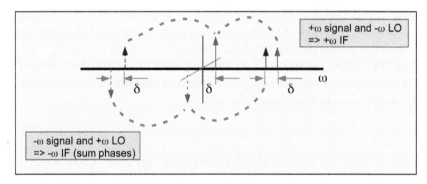

Figure 4-47: The (Low-Frequency) Products of Cosine and Sine

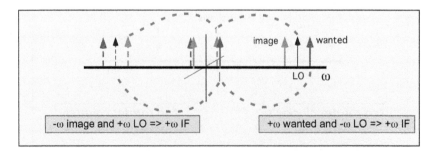

Figure 4-48: I-Channel Mix of Image and Wanted Input

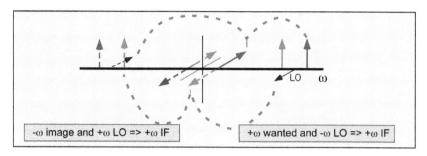

Figure 4-49: Q-Channel Mix of Image and Wanted Input

What happens if we also mix the signal with a sine—the *quadrature* channel? As shown in Figure 4-49, the components at each frequency acquire a different phase depending on whether they originate from the positive- or negative-frequency component of the original signal (as indicated by the dashed lines). Two constituents that end up at the same frequency are not the same in phase: it is this fact that enables us to separate out image and wanted signals.

To exploit the phase difference between the components of the signal, we apply a 90-degree phase shift to one of the branches, for example, the Q branch. This can be done using an R-C network or by simply delaying the branch by a quarter-cycle with a bit of extra transmission line. Recall that a positive phase shift rotates vectors counterclockwise on the positive frequency axis and clockwise on the negative frequency axis. The resulting signal is shown in Figure 4-50, where the rotation operation is explicitly shown on only the negative frequency components for clarity.

The components deriving from the wanted signal are positive at positive frequency and negative at negative frequency. The components deriving from the image are of opposite sign. If we now add this signal to the I branch signal (Figure 4-48), the components from the wanted signal will add, whereas the components which came from the image will subtract and cancel: the image has been rejected. If on the other hand we subtract the two signals, the image will survive and the formerly wanted signal will be

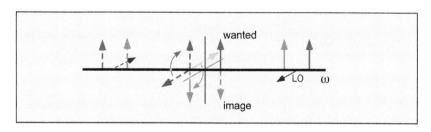

Figure 4-50: Q Branch After Phase Shift

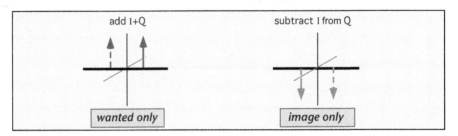

Figure 4-51: Alternatives for IRM Output

rejected, that is, we can choose to accept either the upper or lower signal relative to the LO (Figure 4-51).

A block diagram depicting an IRM is shown in Figure 4-52. The price we pay for achieving rejection without filtering is additional complexity: we not only need two mixers instead of one, but also an accurate means of providing phase shifts both to the LO (to produce the I and Q branches) and the mixed signal to produce the phase-shifted results.

It is worth pausing for a moment to reflect upon Figures 4-48 and 4-49 in the context of a single digital signal. The digital signal generally contains sidebands at frequencies both above and below the carrier, playing the role of the image and wanted frequencies above. If we convert the carrier to DC in only one branch (I here, Figure 4-48), the upper and lower sidebands sum to form the mixed signal and cannot be separated out. If they are the same (an AM or on–off keying signal) we don't care, but if they are not (which is the case with, e.g., QPSK or QAM), we have lost the information needed to demodulate the signal. If we also perform a conversion with a phase-shifted LO signal (Q branch, Figure 4-49), the upper and lower sidebands subtract to produce the mixed result. By having both I and Q branches (the sum and difference of the upper and lower sidebands), we can separately extract the upper and lower sideband amplitudes and

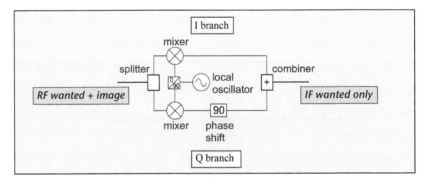

Figure 4-52: Block Diagram of an Image-Reject Mixer

phases and thus the original signal. An I-Q or *complex* mixing operation is necessary to preserve the full information in a signal if the carrier is converted to DC; the same requirement applies in up-converting the transmitted signal. This is why we see I and Q mixers and modulators in our standard block diagram (e.g., Figure 4-5 or 4-37).

To provide an example of how a mixer is actually constructed, we take a quick look at a very popular arrangement, the Gilbert cell mixer. We examine an implementation based on bipolar transistors, but Gilbert cells can also be constructed using FETs as active devices. A schematic diagram of a simple Gilbert cell is shown in Figure 4-53. Figure 4-53 assumes a down-conversion application, but the circuit is equally applicable to a modulator or up-converter.

A cell of this type is considered to be a *triple-balanced* mixer, because the input, voltage-controlled oscillator (VCO), and output connections are all differential. Such an arrangement is very helpful in achieving good isolation. As is apparent with the aid of Figure 4-54, the output voltage is zero if the RF voltage is zero, because in this case equal currents flow in the left and right branches of the mixer (assuming the devices are ideal and symmetric), and the output voltage is the sum of a positive LO contribution (the left branch) and a negative LO contribution (the right branch). Similarly, the output voltage is zero if the LO voltage is zero, because in this case equal currents flow in the upper pairs of the mixer and the left and right branch voltages are individually equal to 0. A fully balanced Gilbert cell will provide good LO-IF and RF-IF isolation if care is taken to ensure symmetric devices and layout.

The bottom pair of transistors converts the RF voltage into a differential current between the left and right branches of the mixer. The upper two pairs of transistors act as switches to direct the differential currents into either the positive or negative output connections (Figure 4-55). When the LO is high, the current is directed to the outermost

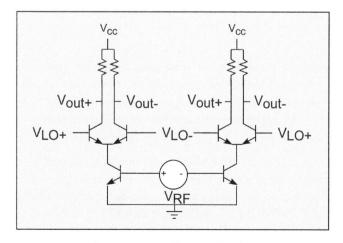

Figure 4-53: Gilbert Cell Mixer

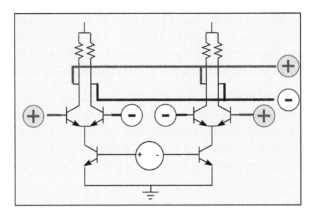

Figure 4-54: Balanced Input and Output Connections

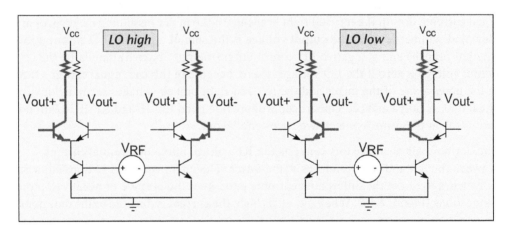

Figure 4-55: Gilbert Cell States

branches and the output is taken in the same sense as the RF voltage. When the LO is low, the current is directed to the inner branches and the output is taken in the opposite sense to the RF current. The mixer in effect multiplies the input signal by a square wave with values of $(+1, -1)$, performing the function of the balanced switch mixer of Figure 4-41.

Reported performance of some representative commercial mixers is summarized in Table 4-7. Performance of mixers incorporated within radio chips is not always reported separately; to focus only on the mixing function we included only discrete mixers in the table. The MESFET and MOSFET are passive mixers and display sizable conversion loss; the SiGe HBT is a Gilbert cell and thus benefits from the gain of the RF (bottom) transistors to provide conversion gain. However, noise figure is comparable between

Table 4-7: Mixer Performance

Technology	Conversion Gain (Loss)	NF (2 GHz)	IIP3 (2 GHz)	LO-RF Isolation	Reference
GaAs MESFET	−9.0 dB	9.5 dB	30 dBm	37 dB	WJ Comm 03
Si MOSFET	−7.5 dB	7.5 dB	31 dBm	30 dB	Sirenza 03
SiGe HBT	+10.4 dB	10.2 dB	4.4 dBm	32 dB	Maxim 03

the three technologies and is quite large. As we discussed in connection with Figure 4-12, in a receiver it is important to precede the first RF mixer with enough gain to ensure that the noise figure of the receiver is preserved despite the mediocre noise performance of the mixing operation.

Mixer linearity, as measured by the input third-order intercept, can vary over a wide range. The difference between the IIP3 and the noise figure is a measure of the mixer's dynamic range. If the IIP3 is low, one must limit the gain before the mixer to avoid the spurious output of a large distorted interferer from blocking reception of the small wanted signal. This isn't going to work very well if the noise figure of the mixer is high. A high noise figure can be dealt with by additional gain before the mixer if the IIP3 of the mixer is large.

Isolation of 30–40 dB is achievable; this is normally sufficient to avoid any serious problems from leakage of the LO signal into the wanted signal. Isolation is dependent on the details of the mixer construction and packaging as much as on the active device technology. Full specification of a mixer is more complex than an amplifier, involving selection of three input frequency bands (RF, IF, LO), three isolations, three power levels, and innumerable possible spurious products.

2.4. Synthesizers

It is perhaps apparent from the discussions of the previous section that mixers don't work very well without a LO. The quality of the LO signal plays an important role in determining several key parameters of the radio performance. As a transmitter, the LO must have sufficient absolute accuracy to deliver the proper transmitted carrier frequency within a precision typically specified in the standards; the receiver needs an accurate LO to convert the incoming RF signal accurately onto the IF bandwidth. The LO must be sufficiently tunable to address all the usable transmit or receive channels, and in the case of frequency-hopping protocols such as Bluetooth, the LO must tune rapidly enough to be ready to receive the next packet at the new frequency. Finally, the random variation in frequency—the *phase noise* of the oscillator—must be small enough to preserve the phase of the transmitted signal and allow accurate detection of the phase of the received signal, because as the reader will recall from Chapter 3 most WLAN modulations are phase sensitive.

Essentially all modern radios use a synthesizer architecture to produce the LO signals. A block diagram of a synthesizer is shown in Figure 4-56. At the heart of a synthesizer is a *voltage-controlled oscillator* (VCO), which produces a signal whose frequency varies monotonically with a control voltage over the desired band. The VCO is embedded in a *phase-locked loop* to accurately set its frequency.

The output of the synthesizer is split, and a portion of the signal drives a divider to produce a signal at much lower frequency, whose zero crossings are accurately synchronized to the LO signal. The divided signal is compared with a reference signal, typically generated by an oscillator whose frequency is set by a quartz crystal resonator. Quartz crystals are piezoelectric; in a resonator this property is exploited to convert an electrical signal to an acoustic displacement within a slice of bulk quartz. The crystal then acts as an acoustic resonator. Quartz resonators have very high quality factors (the reader who is unfamiliar with this concept should refer to section 2.5 below) and, if cut along the appropriate planes, have an acoustic velocity and therefore resonant frequency that is nearly temperature independent. A quartz crystal reference oscillator provides an inexpensive, reliable, accurate reference source. Typical frequencies for these oscillators are around 10 MHz, so the IF LO divisor is a few tens and the RF requires a larger divisor in the hundreds.

A *phase detector* determines the difference between the phase of the divided signal and the reference; this information is used to supply the control voltage for the VCO. A loop filter is provided to optimize the trade-off between responsiveness and stability, as with any feedback system. Roughly speaking, the LO control voltage is adjusted by the feedback loop until the phase error becomes small, ensuring that the synthesizer is locked to an integer multiple of the reference frequency.

The synthesizer in Figure 4-56 is not very flexible: for a fixed N, the output frequency will always be $N \cdot f_{\text{ref}}$. A more versatile synthesizer results if the divisor can be easily varied to allow differing output frequencies. One option is the *integer-N* synthesizer, in which two divisors can be used, differing by 1: for example, 16 and 17. The first modulus N is used for, for example, S cycles and then $(N + 1)$ for $(S - F)$ cycles, for a total of F cycles. After all the cycles are done, the divider outputs one rising or falling edge. The

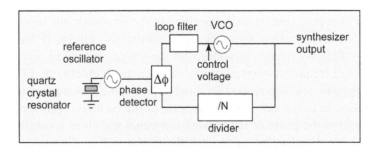

Figure 4-56: Synthesized Local Oscillator Block Diagram

net effect is to divide by $N_{\text{eff}} = (N + 1)S + N(F - S) = NS + S + NF - NS = NF + S$. Thus, by adjusting S (which just involves setting a counter), the overall divisor and thus the output frequency can be adjusted over a wide range, with a resolution of f_{ref}. If higher resolution is desired, the reference frequency can also be divided by some other integer M. However, using larger and larger divisors has a penalty: because the divider outputs an edge only after every N_{eff} cycles of the VCO output, information about the phase of the VCO signal becomes increasingly sparse as the divisor grows. The result is that the variation of phase that can occur without being suppressed by the feedback loop—the phase noise—increases with increasing N.

An integer-N synthesizer is typically adequate for WLAN-type applications, where channels are separated by 5 MHz or some other convenient value, but when better resolution is required a different architecture, the *fractional-N* synthesizer, may be used. In this case the divisor is again dithered between two values differing by 1, but in this case an edge is output after each divide cycle. If the fraction of time spent at divisor N is K, then the output frequencies are $f_{\text{ref}}(N + 1/K)$: the resolution can be made arbitrarily fine by increasing the total number of cycles to achieve accurate control over K. The trade-off is that the VCO frequency is actually varying on an instantaneous basis: it is a frequency-modulated signal. The frequency modulation results in undesired low-level spurious output at a range of frequencies determined in part by the timing of the index dither. Because these spurs are predictable based on the known timing of the divider modulus, they can be corrected for by downstream adjustments. Considerable progress has been made in recent years to minimize spurious outputs of fractional-N synthesizers, but they remain significantly more complex than integer-N architectures.

The absolute accuracy of transmitted signals must meet regulatory restrictions and compliance with standards requirements. For example, the 802.11 classic PHY requires that the transmitted center frequency be accurate to 25 ppm, which works out to ± 61 kHz for ISM channel 6. The receiver typically uses the same synthesizer used on transmit. The difference between the transmitted signal frequency and that to which the receive LO tunes constitutes an effective error in the frequency of the received signal when converted to nominal zero frequency; the error is on the order of 100 kHz. For two successive DQPSK signals at 11 Mchips/second, the error adds a phase difference of $2\pi(10^5)(9 \times 10^{-8}) \approx 0.06$ radians, which is much less than the 1.6 radians between QPSK constellation points.

OFDM signals have special requirements for frequency and phase control. An error in the received frequency results in the displacement of the apparent position of the subcarriers. The orthogonality between subcarriers only applies when the sample time is an integer number of cycles; an error in subcarrier frequency causes inter-subcarrier interference, the mistaken assignment of some of a given subcarrier's signal to its neighbors (Figure 4-57). A 100-kHz frequency offset is obviously a large fraction of the 312.5-kHz separation between subcarriers in 802.11a/g and would cause significant

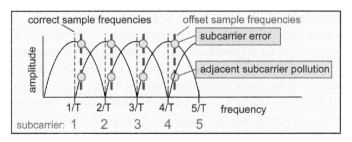

Figure 4-57: Offset Frequency and OFDM Subcarriers

interference between neighboring subcarriers. Fortunately, a simple offset of this type is easily corrected in digital signal processing after demodulation.

Instantaneous random variations in frequency—phase noise—cause the measured constellation points to be smeared out slightly and can be approximately regarded as another noise source in addition to the usual thermal noise. Thus, the total phase noise integrated over a frequency range corresponding to a symbol time degrades the (S/N) of the receiver and the EVM of the transmitter. Phase noise can also have a more subtle deleterious effect by broadening a large interferer on a nearby channel so that part of the interfering signal appears on the wanted channel (Figure 4-58).

An example of synthesizer phase noise performance in a WLAN chipset is given in Figure 4-59. The phase noise is normally measured as power relative to the carrier power (dBc) in a given bandwidth. Recall that one 802.11 OFDM symbol lasts 4 μsec, so phase noise at frequencies less than about 250 kHz corresponds to phase variation from one symbol to another; phase noise at frequencies above about 1 MHz corresponds to phase variation during a single OFDM symbol.

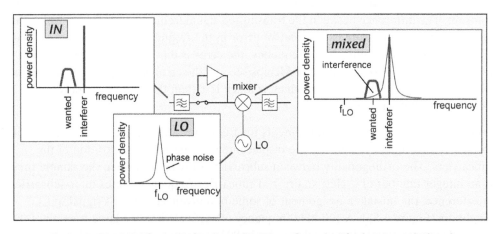

Figure 4-58: LO Phase Noise Broadens Interferer to Block Wanted Signal

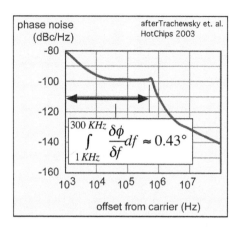

Figure 4-59: Example of Synthesizer Phase Noise for an 802.11a Radio Chip

The total phase error from 1 kHz (corresponding to the length of a complete packet) to 300 kHz is less than 1 degree, so that the radio can use the preamble of the packet to establish frequency offset and phase synchronization and thereafter preserve reasonable phase coherency over the remainder of the packet. Simulations have shown that this level of phase noise corresponds to less than 1 dB degradation in (S/N) required for a given bit error rate.

In early WLAN architectures the VCO was often a separate chip, with the synthesizer logic (the phase-locked loop part) integrated onto the converter chip. Modern chipsets normally include the necessary synthesizers integrated onto the radio chip.

2.5. Filters

Filters are circuits that reject some frequencies and transmit others. They are generally classified into three types: low pass, high pass, and bandpass (Figure 4-60).

Filters play a key role in rejecting undesired signals in a radio (Figure 4-61). The first filter in the receiver, the band or image reject filter, is designed to pass the RF band of interest (e.g., the 2.4- to 2.48-GHz ISM band) and reject signals from nearby bands.

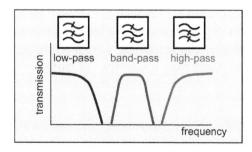

Figure 4-60: Three Types of Filters With Common Schematic Symbol

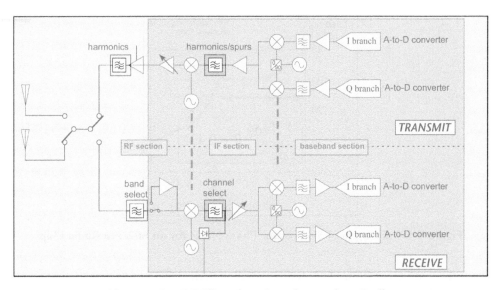

Figure 4-61: RF Filters in a Superheterodyne Radio

After mixing, a superhet radio passes the signal through a channel filter that, in combination with the LO frequency, selects the desired channel (e.g., channel 1 at 2412 MHz). After the demodulation step, or after the conversion step in a direct conversion receiver, additional low-pass filtering is used, although the frequencies of interest in this case are a few megahertz to perhaps 20 MHz, and discrete components or active filters can be used. Filters are also used in the corresponding stages of the transmit operation; in a WLAN superhet radio, a common IF is chosen so that the IF channel filter can be used both for transmit and receive to minimize cost.

The simplest bandpass filter structure is a resonator composed of an inductor and a capacitor (Figure 4-62). The resistor represents unintentional parasitic losses, typically concentrated in the inductor wiring. Resonance occurs at the frequency at which the inductive reactance and capacitive susceptance (i.e., the positive and negative imaginary parts of the conductance) are exactly equal. During each cycle, energy is alternately stored in the inductor (at the moment of peak current flow) and the capacitor (at the moment of peak voltage). An ideal resonator with no loss would appear as a perfect open circuit at resonance, so that all the input current would be transferred to the output. Away from resonance, the current would be shorted through the inductor (low frequency) or capacitor (high frequency). Thus, the resonator acts like a bandpass filter.

The filter can be characterized by a characteristic impedance Z_o and a quality factor Q. The quality factor is the ratio of the characteristic impedance to the parasitic resistance; it expresses the ratio of the energy stored in the inductor and capacitor to the energy

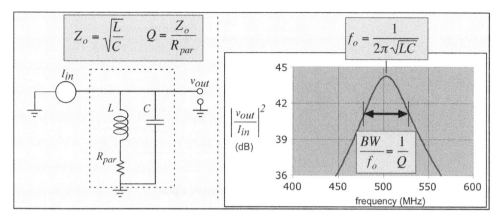

$$Z_o = \sqrt{\frac{L}{C}} \qquad Q = \frac{Z_o}{R_{par}} \qquad f_o = \frac{1}{2\pi\sqrt{LC}}$$

$$\frac{BW}{f_o} = \frac{1}{Q}$$

Figure 4-62: Simple L||C Filter

lost per cycle by the resistor. As shown in the right half of the figure, the quality factor also determines the width of the passband of the filter. Narrow passband filters must have very high Q. For example, an RF band filter for the 2.4-GHz 80-MHz wide ISM band must have a bandwidth on the order of 3% of the center frequency, requiring a Q of at least 30. (In practice, considerably higher Q is needed to make a good filter: the filter should have a fairly flat transmission in the *passband* and a sharp transition to very low transmission in the *stopbands*, rather than the peaked behavior shown in Figure 4-62). On-chip filters constructed with inductors and capacitors in Si CMOS processes are generally limited to Qs of about 10, mostly due to loss in the inductors. Discrete components offer Qs as high as 30 up to a few gigahertz, but complex filters with many elements constructed using discrete components will become physically large and are inappropriate for gigahertz frequencies. Better technologies with high Q, small physical size, and low cost are needed to provide image and channel filtering.

One of the most commonly used technologies for high frequencies is the *surface acoustic wave* (SAW) filter. SAW filters achieve Qs in the hundreds, are available in surface-mount packages, and can pass hundreds of milliwatts without damage. SAW filters are relatively expensive ($1 to $10) and, on the order of 1 cm square, are large enough that the number of filters must be minimized both to conserve board space and minimize cost. The resonant frequency of a SAW filter is somewhat temperature dependent; quartz filters are better in this respect than most other piezoelectric materials but are more expensive to fabricate.

A simplified SAW filter structure is shown in Figure 4-63. The device is constructed on a piezoelectric substrate such as quartz, $LiNbO_4$, or ZnO. Electrical transducers are constructed of a layer of a conductive metal such as aluminum, deposited and patterned on the surface of the piezoelectric using techniques similar to those used in integrated circuit fabrication.

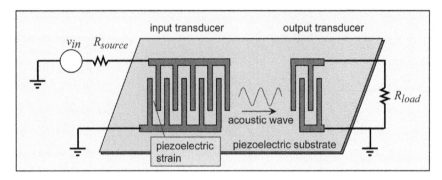

Figure 4-63: Simplified SAW Filter

The input transducer consists of on the order of 100 interdigitated fingers, driven at alternating polarity from an RF source. Between each pair of fingers an electric field is formed within the piezoelectric material, inducing a time-dependent strain that creates an acoustic wave. For an input frequency such that the spacing between fingers is half of the acoustic wavelength, a resonant enhancement of the wave will occur as it propagates along the transducer, as each alternating region of strain will be in phase with the wave and add to the displacement. The resulting strong acoustic wave propagates to the smaller output transducer, where the acoustic strain induces an electric field between the electrodes, resulting in an output voltage. The slice of piezoelectric is often cut at an angle to the propagation axis, as shown in Figure 4-63, so that the acoustic energy that is not converted back to electrical energy is reflected off the edges of the substrate at an odd angle and dissipated before it can interfere with the desired operation of the filter. Because the acoustic wave propagates about 10,000 times more slowly than electromagnetic radiation, wavelengths for microwave frequencies are on the order of $1\,\mu m$, making it possible to create compact high-Q filter designs.

The performance of a fairly typical RF band (or image-reject) filter is shown in Figure 4-64, as the transmission in decibels through the filter versus frequency. Within the ISM band, the loss through the filter is only about $3 \pm 1\,dB$: this transmission is known as *insertion loss*, because it is the loss in band due to insertion of the filter in the circuit. Low insertion loss is important on both transmit and receive. The insertion loss of the transmit filter comes directly out of the signal power, so lossy filters mean bigger transmit power amplifiers that cost more and consume more DC power. On the receive side, the filter loss is essentially equal to its noise figure, and because it is typically placed before the LNA, the filter noise figure must be added directly to the noise figure of the chain.

Other important properties of a filter are the sharpness with which transmission cuts off once the frequency is beyond the edges of the band and the transmission (hopefully small, thus rejection) of out-of-band signals. The bandwidth in this case at a 3-dB decrease in transmission versus the center frequency is about 125 MHz, rather

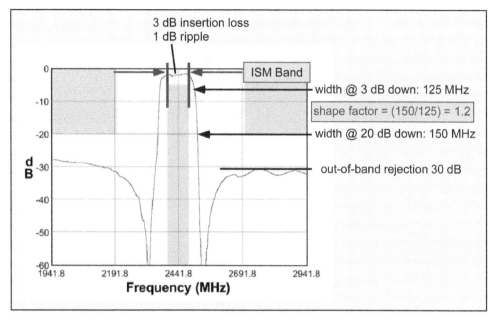

Figure 4-64: SAWTEK 855916 ISM-Band Filter Transmission vs. Frequency (Based on data plot ©2004 TriQuint Semiconductor; used by permission of TriQuint Semiconductor, SAWTEK Division)

noticeably wider than the 83-MHz ISM band: one cannot expect perfect rejection of a signal a couple of megahertz out of band. However, transmission falls quite rapidly thereafter: the *shape factor*, the ratio of bandwidth at 20 dB rejection to 3 dB rejection, is only 1.2. The rejection of signals far from the band edges—for example, the third-generation cell phone (UMTS) downlink frequency at 2170 MHz or the image frequency for a high IF superhet—is a substantial 30 dB.

Performance of an IF (channel) filter is shown in Figure 4-65. Higher insertion loss is tolerable for an IF filter as the transmit power at this point is modest and the receive-side gain before the filter ensures that the loss has little effect on noise figure. The benefit of accepting higher loss is the improved out-of-band rejection: the upper adjacent channel suffers about 43 dB loss and the lower adjacent channel 55 dB. High adjacent channel rejection is necessary for a receive channel filter likely to be exposed to simultaneous transmissions at the wanted and neighboring channels.

A related but distinct filter technology is the *bulk acoustic wave* (BAW) filter, sometimes referred to as a film bulk acoustic resonator. In place of the relatively complex lateral structure of a SAW device, a BAW substitutes a simple sandwich of a piezoelectric thin film between two metal electrodes. If the thickness of the piezoelectric is equal to a half wavelength, a resonant acoustic wave is created within the film. The key problem in fabricating a BAW device based on thin film techniques is to avoid leakage of the acoustic

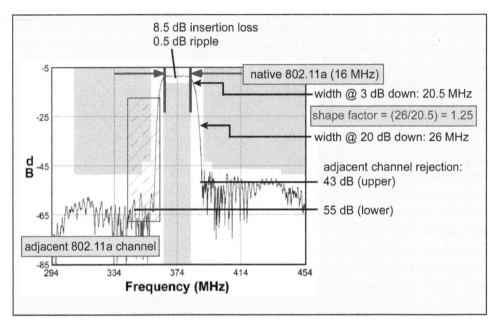

Figure 4-65: SAWTEK 855898 374-MHz IF Filter (Based on data plot ©2004 TriQuint Semiconductor; used by permission of TriQuint Semiconductor, SAWTEK Division).

wave into the substrate, which would represent a loss mechanism and decrease the quality factor of the filter. Two common methods of solving this problem are shown schematically in Figure 4-66. The core BAW device, consisting of a bottom metal electrode, piezoelectric layer such as AlN or ZnO, and a metal top electrode, is the same in both cases. The difference is the means of isolation from the substrate. One approach uses a mirror constructed of layers of alternating high and low acoustic impedance, each one-fourth of an acoustic wavelength thick. The acoustic wave is reflected at each successive interface between the mirror layers with opposite sign. The net phase shift of a half-wave (there and back again through a quarter-wave layer) means that the reflections from all the interfaces add in phase at the bottom electrode to produce a very high reflection coefficient and little transmission to the substrate.

A conceptually simpler but practically challenging method is to fabricate the active part of the piezoelectric layer over a cavity filled with air or vacuum, which conducts little acoustic radiation into the substrate. Such structures are normally formed by subtractive techniques. The cavity is etched into the substrate, filled with a material that can later be selectively etched such as silicon dioxide, and the resulting flat surface used to fabricate the remaining BAW layers. The device can then be immersed in an etchant that removes the cavity fill layer, leaving a void behind.

BAW filters have some advantages over SAW filters. The key dimension is the thickness of the layer rather than the lateral separation of electrode lines as in the SAW

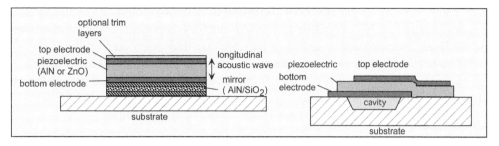

Figure 4-66: Cross-Sectional View of Two Alternative Approaches to BAW Fabrication

device. This thickness can usually be controlled more easily than a linewidth, so that good precision is possible. Typically, piezoelectric thickness are one to several microns for frequencies on the order of one to a few gigahertz. Thin trim layers of metal or dielectric may be added to compensate for any nonuniformity or imprecision in the piezoelectric layers to target precisely the correct resonant frequency. Power handling capability of BAW filters is higher than a similar SAW filter, both because the acoustic wave is distributed uniformly over the piezoelectric material, whereas in a SAW filter the energy is confined within a wavelength of the surface, and because in a SAW filter significant area is consumed by the periphery of the electrodes, edge reflection suppression, and other acoustically inactive regions. BAW filters, being thin-film devices, can in principle be directly integrated in a standard silicon fabrication process, though cost and complexity are significantly increased. High-performance BAW filters are currently commercially available for cellphone duplexers at PCS (1.8 GHz) band in the United States and similar (GSM 1900) frequencies in Europe and Asia.

High-performance filters can also be constructed using ceramic dielectric resonators. The resonator is simply a typically cylindrical slab of high-dielectric-constant material; a coil or other simple electrode structure can excite electrically resonant modes, like those within a metal cavity, which are confined mostly within the dielectric because of the high dielectric susceptibility. The basic challenge is that the size of the filter is on the order of half a wavelength of the electrical radiation in the dielectric; for low frequencies (for example, for IF channel filtering) such a structure is typically impractically large. Ceramic resonator filters become increasingly sensible at higher frequencies: at 5 GHz, the free-space wavelength is only about 5 cm, so a half-wave resonator with a dielectric constant of 10 would require a minimum dimension of about $5/\sqrt{(10)} \approx 1.5$ cm, just at the boundary of a practical surface-mount device. Ceramic resonator construction is fairly simple; the main challenges are in the production of an accurately dimensioned, high-dielectric-constant, low-loss ceramic puck.

Design of a radio, particularly a superhet architecture, is often a trade-off between radio performance and filter capability. A good band filter protects the LNA from out-of-band interferers and thus reduces the requirements for third-order distortion in the LNA and mixer; a cheaper filter will require a better amplifier. On transmit, harmonic filters can remove spurs generated by distortion in the power amplifier, reducing the linearity required of this expensive and power-hungry component. Direct-conversion and NZIF architectures are desirable in part because they place the channel filtering operation at low frequencies where integrated solutions are available. Filter technology influences radio design; as it progresses, it may be expected that optimal design choices for low-cost radios will also evolve.

2.6. Switches

Common WLAN protocols such as all 802.11 standards, Bluetooth and 802.15.3, and HiperLAN are all half-duplex: the radio switches between transmit and receive states but does not occupy both at once. Some client cards and most access points also make use of either one of two *diversity antennas*, depending on which produces the best received signal strength at a given moment. The choice of antenna changes from one packet to the next and thus a switching capability between antennas must be provided. The net result is that several switches are usually interposed between the radio input (filter/LNA) and output (power amplifier/filter) and the antenna(s).

The key performance issues in switch design and selection are as follows:

- *Insertion loss*: Is on on? The signal loss encountered in traversing the switch in the "on" state must be as small as practical. Insertion loss subtracts directly from the output power on transmit and adds directly to noise figure on the receive side. Typical values are 1–2 dB at 5 GHz. To minimize insertion loss, the series devices in the switch are made larger to reduce their on-state resistance.

- *Isolation*: Is off off? The isolation is the amount of power that sneaks through from the input to the output despite the switch being nominally in the "off" state. Typical 802.11 applications are not very demanding in this respect. During transmit, the power to the receive side LNA is typically disabled, so that fairly high input power can be tolerated without damage. The transmitter is also powered down during receive but would not be damaged by the small received RF power in any case: the issue here is residual transmitted power or noise interfering with the received signal. Small amounts of power coupled from the unused diversity antenna have little effect on the received power. Isolation values of 20–30 dB are relatively easy to obtain and are sufficient for WLAN radios.

- *Speed*: The 802.11 standards require switching from transmit to receive in roughly 1 μsec to ensure that the transmitter can respond after the short interframe space of 10 μsec. Choice of the diversity antenna must be made during the preamble, which varies somewhat in the various protocols but is much longer than 1 μsec. A slower switch could be used, but it is usually cheaper and simpler to use the same part for both roles.

- *Power handling*: A switch is ideally a passive part, having no effect on the signal when it is in the "on" state. Real switches are constructed with active devices that can introduce additional distortion to the transmitted signal; the output intercept power and compressed power must be sufficiently in excess of the actual transmitted power to ensure that the transmitted signal remains compliant with the spectral mask and regulatory requirements.

Switches can be implemented in any technology that supports active devices. FETs or BJTs in silicon or compound semiconductors are used in this function. FETs typically provide better isolation than BJTs. Compound semiconductors have superior on resistance due to higher electron mobility and better output power ratings due to higher saturated current and breakdown fields but cannot be integrated into the silicon CMOS radio. Separately packaged switches implemented with GaAs FETs are a common option.

A specialized versatile switching technology commonly encountered in microwave applications is the p-intrinsic-n or PIN diode. PIN diodes have a relatively large intrinsic (not intentionally doped) semiconductor region that makes them very resistive at both low and high frequencies when no DC current is flowing. However, application of DC forward bias causes electrons to be injected from the n region and holes from the p region, increasing the conductivity of the intrinsic region. PIN diode microwave resistance can be varied from $> 1000\Omega$ to $< 5\Omega$ by varying the DC current from 0 to 10–20 mA. PIN diodes are inexpensive and have excellent power handling capability, but their use requires provisions for separating the variable DC bias from the RF signal.

Switches can also be integrated in Si CMOS processing. An example of such an implementation, specifically designed as an 802.11a T/R switch, is shown in Figure 4-67.

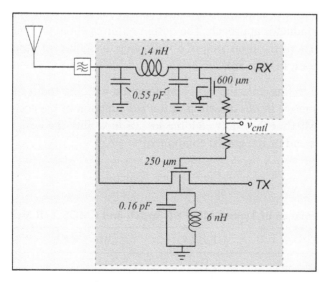

Figure 4-67: Simplified Schematic of Integrated CMOS Switch (After Talwalker et al., ISSCC 2003)

In the TRANSMIT state, the series transistor (250 μm wide) in the TX side is turned on. Note that the substrate side of this transistor is connected to a parallel L‖C filter whose resonant frequency is about 5.2 GHz. Recall that such a filter presents a large input impedance at resonance; thus, the back connection of the MOSFET is essentially isolated from the rest of the (grounded) substrate and is able to follow the imposed RF voltage. The bias resistor on the gate serves a similar purpose; it can be a large value because little charge is required to switch the FET state. The net result is that the FET gate and back gate are free to follow the instantaneous RF potential. Therefore, there is no RF voltage between the channel and gate and consequently little variation in the channel conductance during the RF cycle. Without this provision, the gate and/or substrate voltages would be fixed, causing the channel to be partially pinched off when a large RF voltage is present and leading to signal-dependent conductance—that is, distortion—of the RF signal.

Isolation of the receiver is achieved by turning the large parallel MOSFET (600 μm wide) on as well. This transistor shorts out the second capacitor; the first 0.55 pF capacitor and the 1.4 nH inductor together form a parallel resonator at 5 GHz, which again provides a large impedance and prevents the signal from reaching the FET. The Q of the filter is about 13, providing about 20 dB of isolation. Whatever signal does reach the FET is then shorted to ground by its small impedance, realizing an additional 10 dB of isolation.

In the RECEIVE state, the FETs are both switched off. The C—L—C structure then provides matching of the receiver to the (presumed 50Ω) filter/antenna. The TX FET, being off, provides about 20 dB of isolation. Performance of the switch is compared with a typical GaAs MESFET switch in Table 4-8. The CMOS switch achieves comparable TX output power and isolation, though insertion loss is slightly inferior to the compound semiconductor approach. The design requires about 0.6 mm^2 of silicon area, which is about 10% of the total area of a radio chip: a significant though not catastrophic increase in chip area.

It is thus possible to integrate the switching function with the radio chip. Whether architectures will move in this direction or not is a function of the cost benefits (if any) of doing so and whether they outweigh the loss of flexibility the designer gains from using external instead of integrated components.

Table 4-8: Performance of Typical MESFET Switch and CMOS T/R Switch Compared

Parameter	*MESFET*	*CMOS TX*	*CMOS RX*
Insertion loss (dB)	1.2	1.5	1.4
Isolation (dB)	26	30	15
P1dB (dBm)	31	28	11.5

3. Radio System Design

Now that we are familiar with the various components used in constructing a radio, we can take a look at the overall problem of designing a radio for a given set of requirements.

Because of the half-duplex nature of a WLAN radio and the importance of component cost, it is generally the case that key choices in architecture are the same for transmit and receive. In a receiver, a key choice is the frequency of first conversion. The choice of the conversion frequency or frequencies, and the corresponding LO frequency or frequencies, is generally referred to as *frequency planning*. In a direct conversion architecture frequency planning is simple: there is no IF, and the LO frequency must equal the desired RF channel.

In a superhet design, there are several possible choices of IF (Figure 4-68). In an NZIF radio, the IF is chosen to be comparable with the bandwidth of the baseband signal, typically a few megahertz. Channel filtering is done at megahertz frequencies and can be accomplished with on-chip filters or inexpensive discrete components. However, the image frequency is very close to the wanted frequency and cannot be rejected by filtering. Instead, an IRM design must be used.

Low-IF designs, where the IF is typically chosen between a few 10s of MHz and 200–300 MHz, represent the "classic" approach to superhet radios. The band filter can filter the image frequency effectively if the IF is greater than about 80 MHz (recall from Figure 4-64 that a typical SAW filter has a 20-dB bandwidth of about 160 MHz). Choice of a high value of the IF makes the image filtering easier but requires a high-

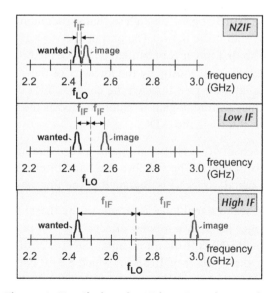

Figure 4-68: Choices for IF in a Superhet Design

performance SAW or similar filter for the IF. In such a design, the IF is common for transmit and receive so that the relatively expensive channel filter can be used in both directions.

In high-IF designs, the IF is several hundred megahertz to a gigahertz. In a high-IF design, image filtering is relatively easy and can be done with discrete components or on chip; however, some of the advantages of a superhet design are vitiated, in that IF gain is more difficult to obtain and channel filtering is difficult.

The designer also needs to choose whether the LO frequency is larger than the wanted RF frequency (*high-side injection*) or less than the wanted frequency (*low-side injection*). High-side injection requires higher absolute LO frequency but less percentage tunability. Low-side injection may make it more difficult to filter spurious outputs of the LO, because they are closer to the RF frequency.

Once the architectural decisions are made, the actual design uses *chain analysis* to follow the signal through the radio, keeping track of the signal strength, noise floor, IM distortion, and maximum unclipped power at each stage. The analysis enables the designer to estimate the receiver sensitivity, gain control requirements, and interferer rejection. On transmit the analysis is simpler, because there are no interferers.

A generic superhet radio chain is shown in Figure 4-69; the performance parameters are loosely based on the Intersil/Virata Prism 2.5, which was a very popular early 802.11b chipset. The radio is shown in the high-gain (best sensitivity) configuration. The signal passes through two switches (diversity antenna select and T/R), each of which contributes 1 dB of loss. The switch IP3 is taken as 50 dBm, so they contribute little distortion for the tiny RF signals likely to be encountered on receive. The band select (image reject) filter is taken to contribute 3 dB additional loss and provide 30 dB image rejection. The LNA has 20 dB of gain, enough so that subsequent stages will contribute only modestly to the overall noise figure. The output IP3 is 11 dBm, so the input IP3 is $(11 - 20) = -9$ dBm. The first mixer is presumed an active mixer with some conversion gain. Interference rejection is set mostly by the linearity of these two stages, because after conversion the signal must pass through the channel filter, which will reject most interferers. After filtering, a variable-gain IF amplifier boosts the signal before final I/Q demodulation, low-pass filtering, and ADC.

The chain analysis for sensitivity is shown in Figure 4-70, assuming the minimum signal required by the 802.11b standard. The switches and band filter do not change the noise floor but decrease the signal, so (S/N) is degraded in the first few stages.

Note that after the LNA, the signal level is far above the thermal noise level: excess noise in subsequent stages has only a modest effect on (S/N). In this maximum gain configuration, the output (S/N) is 16 dB for an input (S/N) of 26 dB, corresponding to an overall noise figure for the receiver of 10 dB, of which 8.5 dB comes from the input switches, filters, and LNA. The resulting signal can be placed roughly in the middle of the ADC dynamic range, allowing for minimal (S/N) degradation due to quantization

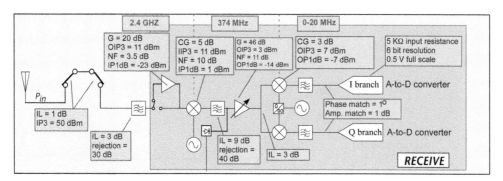

Figure 4-69: Simplified Superheterodyne Receive Chain (IL = Insertion Loss; IP3 = Third-Order Intercept; NF = Noise Figure; P1db = 1-dB-Compressed Power)

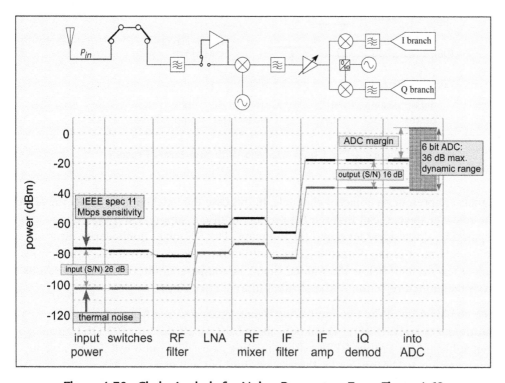

Figure 4-70: Chain Analysis for Noise, Parameters From Figure 4-69

noise in the converter and leaving some room for the signal strength to increase without saturating the ADC. The final (S/N) of 16 dB is quite sufficient to demodulate QPSK with good error rate, in view of the 2 dB gain from complementary code keying.

Now that we have established that the design meets minimum requirements for sensitivity, let's examine selectivity. Figure 4-71 shows what happens when interfering

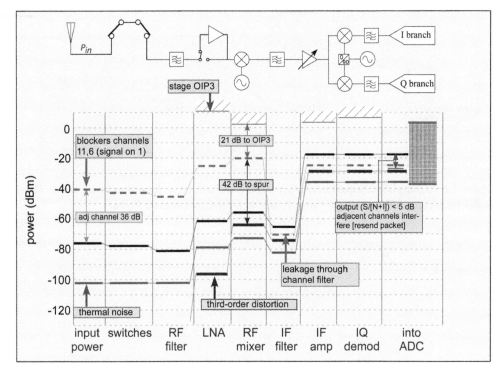

Figure 4-71: Chain Analysis With Interferers Present

signals are present on adjacent channels. In this case the hatched lines depict the IP3 of the relevant stage; recall that the IM product in dBc is suppressed by twice the separation of the signal level and the intercept. The RF mixer nonlinearity turns out to contribute most of the distortion. The resulting IM product (shown as a solid line) is only a few decibels below the wanted signal. A comparable interfering signal results from the limited rejection of the channel filter (dotted lines).

The combination of the two interfering signals gives rise to a (signal/(noise + interference)) ratio (S/(N + I)) of only about 5 dB, insufficient for 802.11b transmission (but sufficient for 1 Mbps BPSK with spreading gain). The radio will not be able to communicate at 11 Mbps under these conditions but will work at 1 Mbps. Retransmission of the packet is also very likely to solve the problem, because the interfering stations are unlikely to both be present at the next transmission slot.

This simple example shows that a good design must trade off sensitivity and linearity. If we had added another 10 dB of LNA gain, the sensitivity would be improved but the distortion of the interferers in the RF mixer would have made the selectivity poor. If the wanted signal is much larger than the sensitivity threshold, significant distortion of the wanted signal could occur in the analog stages or through saturation of the ADC; we need to turn the overall gain down for larger wanted signals. Gain adjustment is usually

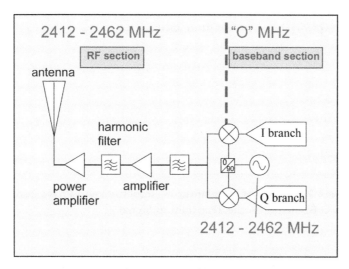

Figure 4-72: Direct-Conversion Transmitter

performed during the preamble to the packet. Improved linearity would give better selectivity but at the cost of increased DC power and chip area. Fortunately, in an 802.11 environment interference is accounted for by retransmission, with graceful degradation in performance. HiperLAN and Bluetooth are time-division multiplexed and somewhat less tolerant of interferers.

We mentioned direct conversion as an alternative to superhet radios. Direct conversion radios use fewer stages and thus fewer components than superhet radios, shown in Figure 4-72. Direct transmission is of particular interest because there are no interfering signals to deal with and thus less need for IF filtering. Because there is no image frequency to remove, transmit filtering need merely be concerned with eliminating harmonics, which is generally much easier.

Direct transmission encounters some special difficulties. The transmitted symbols are points in amplitude-phase space. Errors in amplitude or phase give rise to errors in the transmitted signals, which show up as increased EVM; examples of distorted constellations are depicted in Figure 4-73. Phase match between the I and Q branches is much more difficult when the transmitted signal is at RF frequency rather than IF, as shown in Table 4-9. Direct transmission is only practical with an integrated modulator and requires very careful attention to symmetry in circuit layout to ensure good phase matching. Modern implementations are also incorporating increasing amounts of on-chip calibration of phase and amplitude match.

Direct transmission at high-output power also encounters another somewhat more intractable challenge. Because there is no channel filter, any noise originating in the modulator is converted into broadband noise in the transmitter. The RF band filter has a much sloppier cutoff than an IF channel filter. For example, by reference to

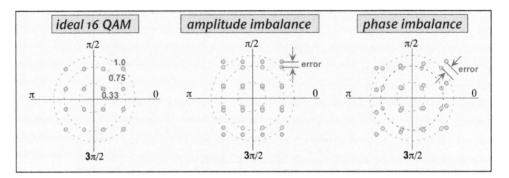

Figure 4-73: Effect of Imbalances in Amplitude and Phase Between I and Q Branches on the Transmitted Signal

Table 4-9: Effect of Increasing Modulation Frequency on Imbalance Requirements

Frequency (MHz)	Phase Offset (degrees)	Time Offset (psec)	Line Offset (μm)	Capacitance Offset (pF)	Implementation
90	1	30	4500	0.15	Hybrid
374	1	7.5	1130	0.04	Multiple ICs
2440	1	1.1	165	0.005	Single-chip

Figures 4-64 and 4-65, the IF filter might achieve 20 dB of rejection by 5 MHz beyond the signal edges, whereas the RF filter requires about 80 MHz outside the band edge to reach the same rejection. Thus, the RF filter doesn't effectively filter any spurious transmitter output near the edges of the band. If the signal power is limited at the modulator to ensure good linearity, the amount of RF gain can be considerable, resulting in broadband noise 20–30 MHz from the nominal carrier exceeding regulatory limits for out-of-band radiation. Broadband noise from the DACs or mixers is mainly a problem for high-output-power (1 W for U.S. ISM band) access points. (Broadband noise is also a major problem for licensed radiators such as cellular telephone base stations, where powers of 10s or even 100s of watts per channel are encountered, and multiple channels may be transmitted with a single radio to reduce component costs.)

Direct conversion receivers present unique challenges to the designer. Because there is no IF gain, the converted signal into the baseband section may be very small: for example, −50 dBm into 200 Ω is 1 mV. If any DC offset is present, it may grow large enough after further amplification to drive the downstream amplifiers to saturation, swamping the tiny wanted signal. Recall (Figure 4-20) that second-order distortion creates IM products at DC. Thus, DC offsets can arise in the radio from second-order distortion of the large LO signal. Offsets can also result from the LO signal escaping into the receive chain due to the finite LO-RF isolation of the mixer (Figure 4-74); the

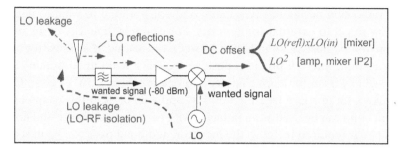

Figure 4-74: Sources of DC Offset in a Direct Conversion Receiver

LO signal may reflect from the filter or antenna back to the mixer, where it is mixed down to DC. These reflections can be time dependent because they depend on the near-antenna environment.

If the signal spectrum is predominantly away from DC, a simple solution to the DC offset problem is to filter the DC out of the signal. For example, the 802.11a and g OFDM signals do not use the innermost pair of subcarriers. A filter that removes DC but passes the third subcarrier frequency would be sufficient to eliminate DC offsets without affecting the signal. However, this approach cannot be used in protocols like Bluetooth, where significant signal power is at DC and very low frequencies. A second approach is to engineer the radio for minimal offset. The mixer must be designed for very good isolation, the amplifiers and filters carefully matched, and all components must have a high second-order intercept. Such an approach is effective but expensive in component cost and power consumption. Finally, the offset may be measured (for example, during the packet preamble) and corrected with an intentional offset from a DAC. Active calibration and correction is becoming more common as digital capability of low-cost radio and baseband chips increases.

Let us conclude with a few general remarks. In deciding how to implement a radio, it is important to note that *analog does not scale*: 1 W is 1 W no matter how small the gate length of the transistor amplifying it. This is in strong contradistinction to digital circuitry, where reductions in lithographic dimensions produce simultaneous increases in speed, density, and reduction in power consumption. Thus, the advantages of integrating many functions in a single chip are much reduced in analog systems compared with digital systems. On the other hand, if we construct a radio of discrete components, we are free to choose the best technology for each application: GaAs pHEMT for the LNA, SiGe or InGaP bipolar transistors for gain and linearity, GaAs for switches and power amplifiers. Discrete passive components also offer significant performance advantages: wirewound inductors can achieve Q of 30 at 2 GHz and don't consume expensive semiconductor area. Very broadband baluns can be fabricated in this fashion. Integrated implementations are ideal for complex functions where branch matching is important. Examples are IRMs, direct modulators, and differential amplifiers with high IP2.

Integrated radio chip design involves many trade-offs related to what to put on the chip and how to get to external components. For example, linearity of an amplifier can be improved by placing an inductor between the source of a transistor and the ground connection, where it provides positive feedback without consuming any DC voltage (a key issue for chips that need to operate from 3.3 V or less). On-chip inductors take a lot of space (see Figures 4-84 and 4-85 later in this chapter) and are lossy. A wire bond can be used as an inductor with better performance and no chip area, but a bond pad is required to get to the inductor, and bond pads can be in short supply, considering the many signals that need to get on and off the chip. The inductor can be eliminated, but then the degraded linearity of the amplifiers must be dealt with, by increasing the size of the amplifier transistors and thus their power consumption, redesigning other parts of the radio chain, or accepting reduced radio performance.

Real radios are a constantly evolving compromise of size, performance, and cost. Today's WLAN radios use an integrated radio chip for core analog functions, which may or may not include the analog-to-digital conversion. A separate predominantly or purely baseband chip is usually though not always assigned the complex tasks of managing the PHY and MAC functions of the protocol. The chips are placed in surface-mount packages and combined with inexpensive, high-performance, discrete components, such as inductors, capacitors, balun transformers, RF switches, and power amplifiers. The components are all mounted with a reflowed solder onto a composite fiberglass-plastic circuit board, which may also contain antennas or antenna connectors for the RF interface, and a bus or wired protocol such as Ethernet for communicating with the host system.

4. Examples of Radio Chips and Chipsets

Radios for WLANs have been undergoing rapid evolution over the last few years. The Harris/Intersil Prism product line (now owned by Globespan/Virata) was a very popular 802.11b chipset, appearing in products from companies such as D-Link and Linksys. The Prism 2 product line featured separate chips for the RF converter, IF converter, baseband including the ADC/DAC and MAC, VCOs, and power amplifier (Figure 4-75). The RF converter and IF converter chips incorporated phase-locked loops to control the VCO frequency. Switches were also external to the chipset. Thus, as many as eight analog active components (not including the reference oscillator) and three SAW filters were required.

A similarly simplified view of a radio based on a more recent Broadcom 802.11g chipset is shown in Figure 4-76. This radio chipset is also capable of dual-band operation using a second radio chip; Figure 4-76 depicts only ISM band operation for clarity. This radio design uses only a single radio chip, which incorporates the complete synthesizer function as well as down-convert/demodulate and direct modulation. Only four active analog components and a single filter are required.

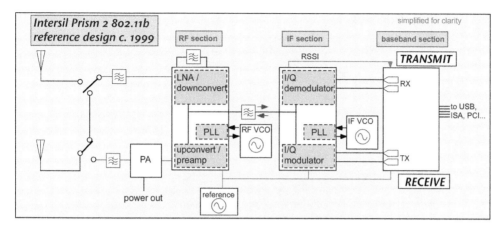

Figure 4-75: Simplified Prism 2 Block Diagram Based on
Published Datasheets and Reference Design

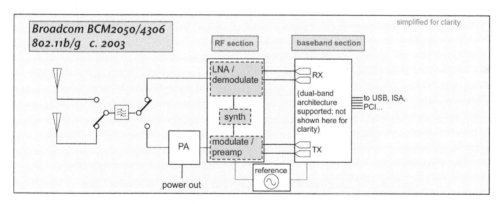

Figure 4-76: Simplified 802.11b/g Radio Block Diagram Using Direct Conversion Chipset

Let us take a closer look at a couple of representative WLAN chipsets. The first is a relatively early 802.11a design from Atheros, described at the 2002 International Solid-State Circuits Conference (ISSCC). A simplified functional diagram of the radio chip is provided in Figure 4-77. The design is a high-IF superhet radio using a 1 GHz IF to make image filtering very easy: simple discrete filters can be used to extract the wanted 5.2-GHz signal from the image at 3.2 GHz. Gain adjustment is provided in the RF, IF, and baseband sections. Active compensation of DC offsets and branch mismatch is provided. The transmit function uses an interesting stacked image-reject up-converting mixer to minimize filtering requirements; an off-chip balun is required to convert the output differential signal to a single-ended signal on the circuit board. An on-chip integer-N synthesizer provides the LO signal for the RF conversions. Digital-analog conversion uses 9-bit ADCs and DACs and oversampling.

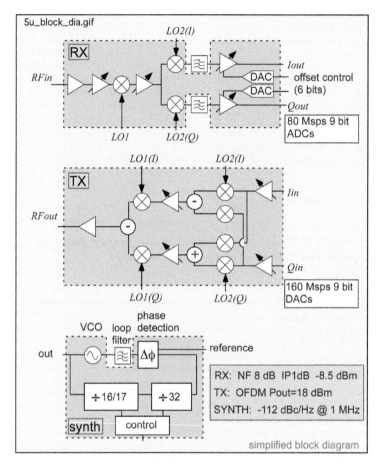

Figure 4-77: Simplified Block Diagram of 802.11a Radio Chip (After Su et al., ISSCC 2002)

A typical transmitted signal is shown in Figure 4-78. The baseband signal is a 6-MHz (BPSK modulation on the subcarriers) signal, which has the highest peak-to-average ratio and is most demanding of the linearity of the amplifiers. The chip is able to produce 17 dBm out while remaining compliant with the 802.11a output mask; this power level may be sufficient for a client card application without an external power amplifier. The price of this very high output power is substantial power consumption: the radio chip consumes 0.8 W in transmit mode.

A more recent radio chip, supporting 802.11b and g, was described by Trachewsky et al. of Broadcom at the 2003 IEEE Hot Chips Conference. A simplified block diagram of the chip is shown in Figure 4-79. This chip uses direct conversion on both receive and transmit. Channel filtering is thus simple low-pass filtering; all the channel filters are active filters whose cutoff frequency can be adjusted to optimize signal fidelity. Two

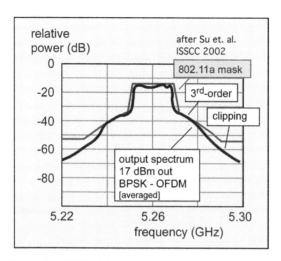

Figure 4-78: Example of 802.11a OFDM Output Spectrum (After Su et al.)

received signal strength indicators are provided, before and after channel filtering, to aid in detection of interferers. Gain adjustment at both RF and baseband is provided; active compensation and calibration is used during idle periods to remove offsets and ensure good I/Q matching. The on-chip synthesizer uses a VCO at two-thirds of the final frequency: the VCO output is divided by 2, and the two complementary outputs of the divider are mixed with the original VCO signal to produce I and Q synthesizer outputs. This displacement of the VCO frequency from the intended RF frequency is a common technique used to ensure that the VCO frequency is not displaced or *pulled* away from its intended value by exposure to a leaked high-amplitude RF signal, such as a fraction of the transmitted signal that may be conducted through the substrate. Receive noise figure is excellent; IP3 is adequate for typical WLAN applications. The transmitted power is modest, and thus an external power amplifier is needed for most applications; however, DC power consumption is correspondingly small.

Performance parameters for these and a number of other reported 802.11 chipsets are summarized in Table 4-10. Certain trends are clear: single-chip radios are the rule, direct conversion is becoming increasingly common, and higher ADC resolution and sampling rate are used as more complex signals become common. Note, on the other hand, that no strong trend in RF performance is evident: receive noise figure is within a range of 4–8 dB, and input intercept varies over the range around −10 to −20 dBm appropriate for WLAN applications. The large transmit power variations are primarily a function of whether an external power amplifier is intended as part of the chipset or not and do not reflect any strong trend in performance scaling. The same can be said for synthesizer performance, save that the complete synthesizer function is integrated in the radio chip in modern chipsets, unlike the older separate VCO solutions.

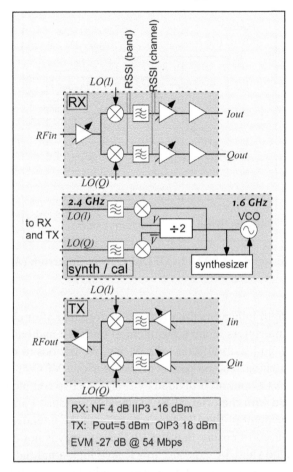

Figure 4-79: Simplified Block Diagram of 802.11b/g Radio Chip (After Trachewsky et al., Hot Chips 2003)

Some recently reported results suggest the direction of future chipset development. At the 2004 ISSCC, Ahola and coworkers described a single-chip radio that supports both the 2.4- and 5-GHz bands. They use a dual-conversion high-IF architecture, with the first fixed LO1 frequency chosen to be roughly between the ISM and UNII band ranges (3840 MHz), so that both bands can be mapped onto about the same IF (1310–1510 MHz) using the same input mixer, saving a bunch of inductors (see Figure 4-84 to get an idea of how much space inductors consume in a chip!). A second but still fixed LO1 of 4320 MHz is used to convert the upper part of the 5-GHz band. The variable second LO2 frequency then directly down-converts the IF signal to baseband. The same approach is used in transmit, although separate output mixers are needed for the two bands. The fixed LO frequencies are relatively easy to provide with low phase noise, and the variable LO is at a reduced frequency where again good

Table 4-10: Summary of Reported WLAN Chipset Performance

Vendor	Intersil – Virata	Atheros	Broadcom	Broadcom	Resonext	Marvell	Athena	AMD	Thomson
Part #(s)	Prism 2	? 2002	BCM2050, 4306	BCM 2060, 4306	Unknown	Unknown	Unknown	Unknown	Unknown
Protocols supported	802.11 classic, b	802.11a	802.11b,g	802.11a	802.11a	802.11b	802.11a	802.11b	802.11a
Chips in radio	5?	1	1	1	1	1	1	1	1
Chips in MAC/baseband	2?	1	1	1		1	1	1	
Radio chip area	?	22 mm^2		11.7 mm^2	13 mm^2	16 mm^2	18.5 mm^2	10 mm^2	17 mm^2
Architecture	Superhet	Superhet	Direct conversion	Direct conversion	Direct conversion	Superhet	Direct conversion	Direct conversion	Superhet dual conv
Technology	0.35 µm SiGe BiCMOS	0.25 µm CMOS	0.18 µm CMOS	0.18 µm CMOS	0.18 µm CMOS	0.25 µm CMOS	0.18 µm CMOS	0.25 µm CMOS	0.5 µm SiGe BiCMOS
IF (MHz)		1000							1225, 60
TX P1dB		22 dBm	18 dBm	19 dBm	15 dBm				15 dBm
TX OIP3						20 dBm	0 dBm	0 dBm	
TX P (CCK)			5 dBm		5 dBm				
TX P (OFDM)		18 dBm	5 dBm	15 dBm					
TX EVM			-27 dB @ 54 Mbps		-28 dB @ 54 Mbps				
RX NF		8 dB	4 dB	4 dB	7 dB		5.5 dB	5 dB	5 dB
RX IP1dB (max gain)		-8.5 dBm	-16 dBm				-20 dBm		
RX IIP3 (max gain)					-18 dBm	-10 dBm	-17 dBm	-8.5 dBm	
RX sensitivity, lowest rate			-97 dBm @ 1 Mbps	-94 dBm @ 6 Mbps		-95 dBm @ 1 Mbps		-96 dBm @ 1 Mbps	
Phase noise		-112 dBc/Hz @ 1 MHz		-100 dBc/Hz @ 30 KHz	-110 dBc/Hz @ 1 MHz	-110 dBc/Hz @ 1 MHz	-115 dBc/Hz @ 1 MHz	-111 dBc/Hz @ 1 MHz	-88 dBc/Hz @10 KHz
Integrated phase noise					1.5° 10 KHz to 10 MHz		-37 dBc 1 KHz-10 MHz (1.6°?)	1 b	
DAC resolution			8 b	8 b		9 b 88 Msps			8 b 160 Msps
ADC resolution			8 b	8 b		6 b 44 Msps			8 b 80 Msps
DC power: TX		0.8 W	144 mW	380 mW	138 mW	1250 mW	302 mW	290 mW	920 mW
DC power: RX		0.4 W	200 mW	150 mW	171 mW	350 mW	248 mW	322 mW	200 mW
Reference	Published datasheets	Su et al. ISSCC 2002 5.4	Trachewsky et al. HotChips 2003	Trachewsky et al. HotChips 2003	Zhang et al. ISSCC 2003 paper 20. 3	Chien et al. ISSCC 2003 paper 20.5	Bouras et al. ISSCC 2003 paper 20.2	Kluge et al. ISSCC 2003 paper 20.6	Schwanenberger et al. ISSCC 2003 paper 20.1

performance is achievable. Because of the 2X separation between the input bands, no special image rejection provisions are needed. The radio chip achieves noise figure of about 5.3 dB, though IIP3 is a rather modest −23 to −26 dBm at maximum gain, apparently due to removal of inductors on the input mixer to allow multiband operation. Transmit output easily meets the EVM and spectral mask requirements at around 0 to −3 dBm out. A similar approach using two separate synthesizers to convert the ISM and UNII bands to a 1.8-GHz IF was described by Zargari et al.

A different approach to a dual-band chip was reported by Perraud and coworkers at this meeting. They used a tunable 9.6- to 11.8-GHz LO, which was then divided either by 2 or by 4 to produce the I and Q signals for direct conversion of the 5-GHz or 2.4-GHz signals. This chip uses a distributed active transformer output amplifier (see section 6) to achieve as much as 12 dBm output from a 1.8-V supply. A Cartesian feedback loop, a linearization technique heretofore only common in high-power amplifiers used for cellular telephone base stations, is used to actively correct distortion in the transmit power amplifier chain, increasing the effective third-order intercept by about 6 dB. The chip achieves less than 5 dB noise figure and good EVM and spectral mask performance.

These reports suggest that it is feasible to create single-chip dual-band radios. Therefore, it seems likely that within 1–2 years, dual-band tri-mode (802.11a/b/g) clients will be common at a cost comparable with today's single-band equipment, at which point the choice of band and protocol will be to some extent transparent to the system user.

IEEE 802.11 chipsets are the most common WLAN technology currently shipping, but by no means the only one. It is worth contrasting a representative Bluetooth design with the 802.11 chips we have examined. Recall that Bluetooth is a low-power low–data rate protocol with a simple Gaussian minimum-shift keying modulation. A simplified block diagram of the chip reported by Chang et al. at ISSCC 2002 is shown in Figure 4-80. The chip is a direct-conversion architecture using an on-chip synthesizer; it is implemented in a silicon-on-insulator technology, which provides good isolation between functional elements on the chip, simplifying such tasks as integrating the synthesizer. The design assumes an external matching circuit and balun to connect the differential interface of the chip to the single-ended antenna.

Recall that Bluetooth's spectrum peaks at zero frequency; DC offsets cannot be solved by filtering in this protocol. Instead, the design uses a wide-dynamic-range ADC to allow some digital DC offset correction. The external balun allows a fully differential chip design with excellent second-order distortion properties. The input second-order intercept IIP2 = 40 dBm. The isolation provided by the insulating substrate minimizes leakage of the LO where it isn't wanted, reducing DC offset generation (Figure 4-74).

The mixers on this chip, like many of the WLAN chips, are implemented as Gilbert cells; a simplified schematic is shown in Figure 4-81. The receive side mixers substitute

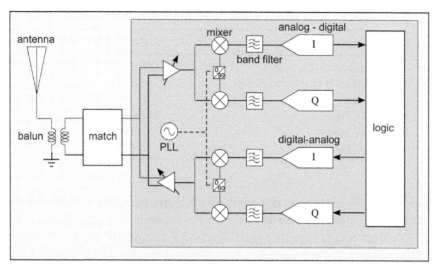

Figure 4-80: Simplified Block Diagram of Direct Conversion Bluetooth Radio Chip (After Chang et al., ISSCC 2002)

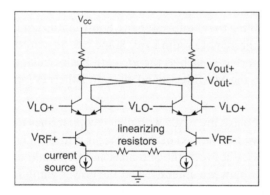

Figure 4-81: Gilbert Cell Mixer Topology (After Chang et al.)

inductors for the resistors to minimize noise generation. An image-reject topology is not needed because the direct-conversion architecture has no image frequency.

Channel filtering is performed digitally. A $\Sigma - \Delta$ converter with 80 dB of dynamic range at 32 Msps is used in this case. Performance of the chip is summarized in Table 4-11. Noise performance is much better than required by the standard. Power consumption is reduced from that required for the 802.11 radio chips, albeit modestly; note, however, that ADC and DAC power are included here. Rejection of the adjacent channel may appear to be modest, but recall that the narrow 1-MHz Bluetooth channels directly abut one another without guard bands, and that the frequency-hopping Bluetooth architecture means that adjacent-channel interference is generally transitory.

Table 4-11: Silicon Wave Bluetooth Radio Performance

Parameter	*Value*
Maximum TX output	3 dBm
Adjacent-channel rejection	−5 dB
RX noise figure	5 dB
RX IIP2	40 dBm
DC power: TX	0.12 W
DC power: RX	0.12 W

A completed radio is more than just a few silicon chips. Inexpensive radios of this type are generally fabricated on a board fabricated of a common fiberglass/plastic composite such as FR4. Most boards have four metal layers: RF ground, RF signal, DC power, and digital data. Along with the key radio chips, the boards carry on the order of 100 surface-mount discrete components. Surface-mount resistors, inductors, and capacitors typically cost from $0.01 to $0.10 each; SAW filters are from roughly $0.50 to $5 each. Power amplifiers and switches are also on the order of $1 to $2 each. A board for a client radio (a *network interface card*) will generally include one or two antennas mounted on the board; an access point will incorporate one or two spring-loaded connectors to allow cabling to remote antennas. Some laptop computers also have built-in antennas cabled to the remote radio card. Some interface to the host or network is also provided. Host interfaces, such as the PC-card bus or universal serial bus (USB), are generally built into the MAC/baseband chip; an Ethernet interface will use a separate chip.

Connections on radio boards are made with plated copper lines a few microns thick and some tens of microns wide. The loss of these lines is critical in ensuring good overall radio performance: losses between the antennas and LNA add directly to the receive noise figure, and losses between antennas and the power amplifier subtract directly from transmitted power. FR4 is the most common substrate used in board manufacture, but it is quite lossy at 5 GHz. A typical 50 Ω transmission line on FR4 has a loss of about 0.2 dB/cm at 5 GHz, so keeping lines short helps. However, on-board antennas need to be reasonably isolated from the rest of the circuit to avoid strong coupling between the antennas and circuit lines, setting a lower limit on the line lengths that can be used in this case. Unintended radiation must also be minimized: lines carrying a high-power LO signal may couple with other lines or radiate, in the case where the VCO is separate from the radio chip.

Some examples of recent radio boards are shown in Figures 4-82 and 4-83. (Note the names are trademarks of the respective vendors.) Figure 4-82 depicts an Airport Xtreme card used in an Apple Computer client card, purchased in 4Q 2003. This card uses a Broadcom 802.11b/g chipset. The high-frequency components are contained within sheet-metal enclosures (the tops have been removed to make the components

Figure 4-82: Apple Airport Xtreme 802.11b/g Radio Card; Plastic Covers and Shielded Enclosure Covers Removed (c. 12/03; Photo by Chuck Koehler)

visible) that ensure good shielding of the receiver from external interference. A separate enclosure is provided for the power amplifier to minimize unwanted coupling of the high-level signal back to the radio chip. The crystal reference oscillator is contained within the main shielding enclosure. Roughly 100 passive components support the operation of the active radio. The use of an off-board antenna and the shielded enclosures allow the connection to the antenna to be kept short.

Figure 4-83 depicts a D-Link DWL G120 802.11b/g USB radio. Again, sheet-metal enclosures surround the radio chip and the front end containing the power amplifier. A Prism chipset is used (descended from Harris Semiconductor, then Intersil, through Globespan/Virata and part of Conexant at the time of this writing). A superhet architecture is used, with separate chips to convert between IF/RF and IF/baseband. To keep the overall structure small, a two-sided board is used; the back side (not shown) contains the modulator and baseband/MAC functions. The radio and front end are again placed in shielded enclosures. The reference oscillator is placed external to the shielded enclosure. Again, a connector to an external antenna is provided, so the transmission line to the antenna is short.

These two boards represent two different approaches to minimization of size and cost. The Apple board uses a small parts count direct-conversion radio, whereas the

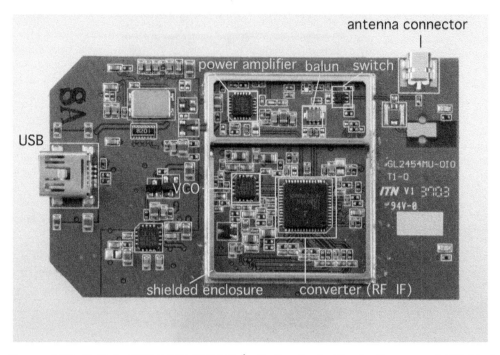

Figure 4-83: D-Link DWLG120 802.11b/g Radio Card; Plastic Covers and Shielded Enclosure Covers Removed; Backside Not Shown (c. 12/03; Photo by Chuck Koehler)

D-Link board uses a well-known superhet radio requiring more parts but with a compact dual-sided packaging arrangement. There's more than one way to accomplish a single task.

In Figure 4-84, we show the radio chip (BCM2050; see also Figure 4-79) from the Airport Xtreme card after removal of the plastic encapsulation. Recall that this is a direct-conversion architecture, so the whole radio function is contained in this chip. It is readily apparent that a great deal of the chip area is taken up by inductors and that although the chip is functionally analog, in practice this is a mixed-signal design with extensive logic area.

In Figure 4-85, the converter chip from the D-Link card shown in Figure 4-83 is similarly depicted after decapsulation. Inductors still consume significant area, though the large areas of logic show even more clearly the mixed-signal requirements of a modern WLAN radio chip.

5. Capsule Summary: Chapter 4

Radios must detect signals of widely varying magnitude in the presence of interferers. A radio is required to minimize excess noise while tolerating large input signals (both wanted and interfering signals) without excessive distortion. Transmit distortion must

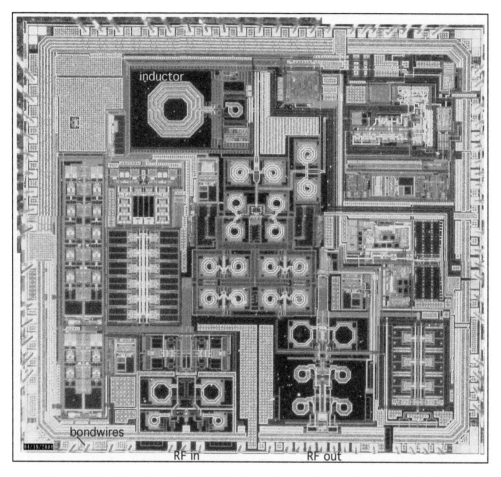

Figure 4-84: Broadcom BCM2050 From Apple Airport Xtreme Card, After Removal Of Plastic Encapsulation; Montage of Four Quadrant Photos (Image Courtesy of WJ Communications)

be minimized to avoid spurious radiation outside the intended channel and band. Radios must provide good frequency selectivity and tunability and adapt to a huge range of input power.

Modern digital radios combine frequency conversion with analog-to-digital conversions. There are three popular architectures: superhet, NZIF, and direct conversion, each with its own advantages and pitfalls. Each analog component has certain key performance parameters. Amplifiers are characterized by gain, distortion, and noise. Mixers add isolation to distortion and noise. Synthesizers must deliver low phase noise and good frequency stability. Filters require narrow bandwidths, low insertion loss, and good rejection of unwanted signals. Switches need good insertion loss, fast actuation, and sufficient isolation. Radio boards are a mixture of discrete and

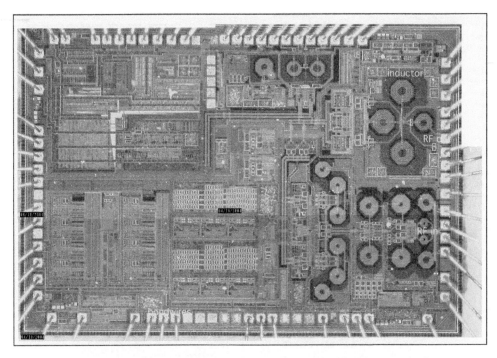

**Figure 4-85: ISL3689 From D-Link USB Card, After Removal of
Plastic Encapsulation; Montage of Four Quadrant Photos
(Image Courtesy of WJ Communications)**

integrated analog and digital components. The final radio chain is a compromise
between gain, noise, distortion, DC power consumption, and cost of component
acquisition and radio manufacture.

6. Further Reading

RFIC Design

The Design of CMOS Radio-Frequency Integrated Circuits, Thomas Lee, Cambridge,
1998: *An encyclopedic introduction to the design of radio components, though the
emphasis is much broader than purely CMOS implementation (which was probably
added to the title to increase sales). Includes treatments of synthesizer operation,
oscillator phase noise, and feedback design.*

Analog-to-Digital Conversion

"Delta-Sigma Data Conversion in Wireless Transceivers," Ian Galton, IEEE
Transactions on Microwave Theory and Techniques, vol. 50, #1, p. 302 (2002)

"Analog-to-Digital Converter Survey and Analysis," R. Walden, IEEE Journal on
Selected Areas in Communications, vol. 17, #4, p. 539 (1999)

Amplifiers

RF Power Amplifiers for Wireless Communications, Steve C. Cripps, Artech House, 1999: *Cripps is bright, opinionated, and brings extensive practical experience to bear on abstruse topics in amplifier design.*

Design of Amplifiers and Oscillators by the S-Parameter Method, George Vendelin, Wiley Interscience, 1982: *Purely microwave-oriented, antedating modern CMOS and SiGe devices, but a useful reference and introduction to matching techniques, low-noise and broadband design.*

"A Fully Integrated Integrated 1.9-GHz CMOS Low-Noise Amplifier," C. Kim et al., IEEE Microwave and Guided Wave Letters, vol. 8, #8, p. 293 (1998)

"On the Use of Multitone Techniques for Assessing RF Component's Intermodulation Distortion," J. Pedro and N. de Carvalho, IEEE Transactions on Microwave Theory and Techniques, vol . 47, p. 2393 (1999)

"Impact of Front-End Non-Idealities on Bit Error Rate Performance of WLAN-OFDM Transceivers," B. Côme et al., RAWCON 2000, p. 91

"Weigh Amplifier Dynamic-Range Requirements," D. Dobkin (that's me!), Walter Strifler, and Gleb Klimovitch, Microwaves and RF, December 2001, p. 59

Mixers

A great deal of useful introductory material on mixers was published over the course of about 15 years by Watkins-Johnson Company as TechNotes. These have been rescued from oblivion (in part by the current author) and are available on the web site of WJ Communications, Inc., www.wj.com. The material is focused on diode mixers but many issues are generic to all mixer designs. Of particular interest are the following:

"Mixers, Part 1: Characteristics and Performance," Bert Henderson, volume 8

"Mixers, Part 2: Theory and Technology," Bert Henderson, volume 8

"Predicting Intermodulation Suppression in Double-Balanced Mixers," Bert Henderson, volume 10

"Image-Reject and Single-Sideband Mixers," Bert Henderson and James Cook, volume 12

"Mixers in Microwave Systems, Part 1," Bert Henderson, volume 17

Switches

"An Integrated 5.2 GHz CMOS T/R Switch with LC-Tuned Substrate Bias," N. Talwalker, C. Yue, and S. Wong, International Solid-State Circuits Conference 2003, paper 20.7, p. 362

Chipsets

"An Integrated 802.11a Baseband and MAC Processor," J. Thomson et al., International Solid-State Circuits Conference 2002, paper 7.2

"A 5 GHz CMOS Transceiver for IEEE 802.11a Wireless LAN," D. Su et al., "An Integrated 802.11a Baseband and MAC Processor," J. Thomson et al. International Solid-State Circuits Conference 2002, paper 5.4

"Broadcom WLAN Chipset for 802.11a/b/g," J. Trachewsky et al., IEEE Hotchips Conference, Stanford University, 2003

"Direct-Conversion CMOS Transceiver with Automatic Frequency Control for 802.11a Wireless LANs," A. Behzad et al., International Solid-State Circuits Conference 2003, paper 20.4, p. 356

"A Multi-Standard Single-Chip Transceiver covering 5.15 to 5.85 GHz," T. Schwanenberger et al., International Solid-State Circuits Conference 2003, paper 20.1, p. 350

"A Digitally Calibrated 5.15–5.825 GHz Transceiver for 802.11a Wireless LANs in 0.18 μm CMOS," I. Bouras et al., International Solid-State Circuits Conference 2003, paper 20.2, p. 352

"A Direct Conversion CMOS Transceiver for IEEE 802.11a WLANs," P. Zhang et al., International Solid-State Circuits Conference 2003, paper 20.3, p. 354

"A 2.4 GHz CMOS Transceiver and Baseband Processor Chipset for 802.11b Wireless LAN Application," G. Chien et al., International Solid-State Circuits Conference 2003, paper 20.5, p. 358

"A Direct-Conversion Single-Chip Radio-Modem for Bluetooth," G. Chang et al., International Solid-State Circuits Conference 2002, paper 5.2

"A Single Chip CMOS Transceiver for 802.11a/b/g WLANs," R. Ahola et al., International Solid-State Circuits Conference 2004, paper 5.2, p. 64

"A Dual-Band 802.11a/b/g Radio in 0.18 μm CMOS," L. Perraud et al., International Solid-State Circuits Conference 2004, paper 5.3, p. 94

"A Single-Chip Dual-Band Tri-Mode CMOS Transceiver for IEEE 802.11a/b/g WLAN," M. Zargari et al., International Solid-State Circuits Conference 2004, paper 5.4, p. 96

Distributed Active Transformers

"Fully Integrated CMOS Power Amplifier Design Using the Distributed Active-Transformer Architecture," I. Aoki, S., Kee, D. Rutledge, and A. Hajimiri, IEEE J. Solid-State Circuits, vol. 37, # 3, p. 371 (2002)

Antennas

1. Not Your Father's E & M

The approach to antenna theory used here, as noted in Chapter 1, is somewhat unconventional in that we eschew all mention of the field quantities **E** and **B** generally regarded as fundamental. All calculations are presented in terms of the vector potential **A** and scalar potential ϕ (which are actually components of the relativistically invariant four-vector version of **A**). In this view, all currents radiate, and there is no physical distinction between the near-and far-field. As a consequence, certain concepts that are widely used in the conventional treatments, such as equivalent currents for fields, are not used here. However, unlike Professor Mead, we adhere to the convention of considering only outgoing radiated fields traveling forward in time; although of great philosophical interest, a purely time-symmetric electromagnetism fits very poorly into the conceptual structure and terminology that antenna engineers depend on.

2. Radiation: The Wireless Wire

We discussed the sorts of signals one might wish to transmit and receive and how the baseband information is placed onto and retrieved from a high-frequency signal. To make this complex exercise useful, it is necessary that the signal actually travel from one radio to another. To convert the output voltage of a transmitter into an electromagnetic wave and reverse the operation at the receiver, we need antennas.

Recall from Chapter 2 that every current induces a vector potential at distant locations (see Figure 2-7). It would at first seem that every signal we create ought to be detectable at a distance. As usual, life is not quite so simple. In most cases, any AC current is balanced by an opposing current (if this were not the case, a DC voltage would be accumulating somewhere). If the two current elements are very close to each other, when viewed from far away the potentials they create will cancel (Figure 5-1). (It can also be shown that the distance is not relevant if countercurrents of equal total magnitude completely surround a subject current, as in, for example, a coaxial cable.) An antenna is a device whose purpose is to *arrange current flows so that the effects do not cancel* at faraway locations.

To treat real systems with more than one infinitesimal current element, it is of course necessary to integrate over the distribution of current within the element. The astute reader, even though warned not to expect to sight electric or magnetic fields, may be somewhat disturbed at the absence of any mention of electric charge or voltages in our discussion so far. The glib answer is that charge is simply the time component of

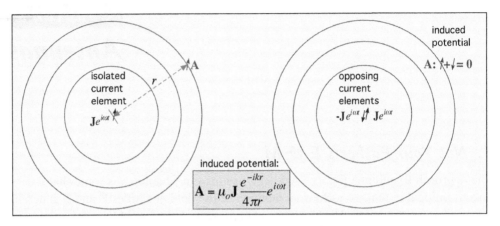

Figure 5-1: Cancellation From Opposing Currents

the relativistic current four-vector and electrostatic potential is correspondingly included in the four-potential, so that everything is contained in a single expression. However, for the reader who, like the author, has quite forgotten how to manage this conversion, we provide the complete separate expressions for vector and scalar potential in equation [5.1].

$$\mathbf{A} = \mu_0 \int \mathbf{J}_s \frac{e^{-ikr}}{4\pi r} e^{i\omega t} d(vol) \quad \phi = \frac{\mu_o c^2}{4\pi} \int \rho \frac{e^{-ikr}}{r} e^{i\omega t} d(vol) \qquad [5.1]$$

The vector potential at some point is due to the sum of all the currents, each weighted by a phase factor and their inverse distance. The electrostatic potential is a similarly weighted sum of the distant charges, with a factor of c^2 to account for the units. The notation here is the same as that introduced in Figure 2-7: μ_o is the magnetic permeability of free space, c the speed of light, $\mathbf{J}$ the (vector) current density, $k = \omega/c$ is the wavenumber, and ρ is the density of charge. As usual, we have assumed a harmonic time dependence for all quantities. (It is always possible to work problems in the time domain, in which case the integral is over all the instantaneous currents and charges at a time delay of (r/c).)

Assume we have accomplished our end and created a potential at some distant point. How does this potential affect the flow of electrons on a wire (which is what it must do if we are to detect the result)? Keeping with the choice of separate consideration of the vector and scalar potential above, we can write the total voltage along some path from one point to another as the sum of a line integral of the vector potential and difference in the scalar potential between the points:

$$v = i\omega \int_1^2 \vec{A} \cdot \vec{dl} \qquad \Delta\phi_{2-1} = \phi_2 - \phi_1 \qquad |V| = v + \Delta\phi \qquad [5.2]$$

magnetic coupling   *electric coupling*   *induced voltage*

In principle, one should integrate over every current in existence to get the correct expressions for the potentials to use in equation [5.2]. Fortunately, in practice we can specialize our attention to the potential due only to currents on the structure of interest—generally a transmitting antenna for the purposes of this chapter.

The operation of an antenna can thus be analyzed by determining how currents and charges are distributed within the antenna, calculating the consequent potentials at the point of interest, and deducing the resulting voltage. To understand the process, we shall undertake such a calculation for a very simple case: the ideal dipole. Like many laborious calculations, the estimation of coupling from current flows is almost entirely avoidable once a certain amount of conceptual travail has been endured; we shall find that most of what we need to know can be couched in terms of the directional properties of a given antenna structure, given the availability of the full calculation for one particular antenna. Also like most simplified approaches to complex problems, this one contains important skeletons concealed in the closet, ready to make a frightening appearance if one is so unwise as to open the door. The skeleton in the antenna closet is accurate calculation of the antenna current distribution and impedance, for which a full and self-consistent accounting of both near and distant currents is required. Needless to say, we shall resist the temptation to such extensive self-abuse in this volume; the reader who can't resist looking must be content to peruse the references provided in section 9.

3. The Ideal Dipole

An ideal dipole is a constant current along a short (but not infinitesimal) length. An ideal dipole can be closely approximated by a bit of wire thin compared with its length and short compared with the wavelength at the frequency of interest. If a uniform electrical current is to flow along the length of the dipole and then terminate, the point of termination must collect the resulting charge; the charge may be reasonably localized to the ends in practice by appending balls or caps to the ends of the wire. Thus, a real structure whose behavior might closely approximate that of an ideal dipole is a short wire ($d \ll \lambda$) with endcaps, as depicted in Figure 5-2. The peak current I_0 is taken to be uniform along the length of the wire, and because the length is short compared with a wavelength, the phase of the current is constant along the wire.

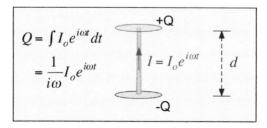

$$Q = \int I_0 e^{i\omega t} dt$$
$$= \frac{1}{i\omega} I_0 e^{i\omega t}$$

$+Q$

$$I = I_0 e^{i\omega t}$$

d

$-Q$

Figure 5-2: Ideal Dipole

To avoid the pain of treating two structures at once, we use an ideal dipole as both the transmitting and receiving antenna. That is, we seek to answer the following question: what voltage results in a distance dipole *dp2* when a current flows in our transmitter *dp1*? We investigate two simplified cases in some detail (Figure 5-3): case I, where the receiver is displaced horizontally with respect to the (vertically oriented) dipole, and case II, where the receiver is displaced vertically along the dipole axis. The separation between dipoles is presumed to be much larger than their size and the wavelength at the frequency of interest, so that one may plausibly assume that the exact shapes of the wire and endcaps are immaterial. In both cases the calculation can be performed with some labor but minimal geometry. We then present without derivation the more general result for a receiver placed at an arbitrary angle with respect to the axis of the transmitting dipole.

Let us first consider case I. The magnetic potential at *dp2* is easy to obtain from equation [5.1], once it is appreciated that the variation of the factor $e^{-ikr}/\mathbf{r}$ is negligible over the short wire because the wire is perpendicular to the separation vector $\mathbf{r}$, and thus this factor can be taken in front of the integral. (This fact arises from expansion of the Pythagorean expression for distance, $r(z) = \sqrt{\mathrm{v}(r^2 + z^2)} \approx r + (z^2/2r)$ for large r, where z is distance along dipole 1. As long as the largest value of the correction to the average distance, $(d^2/2r)$, is small compared with a wavelength, it can be neglected. This happy circumstance is known as the *far-field* approximation; it allows us to treat only the projection of the source distance onto the observation vector and ignore perpendicular distances. When it doesn't apply, we're in the *near-field* and our mathematical life is more complex.) Because the current is also constant over the wire, the integral becomes simply $\int(dz)$, which is the length of the wire, d. The magnetic potential is thus

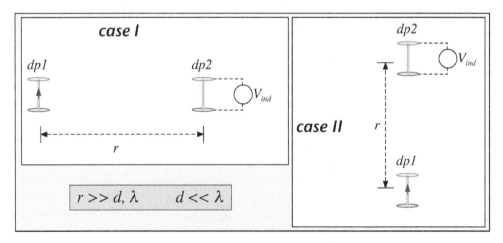

Figure 5-3: Ideal Dipole Coupling, Simple Cases

$$\mathbf{A} = \mu_o \int \mathbf{J}_s \frac{e^{-ikr}}{4\pi r} e^{i\omega t} dz \approx \boxed{\hat{\mathbf{z}} \frac{\mu_o}{4\pi r} e^{i(\omega t - kr)} \mathbf{I}_o d} \qquad [5.3]$$

where now r refers to the overall separation between the dipoles.

The electrostatic potential calculation is even easier. By the same argument advanced previously, the distance between any point on $dp2$ and the top of $dp1$, r_+, is the same as the distance to the bottom of $dp1$, r_- (Figure 5-4).

Because the charge on the top is equal in magnitude and opposite in sign to that on the bottom, their respective contributions cancel: $\phi = 0$ at the second dipole (equation [5.4]). Electric coupling makes no contribution to the voltage in case I.

$$\phi = \frac{\mu_o c^2}{4\pi} \int \rho \frac{e^{-ikr}}{r} e^{i\omega t} dv = \frac{\mu_o c^2}{4\pi} \frac{e^{i(\omega t - kr)}}{r} [Q_+ + Q_-] \boxed{= 0} \qquad [5.4]$$

According to equation [5.2], we need merely integrate the vector potential along $dp2$ to obtain the induced voltage. Because the vector potential is constant with position, the integration again collapses into a multiplication by the dipole length d. The induced voltage is

$$v_o = i\omega \frac{\mu_o \mathbf{I}_o d^2}{4\pi r} e^{-ikr} = V_o \qquad [5.5]$$

The voltage is proportional to the current on $dp1$ and to the square of the dipole length (one factor from each dipole). The magnitude of the induced voltage falls as $(1/r)$ with distance. The induced voltage increases linearly with increasing frequency. This dependence arises not from any change in the potential but from the response of the electrons to the potential (equation [5.2]).

Case II is harder work with a simpler answer. The vector potential calculation proceeds exactly as before, to the same conclusion. However, the calculation of the electric potential does not. In this case, because the axis of the dipole is along the direction of the separation r, the distance to the top of $dp1$ is always smaller by the dipole length d than the distance to the bottom, no matter what the value of r (Figure 5-5). The electric potential is not zero. Although the $(1/r)$ term and thus the magnitude of the electrostatic

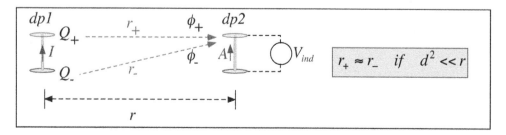

Figure 5-4: Electric Coupling, Ideal Dipole Case I

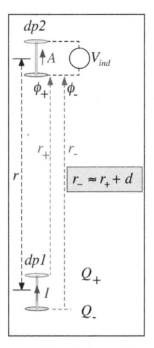

Figure 5-5: Electric Potential, Case II

potential are the same at the top and bottom of *dp2*, the phase is not. The top of dipole 2 is farther away in the direction of propagation and thus receives an additional phase delay. In detail, the calculations are for the bottom of dipole 2:

$$\phi_{bottom} = \frac{\mu_o c^2}{4\pi} \int \rho \frac{e^{-ikr}}{r} e^{i\omega t} dv$$

$$= \frac{\mu_o c^2}{4\pi} \frac{e^{i(\omega t - kr)}}{r} \left[Q_+ e^{ikd} + Q_- \right] \qquad [5.6]$$

and for the top:

$$\phi_{top} = \frac{\mu_o c^2}{4\pi} \frac{e^{i(\omega t - kr)}}{r} \left[Q_+ + Q_- e^{-ikd} \right] \qquad [5.7]$$

The electric coupling is the difference of these voltages:

$$\Delta\phi = \frac{\mu_o c^2}{4\pi} \frac{e^{i(\omega t - kr)}}{r} \left[Q_+ + Q_- e^{-ikd} - Q_+ e^{ikd} - Q_- \right] \qquad [5.8]$$

Because the size of the dipoles is small compared with a wavelength, $kd \ll 1$ and we can expand the exponentials: $e^x \approx 1 + x + x^2/2$ for $x \ll 1$. The quadratic term is needed, because the first-order terms cancel. We obtain

$$\Delta\phi \approx -\frac{\mu_o c^2}{4\pi} \frac{e^{i(\omega t - kr)}}{r} \frac{1}{i\omega} I_o \left[2 + ikd + \left(\frac{(kd)^2}{2!} \right) - 2 - ikd + \left(\frac{(kd)^2}{2!} \right) \right]$$

$$= \frac{\mu_o c^2}{4\pi} \frac{e^{i(\omega t - kr)}}{r} \frac{1}{i\omega} I_o (kd)^2 = -i \frac{\mu_o I_o d^2}{4\pi r} \frac{c^2}{\omega} \left(\frac{\omega}{c} \right)^2 e^{i(\omega t - kr)}$$

[5.9]

where we have made use of the substitution $k = (\omega/c)$. Once we cancel the common factors, we find that this expression is just exactly enough to cancel the magnetic coupling:

$$\Delta\phi_o + v_o = -i\omega \frac{\mu_o I_o d^2}{4\pi r} e^{-ikr} + i\omega \frac{\mu_o I_o d^2}{4\pi r} e^{-ikr} \boxed{= 0}$$

[5.10]

There is no induced voltage in case II: *dipoles do not couple along their axes.* The cancellation is by no means accidental but arises because the amount of charge at the ends of the dipole is determined by the current flow. Each coulomb of charge q flowing from bottom to top in some time dt delivers both a current (q/dt) and a charge $+q$ to the top while subtracting a charge $-q$ from the bottom. This relationship is nothing more than insisting that charge be conserved. Charge conservation can be more generally expressed, for those familiar with vector notations, as

$$\nabla \cdot j = -\partial\rho/\partial t$$

[5.11]

The notation "$\nabla\cdot$" indicates the vector divergence of the current, which in one dimension just collapses to the derivative d/dz. Equation [5.11] is the mathematical generalization of the common sense notion that if current is not constant along a wire, charge must be accumulating somewhere.

It can be shown, though we relegate the demonstration to the appendices, that charge conservation implies a relationship must exist between the vector and scalar potentials for any source, not just a dipole. For plane waves far from the source we can write

$$\hat{\phi}(\vec{r}) = c(A_{long})$$

[5.12]

where A_{long} is the longitudinal part of the vector potential: the part directed along the distance vector **r**. Equation [5.12] implies that coupling along an infinitesimal wire in the direction of the distance vector **r** is *always 0*. To demonstrate this fact, we insert the relationship [5.12] into the expression for total coupling [5.2] and expand the potential using $\Delta\phi = d\phi/dr \cdot \delta r$; recall that at large r, $d\phi/dr = d[e^{-ikr}]/dr = -ik\phi$. We obtain

$$V_{along\ r} = v + \Delta\phi = i\omega A_{long}\delta r - ikc A_{long}\delta r = 0$$

[5.13]

because $kc = \omega$. Only the *transverse portion of the vector potential* leads to net coupling with a distant antenna; the longitudinal portion is always canceled by the scalar potential. This is the real reason we show only currents as sources in Figures 2.7 and 5.1: charges do not make an independent contribution to the coupling at large r.

We can now immediately guess that in the more general case in which the separation between dipoles *dp1* and *dp2* is at an arbitrary angle θ to the axes, the net induced voltage should vary as $\sin^2(\theta)$. One factor of $\sin(\theta)$ comes from the projection of the potential onto the direction perpendicular to the separation **r** and the second from the fact that only a portion of that transverse potential acts along the direction of the axis of *dp2* (Figure 5-6).

This supposition can be verified through a calculation of the electric and magnetic coupling, which proceeds exactly as before except for somewhat more complex geometric considerations. The results are provided in equation [5.14] and depicted in Figure 5-7.

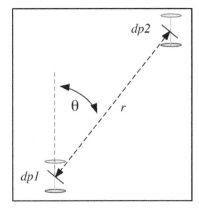

Figure 5-6: General Case of Separation Vector Inclined at an Angle to the Axes

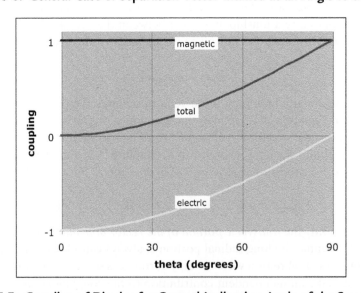

Figure 5-7: Coupling of Dipoles for General Inclination Angle of the Separation r

$$\Delta\phi_o = -i\omega\frac{\mu_o I_o d^2}{4\pi r}e^{i(\omega t - kr)}(\cos{(\theta)})^2 \qquad v_o = i\omega\frac{\mu_o I_o d^2}{4\pi r}e^{i(\omega t - kr)}$$

$$\boxed{\textit{electric coupling}} \qquad\qquad \boxed{\textit{magnetic coupling}}$$

$$V = i\omega\frac{\mu_o I_o d^2}{4\pi r}e^{i(\omega t - kr)}\left[1 - (\cos{(\theta)})^2\right] = i\omega\frac{\mu_o I_o d^2}{4\pi r}e^{i(\omega t - kr)}(\sin{(\theta)})^2$$

$$= \frac{i\omega\mu_o I_o}{4\pi r}e^{i(\omega t - kr)}(d\sin{(\theta)})(d\sin{(\theta)})$$

$$\qquad\qquad dp1 \qquad\qquad dp2$$

$$\boxed{\textit{dipole coupling}}$$

[5.14]

Note that in equation [5.14] we suggested that each dipole contributed a factor of $d\sin{(\theta)}$ to the resulting voltage. One might be tempted to infer that the signal between any pair of antennas might be the product of some characteristic associated only with the transmitting antenna and another similar characteristic of the receiving antenna. In this case, we could characterize each antenna only once and predict its performance as part of a link with any other antenna, even of a differing type, without having to measure or simulate every conceivable pair, just by multiplying together the relevant values. Remarkably, this is exactly what happens once we have constructed the tools needed to define a radiation pattern for each antenna—the topic of the next section.

4. Antenna Radiation Patterns

Because we have imagined that *dp1* produces an outgoing potential, it seems sensible to ask the following: what power is this wave capable of delivering to a unit area at some (far) distant point? This is equivalent to asking what is the largest power that could be delivered to any antenna (normalized to the area thereof)? Although the full derivation of this quantity, known as the Poynting vector, is laborious and consequently has been relegated to Appendix 5, a simple heuristic argument based on the results of section 3 can produce the requisite formula. We established that only the transverse part of the vector potential **A** produces any voltage, and we know from equation [5.2] that the magnitude of this voltage is $\omega A_{tr}dl$ for some infinitesimal length dl. The power this voltage delivers is just $V^2/2R$ (the factor of 2 comes from the harmonic time dependence), the square converting an infinitesimal length into an infinitesimal area. What should we use for the effective resistance R? The obvious candidate is the impedance of free space, $\mu_o c$, which can readily be obtained as the ratio of voltage to current from [5.1] and [5.12]: $\phi = cA = c(\mu_o I) \Rightarrow (\phi/I) = \mu_o c$. Inserting this value in the formula for power and using $\omega = ck$, we obtain an expression for the power delivered by a transverse potential to a unit area, P:

$$P = \frac{\omega k |\vec{A}_{tr}|^2}{2\mu_o}$$

[5.15]

Let us find the power delivered by an ideal dipole. We multiply the expression for the vector potential, equation [5.3], by sin (θ) to obtain the transverse part and then insert the result into equation [5.15], to obtain equation [5.16]:

$$P = \frac{\mu_o \omega k I_o^2 d^2}{32\pi^2 r^2} \left(\sin\left(\theta\right)\right)^2 \hat{r} \qquad [5.16]$$

The expression for P, although perfectly correct, has a $(1/r^2)$ dependence that causes the absolute value to vary wildly with distance, in a fashion that is characteristic of any radiating energy but not particularly informative about this or any other antenna. It is therefore useful to multiply the power density by the square of the distance to obtain a quantity that depends only on angle: the *radiation intensity U* (Figure 5-8), the power delivered to a unit solid angle in the direction (φ, θ).

The radiation intensity for an ideal dipole is given in equation [5.17]. Note that U has no dependence on distance and in this case no dependence on azimuthal angle φ, as one would expect because an ideal dipole is symmetric about its axis.

$$U = \frac{\mu_o \omega k I_o^2 d^2}{32\pi^2} \left(\sin\left(\theta\right)\right)^2 \hat{r} \qquad [5.17]$$

Radiation intensity U as a function of angle is a very important quantity in antenna design and use. In most cases the intensity is normalized to the maximum value with angle to produce a relative measure of the power delivery into a given solid angle: this quantity is known as the *radiation pattern* of an antenna. Many differing ways of depicting this important quantity will be encountered in the literature. Let's look at a few alternatives. In the left half of Figure 5-9 we show a pseudo–three-dimensional view of the pattern of an ideal dipole: it resembles a donut with a small center hole. In this view, the distance from the center of the donut to a given point on the donut surface corresponds to the power delivered along that vector to the surface of a far-distant sphere.

This sort of solid model view is very helpful in gaining intuition about the behavior of the radiation pattern, but it is awkward to use quantitatively (at least on a printed page).

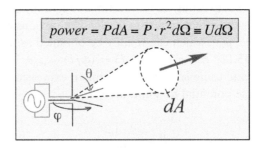

Figure 5-8: Definition of Radiation Intensity *U*

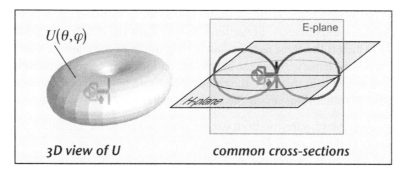

Figure 5-9: Pseudo–Three-Dimensional View of Ideal-Dipole Intensity Pattern (Left) and Definition of Popular Cross-Sectional Views (Right)

Cross-sectional slices of the donut are much easier to plot and display. It is very common to depict cross-sectional images of U in a plane along the direction of current flow in the antenna, known as the E-plane, and a second plane perpendicular to the E-plane, known as the H-plane. (Though we have not made use of the fields, we should note that the nomenclature here arises from the fact that the electric field of a dipole lies in the E-plane and the magnetic field in the H-plane.) An E-plane cross-section of the radiation pattern for the ideal dipole, normalized to the maximum value at $\theta = \pi/2$, is shown versus angle in Figure 5-10. The H-plane for an ideal dipole isn't very interesting: because U is independent of azimuth, it's just a circle of radius 1.

Here we depict the antenna pattern using linear radial coordinates and the distance along the radius to reflect the magnitude of U. Many other formats will also be encountered. It is common to use logarithmic radial scales (so that the pattern is shown in dB). Cartesian

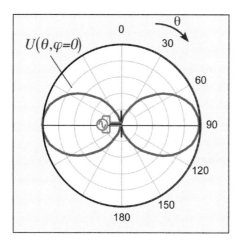

Figure 5-10: Radiation Intensity for an Ideal Dipole, Normalized to U_{max}; Cross-Section Through the Plane $\varphi = 0$

(rectangular) plots with angle as the *x*-axis are also used, using either linear or logarithmic scales for the radiation intensity. Sometimes people show the relative magnitude of the electric field (which is the same as the relative magnitude of the transverse potential *A*) instead of the power density. As we see below, the pattern may be measured relative to an ideal isotropic antenna; in this case, the radial coordinate may be reported relative to an isotropic radiator rather than versus the peak intensity.

Several very important figures of merit can be extracted from the radiation pattern. The *directive gain D* is the value of *U* at any angle divided by the average over all angles (Figure 5-11). The most interesting angle is usually that which provides the maximum value of *D*: this maximum is the *maximum directive gain* D_{max}, also known as the *directivity* of the antenna. An unqualified reference to antenna gain or directivity almost always refers to the direction of maximum directive gain.

The reader should note that the use of the word "gain" does *not* imply that there is any. An antenna is a passive device; it can only radiate up to the energy put into it, no more. The use of the term gain in connection with antennas denotes the ability of an antenna to focus the radiated energy into a particular direction. For a receiver in that favored direction, the effect is the same as if the antenna had added power to the beam, but this advantage is only obtained by removing radiated power from some other direction.

An *equivalent isotropic antenna* pattern is one that radiates the same total energy as the subject pattern but does so uniformly in all directions: the directivity is 1 and the radiation intensity is equal to the average value over the actual pattern. (Note that no

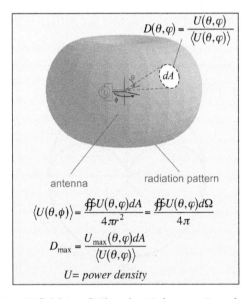

$$D(\theta,\varphi) = \frac{U(\theta,\varphi)}{\langle U(\theta,\varphi)\rangle}$$

$$\langle U(\theta,\phi)\rangle = \frac{\oiint U(\theta,\varphi)dA}{4\pi r^2} = \frac{\oiint U(\theta,\varphi)d\Omega}{4\pi}$$

$$D_{max} = \frac{U_{max}(\theta,\varphi)dA}{\langle U(\theta,\varphi)\rangle}$$

U= power density

Figure 5-11: Definition of Directive Gain as a Function of Angle, and Maximum Directive Gain

real antenna is isotropic, though the ideal dipole is rather close except along the axes.) A closely related figure of merit is the beam width θ_A or *beam solid angle* Ω_A of an antenna. The beam solid angle is an imaginary construct: it is the solid angle required to radiate all the power of the antenna, if the radiation intensity U were uniform and equal to its maximum value within the beam, and 0 outside of it. The beam solid angle (in steradians) is inversely related to the directivity of the antenna (equation [5.18]). The beam width (in radians) is approximately the square root of the beam solid angle when the latter is small (equation [5.19]). A highly directive antenna has a narrow beam.

$$\Omega_A = \frac{4\pi}{D_{max}} \qquad [5.18]$$

$$\theta_A \approx \sqrt{\Omega_A} = \sqrt{\frac{4\pi}{D_{max}}} \qquad [5.19]$$

These concepts are illustrated in Figure 5-12. An equivalent isotropic antenna is a convenient simplified representation of a fairly isotropic antenna like the ideal dipole; a more directional antenna like a parabolic reflector is more sensibly represented in terms of its beam width or beam solid angle.

Real antennas unfortunately can never radiate quite as much power as is put into them, due to ohmic losses in the conductors and dielectric losses in the insulators of which they are made. The ratio of the actual integrated radiation to the input power is the antenna *efficiency e*:

$$e \equiv \frac{P_{rad}}{P_{in}} \qquad [5.20]$$

The product of efficiency and directivity is the ratio of the actual radiation intensity measured in the direction of maximum directive gain to the input power: the antenna *power gain G*. Power gain is the directivity reduced by the efficiency

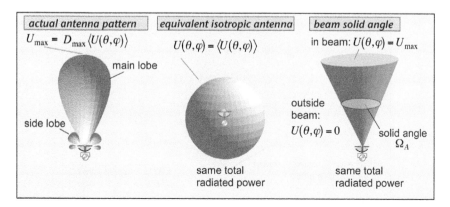

Figure 5-12: Definition of Equivalent Isotropic Pattern and Equivalent Beam Pattern

$$G = eD_{\text{max}} \tag{5.21}$$

When the term gain is used in connection with an antenna with no further qualification, it almost inevitably refers to the power gain. The various gains are summarized in Table 5-1.

One more quantity of considerable practical importance is the *equivalent isotropically radiated power* (EIRP). EIRP is defined as the product of power gain and input power:

$$EIRP \equiv GP_{\text{rad}} \tag{5.22}$$

EIRP is the power that would be radiated by an isotropic antenna with radiation intensity equal to the maximum intensity radiated by the actual antenna in question (Figure 5-13):

$$EIRP = 4\pi U_{\text{max}} \tag{5.23}$$

EIRP is a convenient way of expressing the maximum damage a radio can create when acting as an interferer. Regulatory agencies frequently define limitations on radiated

Table 5-1: Definitions of Various Antenna "Gains"

Parameter	Name	Definition
$D(\theta, \varphi)$	Directive gain	Power in (θ, φ)/average power
D_{max}	Directivity	Maximum value of directive gain
G	Power gain	Maximum power density/input power radiated uniformly

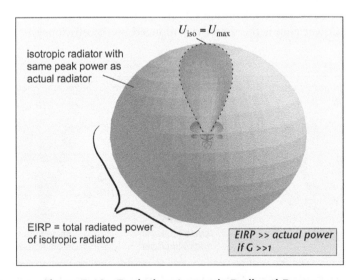

Figure 5-13: Equivalent Isotropic Radiated Power

power in terms of EIRP so as to avoid having to set separate limits for each possible value of antenna gain.

Just to get a feel for the meaning of these various quantities, let us examine some numerical values for the only example we have available, the ideal dipole. The radiation intensity U was provided as a function of angle in equation [5.17]. By averaging over angle we can obtain the normalized radiation intensity of an equivalent isotropic antenna:

$$\langle D(\theta, \varphi) \rangle = \frac{\int\limits_{0}^{\pi} \int\limits_{0}^{2\pi} \sin^2(\theta)\sin(\theta)d\theta d\varphi}{4\pi} = \frac{2}{3} \qquad [5.24]$$

The maximum directive gain, which is equal to the power gain if the efficiency is 1, is

$$D_{\max} = \frac{1}{2/3} = \frac{3}{2} \approx 1.8 \; dB \qquad [5.25]$$

An ideal dipole is not terribly directional. Thus, although we can sensibly define a beam solid angle,

$$\Omega_A = \frac{8\pi}{3} \approx \frac{2}{3}(sphere \; surface) \qquad [5.26]$$

for an azimuthally isotropic antenna, it doesn't make a lot of sense to define a beam width in terms of the square root of Ω but rather to define the beam width in θ, which is $\Omega/2\pi$ or 4/3 of a radian (about 76 degrees). Finally, the EIRP for 1-W input power is

$$EIRP = \frac{3}{2}P_{rad} = 1.5W \qquad [5.27]$$

The total radiated power can be obtained by, for example, multiplying the peak power by the beam solid angle:

$$P_{rad} = \frac{\mu_o \omega k I_o^2 d^2}{32\pi^2}\Omega_A = \frac{\mu_o \omega k I_o^2 d^2}{12\pi} \qquad [5.28]$$

It is worthwhile to define one additional quantity related to the radiated power: the *radiation resistance*. The power that the antenna radiates must have come from the signal source that supplies the driving current. From the point of view of this source, the antenna equivalent circuit must include a resistance, whose dissipation accounts for the power lost to radiation (Figure 5-14). The radiation resistance is obtained by setting the electrical power dissipation from the source equal to the total power radiated by the antenna.

For the case of the ideal dipole, the radiation resistance is easy to calculate, because the current is constant and we already know the relationship between the current and the total radiated power:

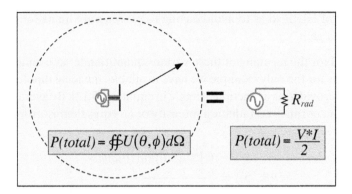

Figure 5-14: Equivalent Circuit of a Radiating Antenna

$$P_{rad} = \frac{1}{2}RI_o^2 = \frac{\mu_o\omega kI_o^2 d^2}{12\pi} \tag{5.29}$$

The radiation resistance is easily found:

$$R_{rad} = \frac{\mu_o\omega kd^2}{6\pi} = 80\pi^2\left(\frac{d}{\lambda}\right)^2 \tag{5.30}$$

where the version on the right arises after a bit of algebra, plugging in the values of μ_o and the velocity of light c. For a 0.6-cm ideal dipole (the biggest thing one might fit inside a PC card) transmitting an ISM band signal ($\lambda = 12.5$ cm), the radiation resistance is about $2\,\Omega$: pretty small compared with the typical line impedance of $50\,\Omega$. The reader may be aware that a generator delivers power most efficiently to a load of the same impedance as that of the generator; thus, a $50\,\Omega$ line will not deliver power very well to a $2\,\Omega$ antenna. It is apparent that some sort of *impedance matching* needs to be used when a small dipole (or any other very small antenna) is to be used. A brief discussion of matching techniques and terminology are provided in Appendix 3. In the cases we discuss in the remainder of the chapter, the antennas used are intrinsically well matched to the generator impedance, and indeed such antenna structures are very popular because of their ease of use.

5. Antennas as Receivers

We now have a prescription for dealing with antennas as transmitters: use the currents on the antenna to find the distant potentials, from which in turn we obtain the radiation pattern of the antenna, conveniently expressed in terms of directivity (or, if ohmic losses are accounted for, power gain). How do we translate that radiation pattern into the quantity that is actually interesting, the received power?

It is certainly possible to carry out a separate calculation of the induced voltage due to the potential on each receiving antenna structure of interest, but fortunately this additional labor can be avoided in almost all cases of practical interest. Let us see how this convenient situation arises.

Without delving into any of the details, it certainly seems sensible to argue, in analogy with our commonplace experience at optical frequencies, that a given antenna structure acting as a receiver could be treated as if it were a sort of window, collecting all the energy falling within its bounds and none elsewhere. In formal terms, let us attempt to treat the receiving antenna as being characterized by an effective aperture A_{RX} (Figure 5-15).

The received power is then the product of the transmitted power density and the effective aperture. For an isotropic antenna, the fraction of the total radiated power P_{TX} received is the ratio of the effective aperture to the total area of a sphere at r, $4\pi r^2$. A real antenna multiplies the power density by the power gain in the direction of the main lobe. The received power P_{RX} (in the direction of maximum directive gain) can thus be written:

$$P_{RX} = P_{TX}\left(\frac{A_{RX}}{4\pi r^2}\right)G_{TX} \qquad [5.31]$$

We now invoke the principle of *reciprocity*: transmitting from antenna 1 and receiving at antenna 2 in general should work the same way as transmitting at antenna 2 and receiving at antenna 1. Although we do not attempt to prove the applicability of this principle, consideration of equation [5.1] indicates that it is certainly plausible: the only quantity that enters into the determination of the potential is the absolute value of the distance. The potential at [2] from a current at [1] is the same as the potential at [1] from a current at [2]. It can be shown that, except in the presence of, for example, a magnetized plasma (a highly unusual application environment for a wireless local area network [WLAN]), reciprocity is a valid assumption.

In our case, reciprocity implies that we can swap the transmit (TX) and receive (RX) tags and get the same result:

$$P_{in}\left(\frac{A_{TX}}{4\pi r^2}\right)G_{RX} = P_{in}\left(\frac{A_{RX}}{4\pi r^2}\right)G_{TX} \qquad [5.32]$$

Obviously the power cancels, as does the sphere area term. We are left with

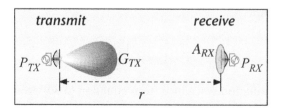

Figure 5-15: Receive Antenna as an Effective Aperture A_{RX}

$$G_{RX}A_{TX} = G_{TX}A_{RX} \longrightarrow \frac{G_{RX}}{G_{TX}} = \frac{A_{RX}}{A_{TX}} \qquad [5.33]$$

The ratio of the effective aperture of two antennas as receivers is the *same as the ratio of their gains when used as transmitters*. This key result means that once we have characterized the gain of a transmitting antenna, we're done. There is no need to separately measure its receiving properties. (One can, of course, also proceed in the other direction, measuring the antenna as a receiver and predicting its behavior as a transmitter.)

However, equation [5.33] only provides us with a relative measure of received power. To obtain the actual values of aperture and thus received power, we need at least one antenna whose effective aperture and gain are both known. The obvious suspect is the only antenna we have examined in detail: the ideal dipole. We have already obtained both the directivity of a single ideal dipole and the voltage induced on a receiving dipole from the current on a transmitting dipole. All we need to do is convert these voltages and currents into power.

The induced voltage in the direction of maximum power gain, from equation [5.5], is

$$|V_o| = \frac{\mu_o \omega I_o d^2}{4\pi r} \qquad [5.34]$$

The power density in the same direction is obtained from equation [5.16]:

$$P = \frac{\mu_o \omega k I_o^2 d^2}{32\pi^2 r^2} \qquad [5.35]$$

The received power is the product of the power density and the effective aperture. This must be the same as the power dissipated by the induced voltage into a matched load. We already established (Figure 5-14) that a transmitting antenna is electrically equivalent to a resistance R_{rad}. By reciprocity, the receiving antenna appears as a voltage source equal to the induced voltage and an equivalent source impedance R_{rad}. (The impedance of an ideal dipole would also have a significant reactive part; we presume this has been removed without loss by appropriate matching.) The received power is thus obtained by attaching this antenna to a load resistance equal to R_{rad} (Figure 5-16).

Because the voltage is equally split by the internal resistance and the load, the power is divided by 4. We obtain

$$P \cdot A_{RX} = \frac{1}{2}\frac{|V_o|^2}{4R_{rad}} \qquad [5.36]$$

Substituting for the radiation resistance from equation [5.30] produces after some cancellations:

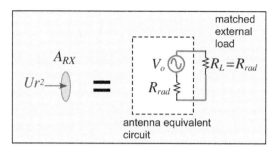

Figure 5-16: Power Density × Effective Aperture = Power Dissipated into Matched Load

$$A_{RX} = \frac{3\pi}{2k^2} = \frac{3\pi}{2}\left(\frac{\lambda}{2\pi}\right)^2 = \frac{3}{8\pi}\lambda^2 \tag{5.37}$$

The effective aperture of an ideal dipole is equivalent to a square about a third of a wavelength on each side, or a circle about 0.4λ in diameter. Note that the effective aperture is *independent of the size of the dipole.* The radiation resistance and power density decrease with decreasing dipole size to cancel the reduction in induced voltage. There is a limit, however, to how small an antenna can be in practice. For small dipoles the radiation resistance will become very tiny, and ohmic losses in the antenna and the matching network needed to match to the small radiation resistance will become significant, reducing antenna efficiency.

Even though no real antenna is isotropic, it is conventional to measure gain relative to an isotropic antenna. It is thus useful to calculate the effective aperture of an isotropic antenna if one existed. Now that we know the directivity and aperture of the ideal dipole, we can use equation [5.33] to obtain the aperture of an isotropic antenna (whose directivity is by definition = 1):

$$A_{iso} = \frac{G_{iso}A_{ideal}}{G_{ideal}} = \frac{(1)\left(\frac{3}{8\pi}\lambda^2\right)}{\left(\frac{3}{2}\right)} = \left(\frac{\lambda^2}{4\pi}\right) \tag{5.38}$$

An isotropic antenna has an effective aperture equivalent to a square a bit more than a quarter wave on a side. For the ISM band, this is about 3 cm. Because the gain of an ideal isotropic antenna is equal to 1, we can now use equation [5.33] to obtain the aperture for any antenna and therefore by [5.31] obtain the received power for any antenna whose gain is known.

$$P_{RX} = P_{TX}\left(\frac{1}{4\pi r^2}\right)G_{TX}\frac{G_{RX}}{G_{iso}}A_{iso} = P_{TX}\left(\frac{1}{4\pi r^2}\right)G_{TX}G_{RX}\left(\frac{\lambda^2}{4\pi}\right) \tag{5.39}$$

After a minor rearrangement, we obtain the celebrated *Friis equation:*

$$P_{RX} = P_{TX} G_{TX} G_{RX} \left(\frac{\lambda}{4\pi r}\right)^2 \qquad\qquad [5.40]$$

The Friis equation allows us to find the received power for any antenna pair at any separation if we know the transmitted power and the power gains of the antennas (assuming of course that the link direction is in the direction of maximum directive gain for both antennas). This equation implies that all we need to do is calculate the radiation pattern and consequently the directivity of a known antenna structure (for efficient antennas) to predict the behavior of a link including this antenna. Empirically, it implies that a single characterization of the link behavior of an antenna using a calibrated reference antenna as the other half (receiver or transmitter) is sufficient to predict the results of any link with any other antenna whose gain is known: a measured radiation pattern tells us everything we need to know to find the link power.

A brief digression is in order to debunk a bit of erroneous physics. The Friis equation is often taken as demonstrating that short wavelength radiation does not propagate readily, because of the factor of λ/r. Reference to equation [5.39] demonstrates that this interpretation is incorrect: the dependence on wavelength arises from the effective aperture of an isotropic antenna, which becomes small as the wavelength shrinks. The received power density is independent of wavelength; a large (and highly directional) receiving antenna will provide the same received power independent of wavelength, though the pointing accuracy requirement becomes increasingly stringent as the antenna gain grows.

We are now equipped with (almost) everything we need to know to engineer a radio link. The interesting quantity is the ratio of received power to transmitted power—the link loss. To find it, we need to know the power gain of the transmitting and receiving antennas. We can calculate the gains using equations [5.1] and [5.15] if the current distribution on the antenna is known or can be guessed, and we can ignore antenna losses. Alternatively, we can measure the gains with a calibrated transmitting or receiving antenna or just look them up in the data sheet if there is one. We then insert the gains in the Friis equation for a known separation to obtain the link loss (at least in outer space; accounting for the inconvenient obstacles encountered in earthbound propagation is the subject of the next chapter).

To make this procedure work, we also need to somehow deliver the input power to the transmitting antenna and get it out of the receiving antenna. Without delving in detail into the subject of matching, it is quite helpful if the radiation resistance of the antenna is close to the impedance of the antenna connection or cable, typically 50 or 75 Ω. An antenna whose impedance is wildly different from the cable impedance will reflect much of the input power back to the generator. Moderate excursions can be accounted for with appropriate matching circuitry, but as we will see, several popular antenna types are chosen because they are well matched to common impedances with minimal additional labor.

The selection of an antenna for an application is thus reduced to the consideration of the power gain, frequency for good impedance match, and real-world performance: size, durability, appearance, mounting, and cost. With these properties in mind, we devote the remainder of this chapter to a survey of commonly encountered antenna types, with the object of enabling the reader to intelligently select the appropriate antenna architecture for a given application.

A brief remark on one last important topic is needed before we proceed to our survey. The reader may recall that the transverse portion of the vector potential is itself still a vector, whose direction is the projection of the source current onto the plane perpendicular to **r**. This direction is the *polarization* of the radiated potential. From equation [5.2] it is apparent that potentials can only act in the direction in which they point: $A \bullet dl = 0$ when the two vectors are perpendicular. An antenna can only transmit to another antenna with the same polarization. For dipoles, this just means that a horizontal transmitter will transmit to a horizontal receiver but not to a vertical receiver. For intermediate angles of rotation about the interantenna axis, a $\cos(\theta_{pol})$ term must be introduced into the Friis equation to account for the loss of received power due to cross-polarization. Keeping track of polarization is of considerable importance in line-of-sight bridging, where it can be reasonably assumed that the received signal has the same polarization as that which was transmitted. However, in indoor or complex outdoor environments, a significant proportion of the total power has undergone one or more reflections, which in general will cause an unpredictable rotation of the axis of polarization due to differences in reflectivity. (See Chapter 6 for more details.) Thus, the polarization of an indoor signal may not closely resemble that of the original transmitted signal and is not well controlled. Furthermore, the user of a mobile client such as a laptop or phone is unlikely to orient their equipment precisely and reproducibly. In summary, in WLAN applications, cross-polarization is usually a modest influence (a few decibels in a 40- to 60-dB link loss budget), and the system user cannot predict or control it. It is for this reason that we haven't devoted much attention to the role of polarization in antenna operation.

6. Survey of Common Antennas

6.1. The Half-Wave Dipole

As we increase the length of an ideal dipole, the radiation resistance increases and the radiated power at a fixed peak current increases, making the antenna easier to use. However, as the length of the antenna becomes comparable with a wavelength, it is apparent that the current along the antenna can no longer be considered constant, because different parts of the antenna must necessarily be at different phases with respect to the exciting voltage or current. We examine one particularly interesting case, a dipole whose length is roughly half the vacuum wavelength, $\lambda/2$.

In the case of a long dipole, there is no longer any option of containing all the charge accumulation at the ends, nor any reason to do so. Therefore, we shall assume the dipole is configured as two pieces of straight wire driven from the center (Figure 5-17). The current must then go to zero at the ends of the wire. It is plausible and very nearly correct to guess that the current on the wire should have a sinusoidal distribution with distance along the antenna, as shown in Figure 5-17. Once we have asserted a form for the current distribution, we need merely integrate equation [5.1] as before to find the potential distribution and radiation pattern. However, the integrals are significantly more complex because there's actually a function that must be integrated rather than a constant. As the exercise is somewhat laborious and not informative, we merely provide the result as equation [5.41]:

$$\mathbf{A} = \hat{\mathbf{z}} \frac{\mu_o I_m d}{2\pi^2} \frac{e^{i(\omega t - kr)}}{r} \frac{1}{(\sin(\theta))^2} \cos\left(\frac{\pi}{2}\cos(\theta)\right) \qquad [5.41]$$

Here the angle θ is as usual defined relative to the axis of the dipole, and the vector potential is in the z-direction. The transverse component of the potential is obtained by multiplying the vector $\mathbf{A}$ by $\sin(\theta)$. From equation [5.15] and [5.41] we can obtain the radiation intensity distribution:

$$U = \left(\frac{\mu_o \omega k I_m^2 d^2}{8\pi^4}\right) \frac{\left(\cos\left(\frac{\pi}{2}\cos(\theta)\right)\right)^2}{(\sin(\theta))^2} \qquad [5.42]$$

The expression for the radiation pattern is of the same form as that for an ideal dipole (equation [5.17]), but with a somewhat different and rather more complex geometry factor. In Figure 5-18 we show a vertical (E-plane) slice through the radiation pattern. The pattern is very similar to that obtained from an ideal dipole (Figure 5-10), albeit somewhat flatter and more horizontally directed. An integration over the distribution produces a directivity of about 1.64 or 2.2 *dBi*. (The terminology dBi denotes dB of gain measured relative to an ideal isotropic antenna. Antennas are also sometimes characterized vs. an ideal half-wave dipole; the resulting gain is denoted dBd.) This is

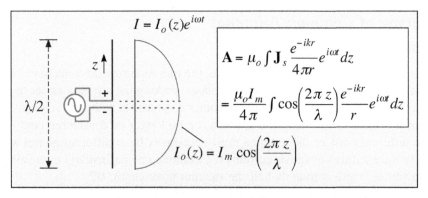

Figure 5-17: Center-Fed Half-Wave Dipole, Showing Current Distribution Along the Antenna; Inset: Integration of the Current to Obtain the Potential

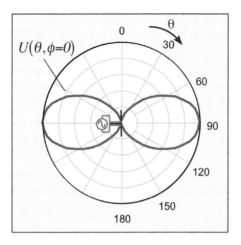

$U(\theta,\phi=0)$

Figure 5-18: E-Plane Radiation Pattern for Ideal Dipole

only slightly in excess of the directivity of the ideal dipole of 1.8 dBi: half-wave dipoles are not appropriate if significant antenna gain is required.

Half-wave dipoles are popular antennas for several reasons. Although it is not obvious from the work we have done here (remember the skeleton in the closet?), a half-wave antenna is nearly resonant, that is, the input inductance and input capacitance approximately cancel to leave a real input impedance. The real input impedance is the radiation resistance, which can be calculated by integrating equation [5.42]; for an infinitely thin wire antenna this value is about 73 Ω, decreasing slightly for wires of finite diameter. Thus, a half-wave dipole is quite well matched to 75 Ω coaxial cable and not too bad for 50 Ω cables. Empirical adjustment of the length of the dipole or a simple matching structure at the feed can be used to tweak the match. If the dipole is fed somewhat asymmetrically, the input impedance changes, because the feed point is moving away from the point of maximum current; this approach can also serve for small adjustments of the match.

Half-wave dipoles are inexpensive and readily constructed, because they consist of two pieces of wire connected together. The driving voltage ought to be differential with respect to whatever ground is present (e.g., the coaxial cable shielding) to avoid current flow through the ground shield and distortion of the antenna pattern. This can be accomplished with a balun transformer of the sort mentioned in Chapter 4 or by the use of a quarter-wave-long piece of coaxial line attached to the center conductor at the antenna. Details of such arrangements can be found in section 9.

Half-wave dipoles also have some significant disadvantages. An ISM or Unlicensed National Information Infrastructure (UNII) half-wave dipole is roughly 6.2- or 2.7-cm long (respectively), much too large to fit two into a PC card or compact flash form factor. Half-wave dipoles are also narrow-band antennas. For example, for reasonable matching, a half-wave dipole with 0.005 λ-diameter wire will achieve a bandwidth of about 13%. This is more than adequate for the 2.4- to 2.483-GHz ISM band but marginal for the full UNII band (5.2–5.85 GHz) and quite insufficient for dual-band use. Half-wave dipoles are also not very directional. A modest increase in gain is obtained by doubling the size of the antenna (a full-wave dipole): $D_{max} \approx 2.4(3.8\,\text{dBi})$. The full-wave dipole must be shortened slightly to achieve resonance (about $0.9\,\lambda$ for a reasonably thin wire), and the input impedance of a center-fed full-wave dipole is infinite in the ideal case and very large in practice; thus it is less convenient to use than a half-wave dipole. Further extension of the length past a full wave does not result in increased directivity: the main lobe of the radiation pattern splits up and additional nodes (directions of zero radiated power) result.

6.2. Quarter-Wave Monopole

A perfectly conducting ground plane acts as a mirror: the net effect of the currents induced by a source above the ground plane is the same as if there were an image source of opposite sign below the ground plane. This effect can be exploited by combining a quarter-wave length of wire with a ground plane to produce an antenna that, for directions above the ground plane, behaves exactly as if it were a half-wave dipole: the quarter-wave monopole (Figure 5-19).

For an ideal infinite ground plane, there is no radiated power below the dipole ($\theta > \pi/2$), and thus the average power is reduced by a factor of 2 for the same peak power. The directivity is thus double that of a half-wave dipole (about 5.3 dB). The input (radiation) resistance is half that of a half-wave dipole, because the total radiated power is reduced by a factor of 2 for the same peak current density. The resulting real impedance of around 37 Ω provides a reasonable match to a 50 Ω coaxial line; slight empirical adjustments in length can be used to optimize the match. An added benefit is

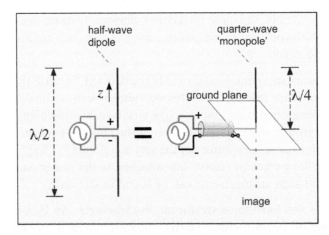

Figure 5-19: Half-Wave Dipole and Equivalent Quarter-Wave Monopole

that the monopole is unbalanced and thus requires no balun transformer: the center conductor of the coax is connected to the antenna (or often just stripped and used as the antenna), and the coax shield is connected to the ground plane.

Real finite-sized ground planes lead to a pattern with reduced radiation in the horizontal direction and some leakage below the ground plane; an example of the pattern for a finite ground plane is contrasted with the ideal case in Figure 5-20. Six wavelengths at UNII band (5.5 GHz) is roughly 33 cm: even the modest ground plane shown in Figure 5-20 may be impractically large for many applications. Smaller ground planes will lead to significant leakage below the plane and reduced directivity, though a good match can still be obtained.

6.3. Folded Dipole

Another popular variant on the half-wave dipole is the folded dipole, shown in Figure 5-21. A folded dipole is constructed of two closely spaced parallel dipoles, shorted together at the ends, forming two quarter-wave lengths of transmission line. Only one side of the line is connected to the signal source. The radiation pattern and directivity are essentially the same as those of a half-wave dipole, but the input impedance is roughly four times higher.

The folded dipole can be analyzed very simply by treating it as the sum of a pair of nonradiating transmission lines and a pair of dipoles (Figure 5-22). We apply half of the signal voltage, with opposing polarities, to both sides of the transmission lines, as shown on the left-hand side of Figure 5-22. Thus, both quarter-wave transmission lines are

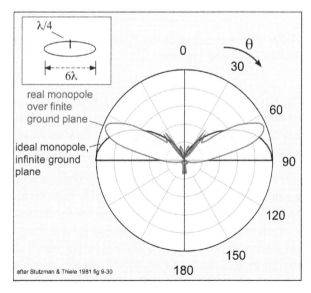

Figure 5-20: Radiation Pattern (Linear Scale) for a Quarter-Wave Monopole Over a Finite and Infinite Ground Plane

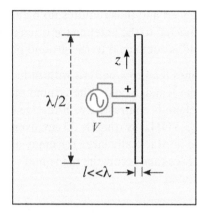

Figure 5-21: Folded Dipole

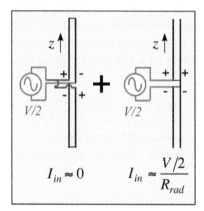

Figure 5-22: Representation of the Folded Dipole as the Sum of Transmission Lines and Dipoles

excited differentially. However, because each line is $\lambda/4$ long and terminated in a short circuit, the apparent impedance at the inputs of the transmission lines is infinite, and they draw no current. In the real case, the currents are very small and cancel in the far field so there is essentially no radiation from the transmission line voltage.

The composite dipole receives the other half, applied to the bases of both quarter-wave wires simultaneously. The composite dipole has essentially the same radiation behavior as a single dipole, because an observer far distant from the antenna cannot tell which of the two dipoles the current is on as long as their separation is much smaller than a wavelength. Thus, the current is that which would exist on a half-wave dipole excited by half the incoming signal voltage. When we add the voltages shown on the left and right sides of Figure 5-22, we get the actual applied voltage of Figure 5-21: the full signal voltage is applied to the left side of the transmission line and none to the right side. Only the current on the left-hand side of the composite dipole is seen by the input: therefore, the current is fourfold less than an equivalent half-wave dipole or, equivalently, the input resistance is fourfold larger: about $280 - 300\,\Omega$. This is a convenient value to match to $300\,\Omega$ twin-lead line, so folded dipoles are often used as part of television or FM radio antennas.

6.4. Waveguide Antennas ("Cantennas")

A waveguide is any conductive channel of uniform cross-section that will support one or more propagating modes—guided waves—along its axis. All hollow waveguides have cutoff frequencies, below which the wavelength is too large to fit in the guide; waves below cutoff are exponentially attenuated and rapidly decrease in amplitude as one proceeds along the waveguide. Different propagating modes are (very roughly) characterized by the number of transverse wavelengths that fit within the guide. Most guides have a range of frequencies in which only one mode can propagate and are normally only used in this range of frequencies. When multiple modes propagate within a single waveguide, undesired variations in amplitude will occur, as the differing modes propagate at slightly different speeds and therefore do not remain in phase.

Waveguide antennas are simply waveguides with one open end. They are a special case of the much broader field of horn antennas, in which the open end is flared to a larger diameter to provide increased power gain. Waveguide antennas have achieved popularity in WLAN circles because they provide reasonable gain and are simple to fabricate using commonly available food storage cans. A simple waveguide antenna is shown in Figure 5-23.

The length of the waveguide should be at least a wavelength to ensure a well-formed single propagating mode but can be longer, though the walls represent a finite source of loss that will affect the performance of a very long guide. The wavelength

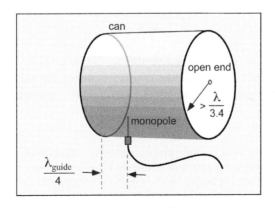

Figure 5-23: Waveguide Antenna

within the guide is longer than the vacuum wavelength (see equation [5.43] below). The diameter must be large enough to allow at least one propagating mode within the guide. The propagating modes of a cylindrical waveguide can be obtained as the roots of various Bessel[1] functions; the calculation is uninformative and we simply quote the results here. For a given wavelength, the smallest diameter guide that will allow a propagating mode is roughly $2\lambda/3.4$: this is about 7.3 cm (3 inches) at 2.4 GHz. Cans that are too small will not work very well, because the desired radiation is nonpropagating and will be rapidly attenuated before reaching the exciting antenna. For best performance the waveguide should be 10–20% larger than the minimum size. The next mode will propagate when the diameter reaches about $2\lambda/2.6 \approx 9.6$ cm (3.8 inches). For a short can, allowing a second propagating mode will cause only a modest degradation in performance. Common can diameters in the United States are 8.7, 10, and 10.5 cm; all will work reasonably well.

An exciting antenna (well, I guess this depends on your point of view) can be readily constructed using a quarter-wave monopole. The length of the monopole is still about one-quarter of the vacuum wavelength, because at this scale the curving walls of the guide are not too different from a flat ground plane. The antenna needs to be offset from the reflecting closed end of the guide by one-quarter of the guide wavelength. For such an offset, the image of the dipole formed by the conducting end of the guide, whose current is oppositely directed from that of the real antenna, is a half of a guide wavelength away. The radiation from the image dipole is phase-shifted by 180 degrees by the time it reaches the real antenna and is thus in phase with it ($-1* -1 = 1$). The dipole and its image add together. If the spacing were a half-wave, the dipole and image would cancel and the antenna would work poorly. The guide wavelength λ_g is longer than the vacuum wavelength by an amount that depends on the ratio of the vacuum wavelength to the smallest allowed wavelength λ_c for the mode

$$\frac{\lambda_g}{\lambda} = \frac{1}{\sqrt{1 - \left(\frac{\lambda}{\lambda_c}\right)^2}}$$
[5.43]

The first-mode cutoff wavelength for a given can diameter D is $3.4/2D$. The guide wavelength becomes infinite right at the cutoff frequency but falls rapidly. For a guide diameter of 10 cm, the guide wavelength is 18 cm and the antenna should therefore be 4.5 cm from the end.

---

[1] Bessel functions are, roughly speaking, the cylindrical analogs of sines and cosines. They are oscillatory functions whose amplitude falls slowly with large arguments. For more details see, for example, **Advanced Engineering Mathematics**, Wylie and Barrett, McGraw-Hill, 1982, Chapter 10.

The power gain of a waveguide antenna can be roughly estimated by equating the physical aperture to the effective aperture, using equation [5.33]. For the smallest allowed aperture we obtain

$$G_{par} \approx \frac{A_{can}}{A_{ideal}} = \frac{\left(\pi R_{can}{}^2\right)}{\left(\frac{\lambda^2}{4\pi}\right)} = \frac{\left(\pi \left(\frac{\lambda}{3.4}\right)^2\right)}{\left(\frac{\lambda^2}{4\pi}\right)} = \left(\frac{4\pi^2}{11.6}\right) \approx 5\,dB \qquad [5.44]$$

The gain of larger apertures will scale linearly in the aperture area; thus a 10-cm guide in this approximation would have a power gain of about 8 dB. Actual power gain will be a dB or two less because of the effects of the finite size of the guide as well as losses in the walls and connection. As noted above, significantly higher gains can be obtained at the cost of mechanical complexity by flaring the open end of the guide; however, for many WLAN applications, an additional 4–5 dB of gain may be quite sufficient, whereas higher gains will impose requirements on pointing accuracy of the antenna that may not be convenient. "Cantennas" can be a useful compromise for achieving improved coverage in homes and small offices at minimal expense.

6.5. Parabolic Reflector Antennas

A parabolic reflector (Figure 5-24) provides a constant distance, and thus a constant phase shift, between a focal point and the aperture plane passing through the focal point. This trick is accomplished by choosing the distance of the reflector from the focal point to compensate the change in distance to the aperture plane at each angle θ. Thus any source placed at the focal point will (approximately) generate a uniform plane wave across the aperture.

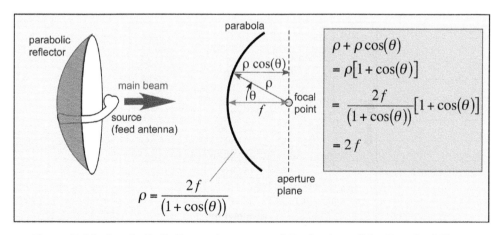

Figure 5-24: Parabolic Reflector Antenna and Derivation of the Required Shape

Just as in the case of the waveguide antenna, the gain of a parabolic antenna can be roughly estimated by setting the effective aperture equal to the physical area of the aperture plane and using equation [5.33]:

$$G_{par} \approx \frac{A_{par}}{A_{ideal}} = \frac{\left(\pi R_{par}{}^2\right)}{\left(\frac{\lambda^2}{4\pi}\right)} = 4\pi^2 \left(\frac{R}{\lambda}\right)^2 \qquad [5.45]$$

However, unlike a waveguide antenna, the size of the aperture is not limited by a mode size but only by practical issues such as available size, manufacturing precision, and cost. Thus, parabolic reflector antennas can achieve very high gains. Figure 5-25 provides a comparison of the predicted gain based on physical apertures from equation [5.45] with measured gain for a few representative ISM and UNII antennas. It is apparent that the simple aperture model is within 2–3 dB of the correct result, becoming more accurate for reflectors which are larger (relative to a wavelength). Gains of 20–30 dB are readily achieved in practice.

The availability of such high antenna gain makes it possible to extend links over long distances (tens of kilometers) even at modest transmitted power levels. However, the price of high gain is narrow beam width (equations [5.18] and [5.19]): a 30-dB gain antenna will have a beam width of only about 6.5 degrees. Excellent pointing accuracy is required to make use of this gain. Mobile applications are not realistic; such high-gain antennas are only practical for fixed point-to-point connections. Setup of the link is awkward at long distances, and very stable antenna mounting is needed. The narrow beams can be deflected by atmospheric conditions. We cover the issues involved in such links in more detail in Chapter 8.

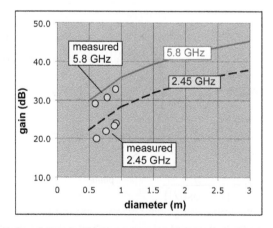

Figure 5-25: Gain of Parabolic Antennas vs. Diameter, Comparing Simple Model Based on Aperture Size (Solid and Dashed Lines) With Representative Measured Values for ISM and UNII Band Antennas

Parabolic antennas are intrinsically wideband, in the sense that the operation of the reflector is only very slightly frequency dependent as long as it is large relative to a wavelength. Practical bandwidth is thus limited by the type of feed antenna and matching used. The polarization of the resulting radiation is not exactly that of the feed antenna but is somewhat modified by the reflector, though this is not usually a major effect. It is desirable that radiation be predominantly in the direction in which the reflector points and not sideways or backward: that is, the antenna should have a large *front-to-back ratio*. A good front-to-back ratio is achieved by arranging the feed antenna to primarily illuminate the center of the reflector but at the cost of reduced gain, because the effective aperture is smaller than the physical aperture. The feed design also affects the relative size of the directive gain outside the main beam, usually expressed as the directive gain at the peak of the largest subsidiary beam relative to that of the main beam.

Gain performance almost equal to that obtained with a continuous reflector is achieved using a reflector made of a grating of wires or rods as long as the spacing is small compared with a wavelength. Such grating antennas provide high gain with much reduced wind resistance and are thus easier to mount reliably in outdoor applications.

An example of a practical UNII band (5.8 GHz) parabolic antenna is shown in Figure 5-26, with the corresponding antenna pattern in Figure 5-27. This antenna is 61 cm in diameter and has a measured power gain of 29 dBi, only slightly below the 31 dB predicted from the simple model of equation [5.45]. The 3-dB beam width (the width measured at directive gain of 3 dB below the maximum value) is just over 5 degrees, in good agreement with that from the simplified width estimate of equation [5.19]. The largest secondary lobe is 18-dB lower in gain than the main beam. The front-to-back ratio is about 35 dB.

**Figure 5-26: MAXRAD MPR58029 5.8-GHz Parabolic Antenna
(Reproduced With Permission of Maxrad Inc.)**

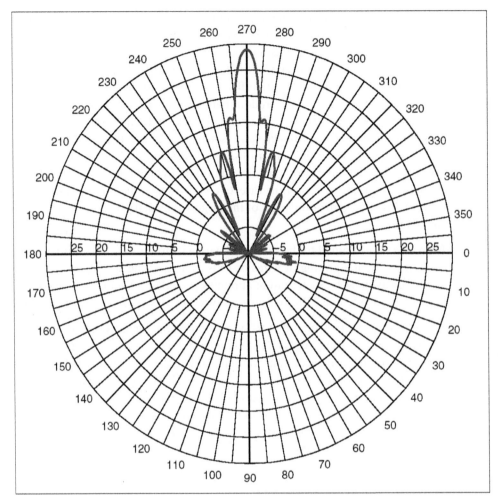

Figure 5-27: Measured Radiation Pattern (in Horizontal Plane) for MAXRAD MPR58029 5.8-GHz Parabolic Antenna; Radial Scale in dBi (Reproduced With Permission of Maxrad Inc.)

6.6. Microstrip (Patch) Antennas

It is very common in microwave engineering to construct simple planar waveguides by placing a narrow conductive strip on a dielectric layer resting on a ground plane. This configuration is known as a microstrip. The thickness of the dielectric is much less than the wavelength of interest.

A microstrip waveguide a half wavelength long is resonant: the inductance and capacitance cancel. In the absence of radiation, the open circuit at the end of the line would appear at the input. In practice, the currents in the top strip and the ground plane do not perfectly cancel because of the termination of current at the ends, and thus there is significant radiation from the resonant piece of microstrip, which acts as an antenna (Figure 5-28). The optimal length is actually slightly less than a half wavelength to compensate for fringing effects at the ends of the patch.

A patch antenna with an infinite ground plane, like an ideal monopole, would not radiate at all below the ground plane; practical patch antennas have very little backside radiation. By symmetry, the radiation of the patch antenna must also go to zero near the ground plane direction, where all the currents in the microstrip and the image are at equal distances from the observer and cancel. Thus, the radiation from a microstrip antenna is mainly directed straight up perpendicular to the plane of the circuit. The detailed radiation pattern is somewhat complex to calculate, but the maximum intensity scales as $\omega k \mu_o I_m^2 t^2$, similar to a dipole but with the patch separation taking the place of the dipole length. Realistic antennas achieve power gain of around 6–8 dB, of which 3 dB is the result of the elimination of the bottom hemisphere.

The input load can be approximated by assigning an effective radiation resistance to each end of the patch of about $120(\lambda/W)$. Because the half-wave line effectively transports the output load to the input, the equivalent input resistance is roughly $60(\lambda/W)$. The width W is generally much less than a wavelength to avoid other resonances; if the patch is square, the input resistance is about $120\,\Omega$. Several simple tricks may be used to match the patch to the incoming microstrip line, with a typical

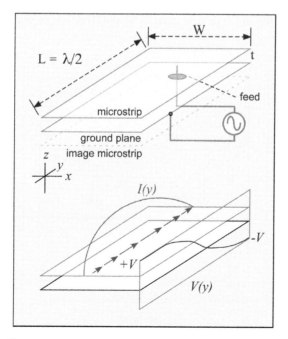

Figure 5-28: Half-Wave Microstrip as an Antenna

impedance of 50 Ω. For example, the feed point may be displaced slightly from the edge of the patch, as shown in Figure 5-28: the input impedance falls as the feed point is moved toward the center of the patch, due to the increase in the local current $I(y)$ relative to the local voltage. Alternately, a bit of line of differing impedance may be used as a quarter-wave transformer. Both techniques may be easily implemented in planar circuits formed on the dielectric. Note that the use of a resonant structure implies that microstrip antennas are generally applicable only to narrow bandwidths.

The special utility of patch antennas arises less because of their characteristics as antennas than because of the ease of their fabrication. Patch antennas can be constructed on planar substrates using the photolithographic techniques used to make printed circuit boards. Batch processes of this type allow an array of patches, with matching structures and feeding connections, to be fabricated as readily (and as inexpensively) as a single patch; a simplified example is depicted in Figure 5-29. Here a bit of widened line serves to split the signal equally without reflections at each bifurcation of the feed line, and a second block serves as a transformer to match the impedance of the patches. The feed lines here have been arranged to ensure that the distance from the input to each patch is equal, so all the patches are driven in phase. Such an array provides an approximation to a uniform plane wave (albeit with perturbations due to the details of the structure), creating a directional antenna with improved gain.

To fabricate such large arrays, it is imperative that losses in the feed lines be minimized. Highly conductive aluminum ground planes, copper lines, and a low-loss dielectric such as Teflon or its variants are typically used. The polarization of the antenna can be

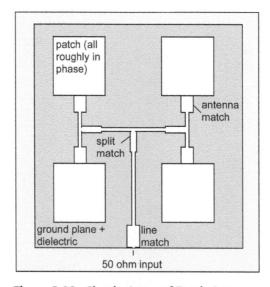

Figure 5-29: Simple Array of Patch Antennas

modified by tilting the patches; square patches can also be driven from both sides, using a bit of line to create a 90-degree phase shift, to produce a circularly polarized output. Small perturbations from uniform phase are easily created by meandering feed lines, so that directivity and the shape and direction of subsidiary beams can be traded off if desired. All these variations have little or no impact on the fabrication cost.

Microstrip array antennas with tens of patches can be readily and inexpensively fabricated; power gains of 10–20 dB are practical. Arrays can be packaged in plastic radomes and are robust to handling and easy to mount. Secondary beam suppression and front-to-back ratios (typical values are around 10 dB and 20 dB, respectively) are not as good as the corresponding parabolic reflector antennas, but the low cost, versatility, and simplicity of patch antenna arrays make them an appealing choice for indoor applications.

6.7. Phased Arrays

The array of patch antennas depicted in Figure 5-29 is an example of an important class of antennas known as phased arrays. A phased array is a group of identical or nearly identical antennas excited with known phase relationships, usually at about equal amplitude. The radiation pattern of the antenna is the product of the radiation pattern of the individual elements and a term due to their relative phases and amplitudes, known as the *array factor*. The array factor is defined for the simple example of an equally spaced linear array in Figure 5-30. The array factor is usually the dominant influence on the antenna behavior when many individual antennas are present.

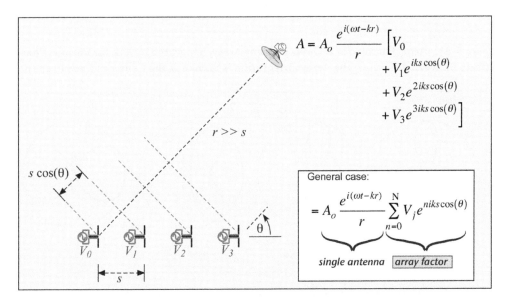

Figure 5-30: Radiation From an Array of Identical Antennas Can Be Treated as the Product of the Individual Antenna Radiation and an Array Factor

The simplest case (and the only one we treat in detail) occurs when all the amplitudes $|V_i|$ are the same and the phase of excitation differs by a constant from one antenna to the next: $V_{i+1} = e^{ik\alpha}V_i$. In this case the excitation voltage can be extracted from the sum, and the array factor takes on the more tractable form:

$$AF(\theta) = |V_j| \sum_{n=0}^{N-1} e^{in(ks\cos(\theta)+\alpha)} = |V_j| \sum_{n=0}^{N-1} e^{in(\psi)} \text{ where } \psi = ks\cos(\theta) + \alpha \qquad [5.46]$$

The array factor is determined by the distance between antennas (normalized to the wavelength), the angle of view, and the phase offset between antennas. Some examples of array factors for a simple three-element array separated by a quarter-wave and a half-wave are depicted in Figures 5-31 and 5-32, respectively. When the pattern is predominantly perpendicular to the axis of the array (e.g., Figure 5-31 or 5.32, $\alpha = 0$), it

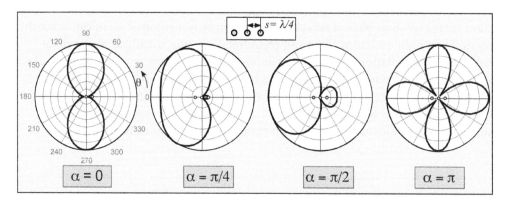

Figure 5-31: Array Factors for a Three-Element Array, Separated by λ/4, for Various Phase Offsets Between Antennas; the Array Orientation Is Shown at the Center of Each Pattern

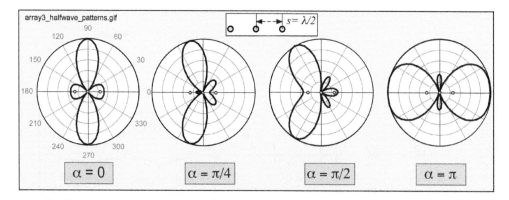

Figure 5-32: Array Factors for a Three-Element Array, Separated by λ/2, for Various Phase Offsets Between Antennas; the Array Orientation Is Shown at the Center of Each Pattern

is commonly referred to as a *broadside* pattern. When the radiation is along the axis of the array (Figure 5-32, $\alpha = \pi$), the pattern is known as an *endfire* array. By adjusting the relative phase and separation of the array elements, a wide variety of patterns can be produced, with varying beam widths and orientation with respect to the array.

The possibilities in an array appear powerful but also perhaps bewildering. It may seem that an independent numerical calculation of every array spacing and phase shift for every number of elements might be needed to figure out what can be achieved and how to achieve it. Fortunately, a systematic investigation of the behavior of the array factor may be accomplished graphically with the aid of a bit of intermediate construction.

First, we note that the array factor written in the form of equation [5.46] is just a geometric series, that is, a series of terms of the form $(1 + x + x^2 + x^3 + \ldots)$. Such a series can be readily summed:

$$\sum_{n=0}^{N-1} x^n = \frac{x^N - 1}{x - 1} \tag{5.47}$$

Applying this formula to the array factor, we obtain

$$AF(\psi) = V \sum_{n=0}^{N-1} \left(e^{i(\psi)}\right)^n = V \frac{e^{iN\psi} - 1}{e^{i\psi} - 1} = V e^{i[N-1]\frac{\psi}{2}} \frac{\sin\left(\frac{N\psi}{2}\right)}{\sin\left(\frac{\psi}{2}\right)} \tag{5.48}$$

Because the absolute phase of the array factor is not usually important (unless we are combining multiple arrays), we can ignore it, obtaining the simplified expression:

$$AF(\psi) = V \frac{\sin\left(\frac{N\psi}{2}\right)}{\sin\left(\frac{\psi}{2}\right)} \tag{5.49}$$

Some examples of this generalized array factor, plotted versus the variable ψ, are given in Figure 5-33 for 3, 5, and 10 array elements.

The array factor is periodic in ψ with period 2π. There are $(N - 1)$ lobes in each period: a big one and $(N - 1)$ smaller ones. The big lobe is $4\pi/N$ wide; the minor lobes are half as wide. The side lobes decrease relative to the main lobe for increasing N, but the dependence is weak for large numbers of elements: the first side lobe is down 12 dB for $N = 5$ and only about 13 dB for very large N.

Now we are somewhat familiar with the array factor and its behavior, but only in terms of the rather obscure variable ψ rather than the physical quantities array separation s, angle θ, and phase shift α. Although the numerical conversion can be accomplished using equation [5.46], a better intuitive grasp is obtained by performing the transformation graphically.

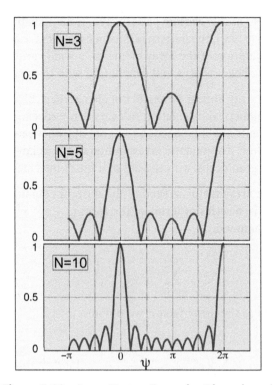

Figure 5-33: Array Factor Examples Plotted vs. ψ

The construction is shown in Figure 5-34. Below the array factor plot, place a circle of radius equal to the propagation phase shift between elements, ks. The center of the circle is offset from the $\psi = 0$ axis by the excitation phase offset α. The array factor as a function of the view angle θ is obtained by drawing a radius of the circle at θ from the horizontal to determine the value of ψ.

Some general conclusions about the behavior of the pattern can now be obtained. When the normalized element separation ks is increased, the circle gets bigger. This means more lobes are included in the pattern and each lobe becomes narrower in θ. If the separation becomes too big, the circle encompasses more than one main beam, and directivity is reduced. Adjusting the offset α slides the circle to the left or right, thus changing the angle of the main lobe. The phase offset α steers the main beam.

Figure 5-35 provides some examples of how the construction works, demonstrating the origin of some of the patterns previously depicted in Figures 5-31 and 5-32.

With this preparation completed, we can now consider several antenna types in which an array of individual antennas is exploited.

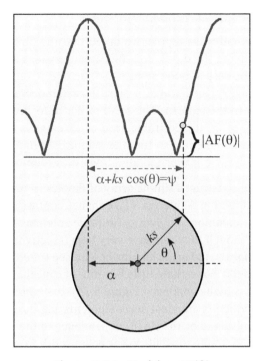

Figure 5-34: Deriving AF(θ)

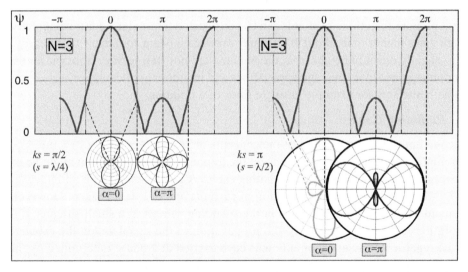

Figure 5-35: Examples of AF(θ) From AF(ψ) for N = 3, Varying Spacing and Phase Offset

6.8. *Azimuthally Isotropic Directional Antennas*

In our discussion of the half-wave dipole, we noted that the directivity is a rather modest 2.2 dBi. Lengthening the dipole to a full wave provides a few decibels of additional gain but at the cost of poor input matching and strong frequency dependence; lengths longer than 1 wavelength suffer from decreased directivity as additional lobes appear in the pattern. This is all rather unfortunate, because the earth is (locally) flat: in many cases, the desirable area of coverage is essentially in a horizontal plane around an antenna, so that the azimuthal symmetry of a dipole is desirable, but the considerable power radiated toward the zenith and nadir is wasted.

To fill this need, one can construct a simple array of dipoles to form an azimuthally isotropic but vertically directional antenna. (To enhance confusion for folks entering the field, these antennas are commonly known as *isotropic* antennas—referring only to the H-plane pattern—even though they can have very high directivity and are not at all isotropic.) The array is fabricated using lengths of wire separated by short inductors, which provide a phase delay but radiate little because of cancellation of the opposed currents on opposite sides of the windings (Figure 5-36). A simple model of the radiation pattern, assuming the inductors are ideal delay elements and the dipoles radiate as a three-element array (multiplied by the $\sin (\theta)$ dependence of the individual dipoles), is also shown in Figure 5-36. Using the half-power beam width from the pattern, we can estimate the directivity to be about 6.6 dBi. The measured half-power beam width for this antenna is 20 degrees, corresponding to a power gain of about 7.5 dBi.

Arrays of more than three elements can be constructed to provide additional directivity, though gains higher than about 10 dBi may produce a beam so narrow that nearby clients slightly offset in height from the antenna are not well served. Isotropic antennas are popular in outdoor coverage applications and in indoor applications where coverage is to be limited to, for example, a single floor of a building.

6.9. *Radiating Cable*

A second interesting application of the concepts of array antennas is a structure known as a radiating cable, though depending on the frequency of application, it might be more appropriate to call the design nonradiating cable. A coaxial cable with an intact shield provides essentially perfect cancellation of currents and does not radiate. However, if the shield is perforated, each defect in the shielding will act as a small antenna; if the perforations do not overly perturb the propagation of the signal within the cable, they may be regarded as an array of antennas each excited at a phase determined by the phase velocity of the wave within the cable, which in a dielectric-filled cable is always less than the velocity of light. The situation is schematically depicted in Figure 5-37.

The radiation pattern of such an array will have a maximum at a value of the angle θ at which the radiation from all the perforations arrives has the same phase delay to the receiver. However, for cable phase velocities slower than the velocity of light and sufficiently short spacings s, there is no angle at which this can occur:

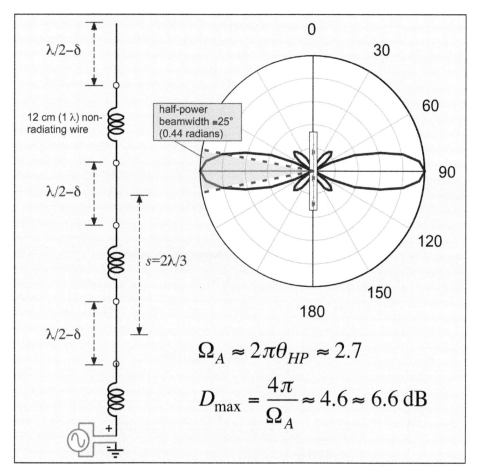

Figure 5-36: Example of Vertical Array Used to Provide "Isotropic" Antenna With High Directivity Into the Horizontal Plane; Modeled Pattern Is Based on Three-Element Array Factor

$$\text{from slot n: } \overbrace{e^{-ik(R-ns\sin(\theta))}}^{\text{cable-to-RX}}\;\overbrace{e^{-i\beta ns}}^{\text{in cable}}$$

$$= e^{-ikR}\left[e^{-is(\beta-k\sin(\theta))}\right]^{n} \qquad [5.50]$$

$$(\beta - k\sin(\theta)) = 0 \Rightarrow \text{all slots in phase}$$

$$\boxed{impossible \text{ for } \beta > k}$$

(A phase shift of 2π between neighboring perforations, which would produce the same effect as a phase shift of 0, is possible at some angle if s is large enough or, equivalently, if the wavelength is short enough.) The same conclusion can be obtained using the array factor construction, as shown in Figure 5-38. If the phase velocity in the cable is slower than in vacuum, $\beta > k$. Therefore, the offset of the

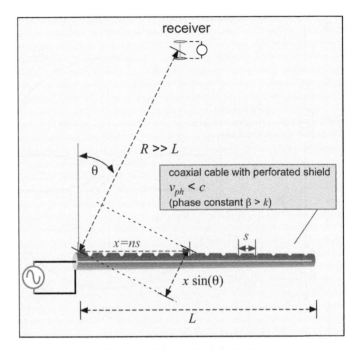

Figure 5-37: Radiating Cable

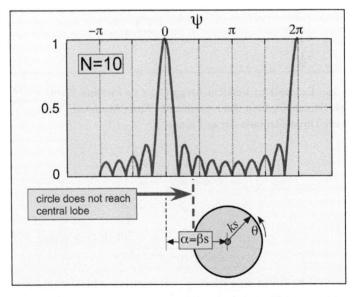

Figure 5-38: Array Factor Construction for Radiating Cable

circle center βs is always larger than the radius of the circle ks. For large numbers of elements N, the main lobe is very narrow and the edge of the circle will always be outside of it for any angle, so there is no main beam.

When the observer is close to the cable, on the other hand, the distance to each perforation varies significantly. The $(1/r)$ factor in the potential calculation must be retained within the integral and suppresses radiation from distant sources, thus preventing cancellation once the receiving antenna is close enough to the cable. In practice, the field is reasonably confined to within about a meter of the cable for typical geometries.

If the frequency is increased until the phase shift between perforations reaches 2π, a backward-propagating radiated wave will appear. This mode of operation produces higher power at the receiver, because the power from many perforations is combined, but radiated power in the backward direction may appear in undesired locations. This mode of operation is used for coverage in tunnels or pipelines.

Radiating cable can be useful in indoor applications, where it can be run along corridors or within tunnels where coverage would otherwise be poor. Radiating cable can be used along the perimeter walls of buildings to provide local coverage in corridors, offices, or conference rooms while radiating very little signal to the outside world, improving security. However, the cable is relatively expensive and laborious to install and does not replace traditional antennas in most applications.

6.10. Yagi-Uda Antennas

Yagi-Uda antennas, often known simply as Yagis, are wire antennas that are simple to construct (though not simple to design) and can achieve quite high directivity. Figure 5-39 depicts a simplified Yagi-Uda structure. Only one element, the *driver*, is actually excited by an external signal source. The driver is often a dipole or folded dipole. A *reflector*, close to and slightly longer than the driver, offsets the pattern of the driver modestly toward the *director array*. Larger numbers of directors produce higher gain if their lengths and locations are properly chosen; gains up to 17–20 dB are achievable. The director array acts as a sort of waveguide, supporting a current wave that is slow relative to the velocity of light in free space. Like radiating cable, Yagi-Uda antennas exploit the slow-moving wave as an array excitation to avoid broadside radiation, resulting in a narrow endfire beam.

The general mechanism of operation is depicted in Figure 5-40. Viewed perpendicular to the long axis of the array, each element constitutes a source driven with relative phase βs, where the phase constant $\beta > k$ is characteristic of the slow wave propagating along the array. Just as in the case of radiating cable, the individual contributions from the array elements will not add in phase in any direction far from the axis, so little power is radiated off axis. (If the antenna is constructed of wires as shown, the pattern of each wire will also have a node in along its axis, contributing a couple of decibels of directivity. However, a Yagi-Uda

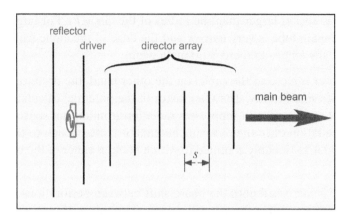

Figure 5-39: Yagi-Uda Antenna

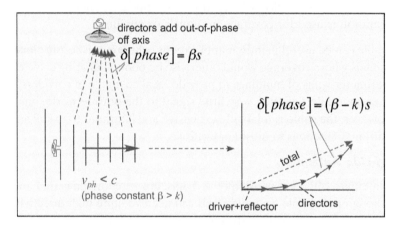

Figure 5-40: Yagi-Uda Operating Principles

antenna may also be constructed of ring-shaped elements that have less directionality than a long dipole.) On-axis, the phase shift from one array element to the next is proportional to the difference between the phase constant of the slow wave and the unperturbed wave. Thus, though all the contributions are not perfectly in phase, the total potential will continue to grow as directors are added until enough phase shift has been accumulated to turn the potential from the next reflector perpendicular to the sum of the previous contributions.

The design of Yagi-Uda antennas is complex, because there is no simple analytic method of predicting the phase velocity of the array wave, as it arises from the interactions of all the currents on the array elements. Numerical simulation of the self-consistent current distribution, taking explicit account of the thicknesses of the elements and perturbations due to conductive mounting bars and masts, is necessary. However, the results of such calculations are widely available as design charts specifying optimal parameters for particular element configurations and frequencies, so that a design from scratch is rarely required.

Yagi-Uda antennas are commercially available from many vendors. Antennas for ISM and UNII bands are of a reasonably convenient size for outdoor applications and are inexpensive to construct and robust in use. However, the bandwidth of a Yagi-Uda antenna is typically only a few percent; the structure is not appropriate for broadband or multiband applications. Yagis are often enclosed in a plastic shroud that reduces wind resistance (and serves to provide some aesthetic benefit, because most folks would not regard a Yagi's spindly structure as an attractive addition to their neighborhood).

6.11. Adaptive Antenna Arrays

All the antennas we examined so far can be regarded as having a fixed radiation pattern: for example, a parabolic antenna radiates along its axis, and if a different axis is desirable the antenna must be moved. However, it should be apparent from the discussion of antenna arrays in section 6.7 that if the phase shift between array elements α could be adjusted electronically, the resulting antenna pattern may be modified in direction and width even if the array elements are physically fixed.

A number of methods can be used to provide multiple values of the phase offset. Some approaches provide a finite number of phase shifts and thus of beam directions; a particular input or output signal can be switched from one state to another electrically. Other methods use analog or digital phase shifting and are capable of continuous modifications of the antenna radiation pattern but are more complex to implement and calibrate.

An example of the first approach uses a phase-shifting network between the radio or radios and an antenna array. One convenient phase-shifting network is known as a *Butler matrix*. The Butler matrix is constructed of two elements: a fixed phase shift and a *hybrid-T* coupler (Figure 5-41). The phase shifter simply introduces a phase shift in the signal and can be implemented for a narrowband application by adding a bit of length to a microstrip transmission line. The coupler is a bit more subtle: a signal placed

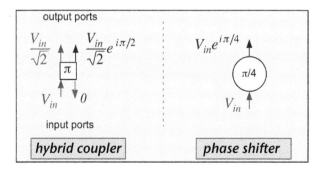

Figure 5-41: Hybrid-T Coupler and Phase Shifter

on one port is split into two output ports, with a phase shift of $\pi/2$ between them. No signal is reflected or appears on the other input port. The coupler's operation is symmetric and reversible, so that any port used as the input couples to two of the three other ports, with the same phase shift between the output signals. Couplers can be constructed on a circuit board using microstrip structures or can be based on hollow waveguides.

A simple Butler matrix constructed using these elements is shown in Figure 5-42. To drive four antennas, two pairs of two couplers are used. The first stage consists of two hybrid-T couplers. Their inner outputs are cross-directed to a second pair of couplers; the outer outputs are phase-shifted by $\pi/4$ before being forwarded. (Note that for this scheme to work, the vertical and crosswise connections in the matrix must actually be of the same length when constructed on a circuit board; thus, careful attention to layout is required for a successful implementation.)

Each coupler in the second pair provides 0 and $\pi/2$ additional phase shifts to its two output ports. By tracing the phase shifts in the diagram, the reader can verify that an input signal at port 1 suffers an additional phase shift of $\pi/4$ between each successive output antenna. Thus, from port 1, the array appears to provide a phase shift $\alpha = \pi/4$. However, an input to the adjacent port 2 will encounter a phase shift of $5\pi/4$ between antennas. In fact, all odd multiples of $\pi/4$ up to $7\pi/4$ are available. This simple Butler matrix provides four phase shifts and consequently four pointing directions from an array of four antennas. The resultant patterns can be estimated using the array factor construction of section 6.7. Larger arrays, using more elements and stages, can be

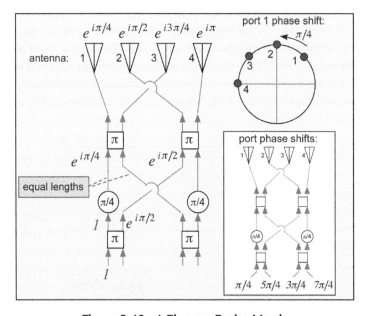

Figure 5-42: 4-Element Butler Matrix

constructed; in general, phase shifts of $(2m - 1)\pi/N$, for $m = 1$ to (N), are available for N antennas. Thus, for large arrays, essentially all phase shifts between 0 and 2π are accessible. We examine an implementation using $N = 16$ for an adaptive WLAN antenna in Chapter 8.

A fully configurable array can be constructed with the use of analog phase shifting networks (which may also include an amplitude adjustment; in this case the whole is known as a *vector modulator*); such a scheme is depicted on the left in Figure 5-43. The relative phase of a physically large array of such antennas is challenging to maintain at high frequencies, over variations in temperature and components. An alternative approach to constructing an adaptive array, which is becoming more practical as the cost of radio chips decreases, is to perform the phase shifts digitally and provide baseband signals (which are near zero frequency and easy to ship from place to place) to a separate radio for each antenna. The synthesizers in these radios must be phase locked to a single reference oscillator, which is again practical because only signals in the megahertz range must be provided across the array.

Adaptive arrays, either the switched variety of Figure 5-42 or the fully flexible version in Figure 5-43, are potentially of great value in WLAN (and other wireless) systems. An adaptive antenna can have a large directivity in applications where a fixed high-gain antenna cannot be used, such as portable or mobile devices. Errors in pointing direction are subject to correction by the adaptation algorithm, so the user has no need to maintain precise orientation. The resulting power gain can extend range or reduce

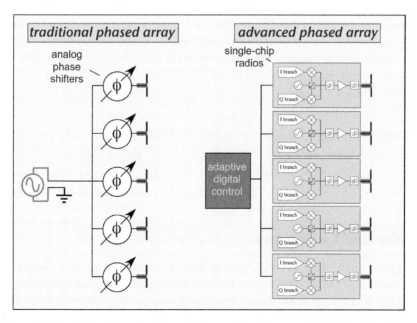

Figure 5-43: Fully Adaptive Phased Arrays Implemented With Analog Phase Shifters or Digital Control

transmit power (or both). The use of higher gain antennas puts more radiation where it is wanted and less where it is not, reducing the overall interference burden each user places on other neighboring users. Adaptive arrays can also be used to defeat interference, by directing an angle of near-zero directive gain (a node) toward the interferer while preserving high directive gain toward the desired client or access point. An adaptive array can increase total WLAN capacity in a given area by partitioning the area into directions, each served by a separate radio link. (This function can also be provided using a fixed set of directional sector antennas and adaptive digital back-end processing to handle handoffs between the sectors. Such a scheme is widely used in cellular telephony.) Adaptive arrays can eliminate dead spots and fading that result from cancellation of signals traveling along multiple paths, by transmitting or receiving in only one optimal direction. Fully adaptive arrays of the type depicted in Figure 5-43 can even provide an optimal pattern adapted to the propagation conditions of each user, the *spatial signature*. (More about indoor propagation is found in Chapter 7). Finally, if multiple antennas are used on both transmit and receive in an indoor environment (*multiple input multiple output* or MIMO architectures), multiple effective beams can be synthesized, increasing the total capacity of the link above Shannon's limit (though no new physics has been introduced: this is just a different way of achieving spatial partitioning).

As with any technology filled with such enticing possibilities, there are important caveats. Adaptive array systems are complex and relatively expensive compared with fixed antennas, though the added expense decreases with time. Expense and complexity increase with the number of elements, but so do the benefits. To provide a significant benefit, the array elements must be spaced by at least a quarter wavelength and preferably a bit more. An array of five or six antennas a half-wave apart at 2.4 GHz will be on the order of 30 cm across, hardly appropriate for a PC-card or personal digital assistant form factor, though perhaps applicable to built-in antennas in a laptop. Control algorithms are complex and in many cases the best approaches are not compatible with existing standards.

It is very likely that adaptive antenna arrays and MIMO architectures will play an increasing role in commercial wireless technology over the next decade, including WLANs and wireless personal area networks, but the exact way in which proliferation of the technology occurs will be complex and uneven due to the practical difficulties noted above.

7. Cables and Connectors

The best antenna isn't very useful if signals can't be delivered to it or extracted from it. Some PC cards and other small radios may incorporate an antenna integrated into the card, driven directly from a microstrip waveguide, so that no connectors or cables are needed. However, the performance of such small antennas (particularly at 2.4 GHz) is limited, and a separate antenna is often used even for card applications, necessitating a connecting cable. External and remote antennas obviously require cables to bridge between the radio and the antenna. Although it is possible to permanently attach the subject cable at each end by soldering, it is more common and much more convenient to use connectors so that cables, antennas, and radios can be readily swapped. Cables and connectors play an important role in real radios.

Losses in these cables and connectors generally precede any amplification. As a consequence, loss in cables and connectors adds directly to the noise figure of a radio receiver (Figure 5-44) and subtracts directly from the transmitted power. Cable losses should be minimized for high-performance links. Minimizing loss often militates directly against the most convenient and inexpensive methods of antenna mounting. Short cables have less loss but detract from the freedom of the system designer to place the antenna and radio where each can most easily be used and accessed. Low-loss cabling is large and expensive, increasingly so at higher frequencies. A separate tower-mounted amplifier can be provided for antennas that are at a considerable distance from the transceiver, but the complexity of this arrangement may exceed that of simply remoting the radio to the antenna site. Designers and users need to know something about cables and connectors.

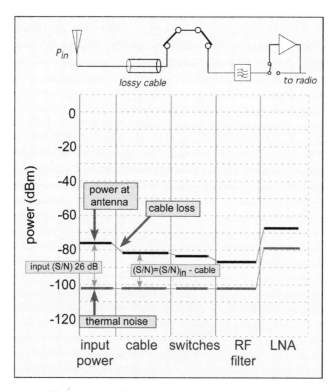

Figure 5-44: Cable Losses Before the Low-Noise Amplifier Subtract Directly From Input S/N and Thus Increase Equivalent Input Noise Figure

Almost all remakeable connections use coaxial cabling. Coaxial cables have a center conductor completely surrounded by a conductive ground shield. Signal currents traveling along the center conductor are compensated by countercurrents in the shield, so radiation from the cable is very small (Figure 5-45).

The space between the center conductor and the shield is generally filled with a low-loss dielectric, although in some cable designs spacers are distributed periodically along the cable with most of the interstitial space filled with air. The ratio of the current in the cable to the voltage between the center conductor and the shield is a real-valued constant: the *characteristic impedance* of the cable. The characteristic impedance is determined by the shield radius b, the center conductor radius a, and the relative dielectric constant ε of the fill material:

$$Z_o = \frac{138}{\sqrt{\varepsilon}} \log_{10}\left(\frac{b}{a}\right) \qquad [5.51]$$

Common values of Z_o are 50 or 75 Ω. Assuming a typical dielectric constant of around 2, a 50 Ω line requires a ratio $(b/a) \approx 3{:}1$. For any given outer cable diameter, the target impedance constrains the size of the center conductor.

Unlike hollow waveguides, coaxial cables work fine with signals down to DC. The highest frequency that can be used is that at which additional propagating modes, such as hollow-waveguide modes (briefly noted in section 6.4 above), become possible within the cable. The *cutoff frequency* is approximately

$$f_c = \frac{c}{\pi\sqrt{\varepsilon}(a+b)} \qquad [5.52]$$

For a 50 Ω line with an outer diameter of 1 cm, the cutoff frequency is about 10 GHz. However, *cable loss* and connector leakage may become prohibitive long before this limit is reached, depending on the application. Loss results from the finite conductivity of the center conductor and shield as well as from within the dielectric material if

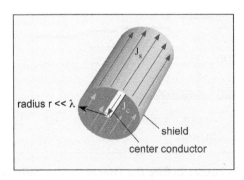

Figure 5-45: Coaxial Cable Configuration

present. The current flows in a narrow region near the surface of the metal, the *skin depth*. Typical skin depths are a few microns at gigahertz frequencies: losses are much larger than they would be if the current flowed through the whole thickness of the metal. The total area for current is roughly the perimeter of the wires multiplied by the skin depth, so loss increases with decreasing cable diameter. The skin depth decreases as the square root of frequency, so cable loss increases with frequency. Cable loss due to conductor resistance is approximately

$$\alpha_c = \{1.5 \times 10^{-4}\} \sqrt{\rho f} \left(\frac{1}{a} + \frac{1}{b}\right) \frac{1}{Z_o} \qquad [5.53]$$

(This loss is expressed in *Nepers/meter*. A Neper is $(1/e)$ of loss, where e is the natural logarithm base, $2.718\ldots$. To convert to dB/m, multiply by 8.68.) For our 1-cm diameter $50\,\Omega$ line, the ohmic loss would be about $9\,\text{dB}/30\,\text{m}$ at $5.8\,\text{GHz}$. Flexible coaxial cables generally use some sort of multistrand ribbon material for the shield and are subject to higher losses than would have been the case for a continuous sheath.

A bewildering variety of coaxial cables are available, differing in characteristic impedance, loss, frequency range, flexibility, and cost. *Semirigid* coaxial cables provide the best performance in most respects but at the cost of—surprise!—flexibility. Semirigid cables have a solid copper shield and solid inner conductor, separated by a Teflon variant or other low-loss insulator. The solid shield ensures that essentially no signal leaks out of the cable if it is solidly grounded at the ends; the high conductivity of the solid copper provides low loss. Semirigid cables and cable stock are available in a range of diameters and impedances. Semirigid cable can be bent by hand or with tube bending tools and is sufficiently flexible to allow many centimeters of elastic adjustment over a 10-or 20-cm length of cable, but it is best suited for applications where the connection geometry is essentially fixed.

Flexible coaxial cables come in a wide variety of sizes and construction. They are usually fabricated with a copper or copper-clad steel inner conductor and copper or aluminum ribbon or braid for the outer conductor. Polyethylene and Teflon are common insulators. Foamed insulation may also be used to reduce dielectric constant. From equation [5-53] we see that large diameters are preferred for low loss, but large-diameter cables are relatively inflexible, awkward, and expensive. Small cables are easier to work with, particularly in tight spaces, if the excess loss can be tolerated.

Popular cable nomenclature is based on U.S. military standards. *Radio Guide* (RG) designations dating from the 1940s are still widely used. *M17* names (from military standard MIL-C-17, promulgated in the 1970s) are also used. The nomenclature is arranged in a quite arbitrary fashion; there is no simple correlation between the RG or M17 number and cable size, loss, or frequency rating. Performance specifications required in these standards are based on very old technology, and modern cables can easily outperform the RG specs within the same physical dimensions. The nomenclature is often used in a generic fashion to denote cables that are similar to but not necessarily in compliance with a given specification.

Measured loss for some common RG-designated cables is depicted in Figure 5-46, along with an empirical fit of the data to an expression of the form of equation [5-53]: smaller cable diameters mean bigger losses! (Note that the diameter here is the outer diameter of the cable as provided, including the insulating cover; the diameter of the outer conductor *b* of equation [5-53] is significantly smaller.) The magnitude of the loss shows that 20 or 30 cm of cable is generally not a major problem, but 10 m of, for example, RG58 cable will consume half of the available link budget of a WLAN radio. Long cable runs between the radio and antenna should be studiously avoided and if truly inevitable should use large-diameter cables.

Connectors must satisfy a number of stringent constraints. The size of the largest open area within the connector sets the highest frequency of use; internal resonances show up as frequency-dependent reflection and loss. The ground-to-ground and center-to-center contact resistance between the male and female mating connectors must be reproducibly low even after multiple make-and-break cycles. Spring-loaded female connectors simplify alignment of the center conductor, but the springs can undergo cold working and degrade with repeated usage cycles. Low electrical resistance materials must be used, but good mechanical properties are also required. Gold plating is often used to ensure excellent surface properties and corrosion resistance but of course adds cost to the resulting connector. In modern WLAN applications, such as connecting a radio card to a built-in antenna in a laptop, small diameter cables must be used, so very small connectors are needed.

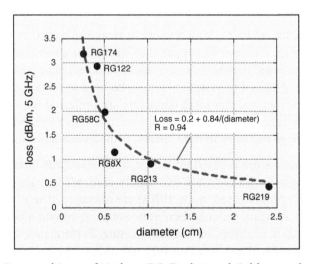

Figure 5-46: Reported Loss of Various RG-Designated Cables vs. the Actual Outer Diameter of the Cable; the Dashed Line Is a Fit to the Data

A wide variety of connector types (though fewer than there are of cables) is available to fill these needs. We shall survey a few common types here, proceeding generally from large to small. The highest performance connectors, such as APC-7 or APC-3.5 types, rely on a face-to-face mating of center conductors produced by precision alignment of high-quality outer shells, avoiding the use of springs or clip arrangements. These connectors are used in instrumentation but are too large, expensive, and slow to be used in most other applications.

Type-N connectors are shown in cross-section in Figure 5-47. N-connectors are versatile and can be used for high power applications. They are physically fairly large and often used to terminate large-diameter low-loss cabling. N-connectors must be aligned and screwed together, using noticeable force. They are mechanically robust and well suited for outdoor use (with rain protection). They can be used to 10 GHz. Note that both 50 Ω and 75 Ω types exist. The two are visually almost identical except for the center conductor diameter, and will mate with each other (to the extent of being screwed together), but a 75 Ω male will make only intermittent contact with a 50 Ω female and a 50 Ω male will bend the springs of a 75 Ω female. The resulting connection problems are frustrating to debug. You have been warned!

BNC[2] connectors are depicted in Figure 5-48. BNCs are extremely common in low-frequency work, and cables thus equipped are often known generically as coaxial cables. The connectors use a convenient twist-on, twist-off attachment approach and are mechanically tough and easy to use, though the connection to the cable is sometimes made sloppily and will fail in the field. BNC connectors are theoretically suitable to 4 GHz, but their use above 1 GHz is inadvisable in the author's experience.

F connectors are shown in Figure 5-49. These are widely used in conjunction with 75 Ω cables for television distribution, both locally within the home and in cable television

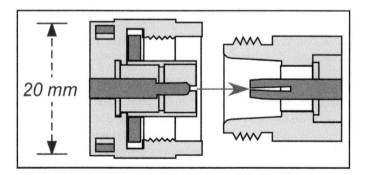

Figure 5-47: Type-N Connectors; in This and Subsequent Diagrams the Male Connector Is at Left and the Female at Right

---

[2] It had been the author's understanding that BNC is an acronym for bayonet Naval connector, but recent posts in discussions groups suggest alternative interpretations. The author is not aware of an authoritative documented origin for the term.

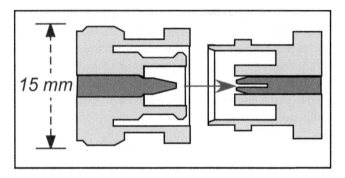

Figure 5-48: BNC Connectors

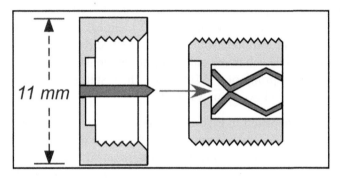

Figure 5-49: F Connectors

systems. They are nominally suitable to 3 GHz but normally used only to 1 GHz (and the author has characterized some inexpensive adaptors that had very poor performance above 600 MHz). Because of their popularity, many low-cost (and often low-quality) connectors are available: some of these are susceptible to problems with grounding, corrosion, and intermodulation distortion (the latter arising from unintended diodes formed by oxidized contacting surfaces).

SMA connectors, shown in Figure 5-50, are very commonly used with semirigid cabling as well as flexible coax. These screw-on connectors are usable to about 18 GHz. SMAs are reliable and provide low loss and reflection but are not intended for an unlimited number of disconnects and are mechanically delicate compared with, for example, N-connectors. SMA connectors should be kept clean and snugged carefully; use of a torque wrench is not indispensable but will improve the reliability of a connection. The very similar "2.4-mm" connector allows for operation to somewhat higher frequencies.

The MCX and MMCX connectors (Figures 5-51 and 5-52) are more recent additions to the stable. The small size and press-fit convenience of these connectors makes them popular for mounting directly onto printed circuit boards, providing local interconnections within a laptop or other portable system. These connectors are rated

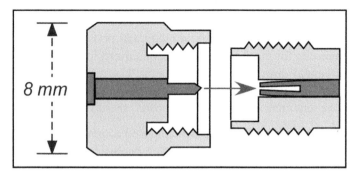

Figure 5-50: SMA Connectors

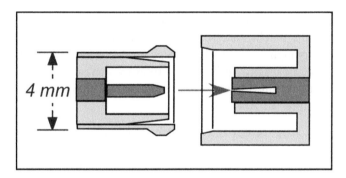

Figure 5-51: MCX Connectors

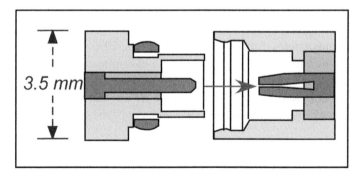

Figure 5-52: MMCX Connectors

for use to 4 GHz. MCX connectors are recommended for use with small cables such as RG174 (M17/119) or RG188A (M17/138). These small cables can have very high losses even if the connector works fine. RG173 is rated to 1 GHz and presents approximately 1.4 dB/m loss if used at 2.4 GHz or 2.1 dB/m at 5 GHz. RG188A is rated to 3 GHz and similarly has about 1.4 dB/m of loss at 2.4 GHz. Very small cables such as these are only suitable for use in lengths of a few tens of centimeters. Other less common very small connectors include the GPO, U.F.L, and AMC types.

8. Capsule Summary: Chapter 5

Antennas are devices to arrange currents whose effects do not cancel at large distances. Their performance can for most purposes be summarized in terms of their radiation pattern and consequent directivity and power gain. The received power in a link between two antennas can be estimated from the Friis equation, requiring only the power gains of the antennas, the transmitted power, the distance, and the wavelength.

A wide variety of antennas has been studied and used commercially. Practical selection must take into account the impedance properties, bandwidth, cost, and mechanical robustness of a particular design as well as its radiation pattern. Wire antennas, such as dipoles, folded dipoles, and Yagi-Uda antennas, are popular because of their simple fabrication and low cost. Configurations from nearly isotropic to highly directive are available. Microstrip (patch) antennas are widely used when integrated into arrays due to the versatility of the batch processes used for their manufacture. Parabolic reflector antennas provide very high gains and are indispensable for long-range outdoor point-to-point links. Adaptive arrays provide powerful tools for increasing range, reliability, and data rate while reducing interference and power consumption; such arrays will play an increasing role in future WLAN systems.

Most antennas use coaxial cables and removable connectors to deliver signals to and from the radio. Appropriate selection and use of cables is necessary to obtain a wireless system whose performance reflects the radio and antenna rather than the cable.

Further Reading

Electromagnetism

Collective Electrodynamics, Carver Mead, MIT Press, 2000: *A polemic for the eradication of fields and elevation of potentials in electromagnetic discourse. A fascinating book with entertaining digressions on the history of electromagnetism and quantum theory but of questionable utility for the RF engineer. (This book is, of course, an attempt to partially repair that deficiency, judgment of the success thereof being left to the reader.)*

All other references cited below use conventional field-based computational techniques. The reader who works through some of the calculations will hopefully agree that the potentials-only approach we have taken is simpler and more direct.

Antenna Theory

Antenna Theory and Design, W. Stutzman and G. Thiele, Wiley, 1981: *Old but still good.*

Antenna Theory, C. Balanis, Wiley, 1997: *More recent and thus includes treatment of newer varieties of antennas, e.g., microstrip arrays.*

Antenna Handbooks

Antenna Engineering Handbook, Richard Johnson, editor (3rd edition), McGraw-Hill, 1993: *This is a very thorough reference book, covering nearly every imaginable type of antenna and antenna application. The treatments are necessarily terse and provide key results with minimal or no derivation, but the quality of the chapters is uniformly high, not always the case in a compendium of this type, and some effort has obviously been made to keep notation consistent as well. Nice supplements on transmission lines and matching and propagation. Anyone who is seriously involved in antennas will find this book useful.*

Practical Antenna Handbook, Joseph Carr, McGraw-Hill, 1994: *This book is of much more limited scope than Johnson above, primarily focused on large wire antennas for long-wave applications, but it has a nice introduction to impedance matching and some practical information on measurement techniques.*

Yagi-Uda Antennas

"A New Method for Obtaining Maximum Gain from Yagi Antennas," H. Ehrenspeck and H. Poehler, IRE Transactions on Antennas and Propagation, October 1959, p. 379. *While there is a vast literature on Yagi-Uda antennas, this is the only article I found that went beyond empirical investigation or numerical modeling to provide some sensible explanation of how they work.*

Cables and Connectors

RG cable designations: http://www.bluejeanscable.com/articles/rg6.htm

Connector info: http://www.wa1mba.org/rfconn.htm

Frequency chart for connectors: http://www.amphenolrf.com/rf_made_simple/freqrange.asp

Propagation

1. Propagation in Free Space

Imagine a wireless local area network (WLAN) somewhere in interstellar space. (Ignore for the moment the question of how we got there and what we're doing with a WLAN so far from home, and for that matter the magnetized plasma background likely to be present in an interstellar environment.) The received signal is just a time-delayed attenuated version of the transmitted signal. The received power is described by the Friis equation 5.40 in terms of the transmitted power, the distance between transmitting and receiving antennas, and the directive gains of the respective antennas (Figure 6-1). Propagation is independent of wavelength, direction, and polarization (recall that the factor of λ^2 arises from the antenna size, not the properties of electromagnetic waves).

The reader who is only responsible for interstellar or perhaps circumstellar WLAN installations is done and can skip the remainder of this chapter. For the rest of us, complex environments with obstacles are a part of life.

2. Propagation in Earthbound Environments

Real indoor and outdoor environments introduce a multitude of propagation paths; for example, Figure 6-2 depicts in cartoon style a representative indoor propagation environment.

Each path interacts with differing objects and obstacles. The resulting potential at the antenna is the sum of all the paths with their unique phase, amplitude, and polarization.

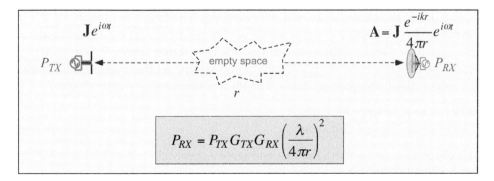

Figure 6-1: Propagation Between Antennas in Empty Space

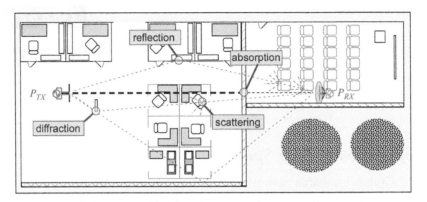

Figure 6-2: Multipath Propagation in an Indoor Environment

The resulting received signal is attenuated, delayed, and dispersed in time in a complicated fashion. The effects are large and can significantly affect the ability of a radio link to function. For example, in Figure 6-3 we show some real data for indoor propagation in a typical commercial building. Signal strength varies by as much as 50 dB from that expected in free space.

Indoor and outdoor environments are also significantly different from the perspective of a WLAN signal. The typical obstacles in an indoor environment vary from sub-wavelength to tens of wavelengths in size; propagation distances are tens of meters. Outdoor obstacles and paths are one to two orders of magnitude larger. Indoor environments are delineated by walls and filled with small metal and wooden objects, as well as water-filled human beings running around, but have no cars and only artificial trees. Outdoor environments are delineated by buildings (if present) and are usually filled with cars, trucks, plants, and trees, with optional wind, rain, and snow; large bodies of water may be present. Long outdoor links become sensitive to gradients in atmospheric temperature and humidity that lead to refraction of the otherwise direct beam.

There are three responses to this sort of complexity. The first is to attempt to understand each interaction in some detail, so that the effects on the transmitted wave can be understood and if appropriate mitigated. The second approach is to treat propagation on a statistical basis and build radio systems so that reception is robust to variations in signal strength and delay. The third is summarized in the rhyme

> *When in trouble, when in doubt,*
> *Run in circles, scream and shout.*

In this chapter we pursue both the first and second approaches. We provide some basis for understanding the sort of absorption, reflection, refraction, diffraction, and scattering effects that are important for radio propagation. We also discuss some

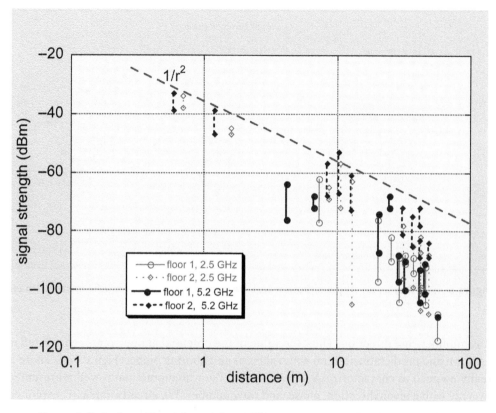

Figure 6-3: Indoor Signal Strength vs. Distance at 2.5 and 5.2 GHz (Single-Frequency Signal) (Image Courtesy of WJ Communications Inc.)

approaches to mitigating the effects of propagation variations through the use of robust modulation schemes and multiple or smart antennas. When both these approaches fail, irrational panic remains an alternative course of action, though the author recommends the more classic approaches of blaming the uninvolved or flight to a foreign jurisdiction.

2.1. Background: Rays and Geometric Optics

Before we proceed with the main thrust of the discussion, it is worth a moment to introduce the concept of a ray as the geometric limit of a wave. So far we have treated propagation as taking place through the vector potential **A**. The vector potential is a vector field, an entity that associates a unique direction and amplitude with every point in three-dimensional space. In the most general case, no further simplification is possible: an arbitrary vector field must be described in terms of its value at every point. However, in most cases of interest for WLAN users, the values of the potential are primarily of interest at long distances from the transmitting antenna and other obstacles. The vector potential is described by an exponential with a known frequency,

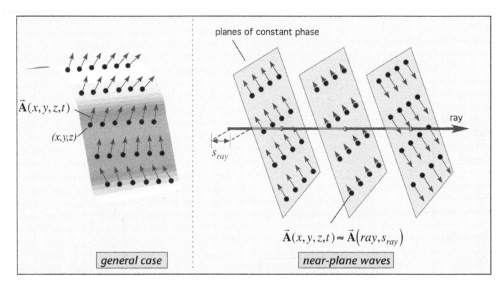

Figure 6-4: A Ray as an Approximation to a Vector Field With Planes of Constant Phase

wavenumber, and direction. Thus, the value of the potential at one point is correlated in a known and predictable fashion with values at neighboring points (Figure 6-4). There is really no need to characterize **A** at every point: the whole potential is well represented by a ray with a given direction, phase, and wavenumber. We already implicitly exploited this simplification in Figure 6-1 by drawing the individual propagation paths as independent lines with independent interactions with localized obstacles.

Rays are a valid approximation to the vector field as long as the phase and direction of neighboring rays doesn't change much over a lateral distance comparable with a wavelength. In that case, we can divide the propagation problem into distinct rays following identifiable paths and add the resulting phase and intensity at the end to estimate the received signal: this is known as the *geometric optics* approximation. This approximation enables us to treat a complex environment as composed of distinct segments and combine them at the end. Interactions between obstacles can be accounted for to lowest order as the result of multiply-reflected or -diffracted rays. In this fashion we avoid needing to solve simultaneously and self-consistently for all currents and potentials.

Note, however, that merely adding amplitudes cannot be expected to produce an accurate estimate of the signal: it is necessary to account for the variation in phase encountered by the ray along its path. Accurate estimation of the relative phase of different paths requires knowledge of path length to better than 1 cm accuracy (out of total path lengths of tens to hundreds of meters) as well as detailed understanding of obstacles to similar precision. Such accuracy is unlikely to be obtainable in real environments. Practical analysis of propagation environments with multiple paths therefore cannot realistically predict actual instantaneous signal amplitude but merely set bounds on possible amplitudes. As we discuss in somewhat more detail in section 3, real signals do exhibit large variations in amplitude with small displacements in time and space; this effect, known as *fading*, is the result of the sensitivity of signal power to small variations in relative phase along the various ray paths.

2.2. Refractive Index

The vector potential at a point is the sum of the potential induced by all relevant currents. The presence of a material object can always be accounted for by estimating the currents induced in the material by an external potential and adding their effects to those of the external current to arrive at the total potential. However, in the case of dielectric materials (those which have negligible DC conductivity but whose charges can be polarized in the presence of an external potential), a much simpler approach is possible. For most dielectrics, the induced current can be taken to be proportional to the impinging (transverse) potential, resulting in a change in the effective phase constant of the propagating wave, that is, a change in the velocity. Instead of writing a complex integral over induced currents, we just introduce a new phase constant β to replace k. The potential within the medium can then be written as

$$A = A_o e^{i(\omega t - \beta r)} \quad v_{ph} = \frac{\omega}{\beta} \qquad [6.1]$$

The reduction of the phase velocity within the medium, v_{ph}, relative to that in vacuum (c, the velocity of light) is the *refractive index* of the material, n:

$$n = \frac{c}{v_{ph}} \qquad [6.2]$$

The refractive index is usually > 1, signifying a phase velocity less than that of light, though the opposite result is possible in the presence of glow-discharge plasmas or other resonant media. Typical values for solid materials encountered in ordinary environments vary from $n \approx 1.3$ for foams or pressed paper ceiling panels to $n \approx 3$ for wood or concrete. The local effect of the external potential depends on the movement of charge within the material; it is generally weakly dependent on frequency except in the vicinity of a resonant response. Molecular vibrational resonances in liquids and solids are at infrared frequencies; unhindered rotation of molecules takes place at high microwave frequencies and may occur in liquids but is less likely in solids. The refractive index of most solid dielectrics doesn't vary strongly with frequency.

2.3. Absorption

In conductive materials, the induced current is 90 degrees out of phase with the external potential. It can be shown using the differential formulation of the propagation equations that such a current leads to a decreasing amplitude with distance (even in the case where the source is effectively at infinite distance and thus the ($1/r$) term is constant): the material *absorbs* the radiation (see Appendix 5 for more details). If absorption is weak enough so that little change in amplitude takes place in a wavelength, we can simply tack on an attenuation term to the previous expression for the potential

$$\mathbf{A} = \mathbf{A}_o e^{i(\omega t - kr)} e^{-\alpha r} \qquad [6.3]$$

The real exponential has the effect of attenuating the wave as it propagates (assuming $\alpha > 0$), with a characteristic length of $(1/\alpha)$; an example is shown in Figure 6-5.

The attenuation constant is related to the conductivity σ of the medium

$$\alpha = \frac{\sigma \mu_o c}{2n} \qquad [6.4]$$

where n is the refractive index (see section 2.2). This expression can be understood if we recall that $\mu_o c$ is the impedance (i.e., 1/(conductance)) of free space (see Chapter 5, section 4). The attenuation relative to a wavelength is then essentially proportional to the ratio between the conductivity of the absorbing medium and that of the space in which it resides. However, it is most often the attenuation constant α that is available, rather than the conductivity.

The attenuation constant has dimensions of (1/length), typically (1/m). Though in formal terms such a description is quite sufficient, the fact that α represents an exponential attenuation is often expressed by considering it to be in units of *Nepers/meter*, where a Neper represents a decrease in amplitude of $(1/e) \approx 0.37$, or a reduction in power of $(1/e^2) \approx 0.135$. Attenuation can also be expressed in more convenient units of dB/m:

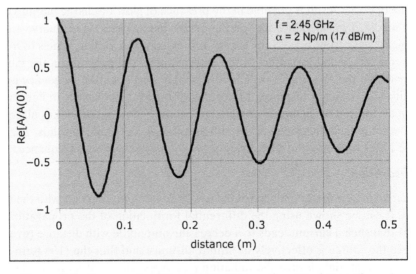

Figure 6-5: Propagating Wave With Attenuation (Real Part Shown)

$$a\left(\frac{dB}{m}\right) = 8.68\alpha\left(\frac{Nepers}{m}\right) \qquad [6.5]$$

Conductivity in materials can arise from the motion of either electrons or ions in response to the applied potential. The low conductivities associated with mobile ionic species in liquids and porous solids will lead to microwave absorption. Conductivity in metals is usually so large that the simplistic treatment of equation [6.1] is not valid, and the impinging wave is instead primarily reflected rather than absorbed; we examine reflection in section 2.4.

In addition to the motion of "free" charges in response to the imposed potential, absorption can also result from local interaction of the potential with individual atoms or molecules. Mathematically, the effect is accounted for in exactly the fashion of equation [6.3] by incorporating the additional mechanisms as a different value of the attenuation constant instead of that arising from equation [6.4]. Not all such absorption mechanisms are active at microwave frequencies. The excitation of bound electrons within a molecule to higher electronic states requires frequencies of hundreds to thousands of THz—that is, visible or ultraviolet light. Loss of energy to the vibrations of atoms relative to their bonded neighbors only occurs at frequencies of tens to hundreds of THz, thus in the far to near infrared. On the other hand, rotational excitations of free molecules and rotations within a liquid or solid have characteristic frequencies in the tens to hundreds of GHz and can be significant at WLAN frequencies.

Let us examine as practical examples the absorption of two important media in the real world: atmospheric air and liquid water. The absorption coefficient of air at microwave frequencies is so small it is more convenient to depict as dB/km than dB/m. The absorption is primarily due to oxygen and water vapor; water vapor is a much stronger absorber than oxygen because of its permanent dipole moment but is present in much smaller concentration than oxygen, so oxygen dominates total absorption at low frequencies. These absorptions are actually the collisional tails of resonant absorptions at much higher frequencies (about 22 GHz for water and 60 GHz for oxygen) due to rotational transitions. The left side of Figure 6-6 shows the absorption coefficients of these two species versus frequency for 25% relative humidity at room temperature; the absorption of water vapor can be scaled linearly in concentration to account for changes in humidity and temperature using the graph at the right of the figure. Note that even for high humidity and temperature, the total absorption coefficient at 5 GHz is of the order of 0.01 dB/km: air absorption is completely negligible indoors and very small even for a long outdoor link.

Liquid water is about 130,000 times denser than the ambient water vapor in Figure 6-6; it would hardly be surprising to find a stronger effect on microwave propagation. Liquid water is also qualitatively different from water vapor: individual molecules are surrounded by other molecules at close range, and the positively charged hydrogen atoms form transient *hydrogen bonds* with their negatively charged neighboring oxygens. These strong interactions hinder the free rotation of the individual water

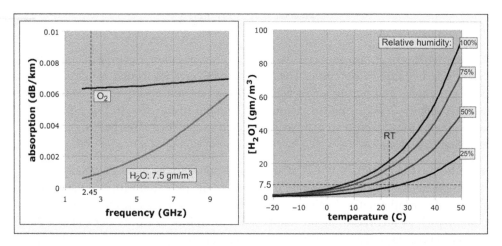

Figure 6-6: (Left) Absorption Due to Oxygen and Water Vapor in Air (From Hyde, In *Antenna Engineering Handbook*, R. Johnson, McGraw-Hill, 1993, Figure 45-2) and (Right) Conversion of Relative Humidity and Temperature Into Water Vapor Concentration (From *CRC Handbook of Chemistry and Physics*)

molecules, but the bonds are constantly broken by thermal agitation. The net change in orientation of the molecular dipoles produces a polarization current and thus a large refractive index; the dissipation that results from interactions and collisions with neighboring molecules appears as absorption. The result is that the refractive index is fairly constant at around 8–9 in the range of frequencies of interest for WLAN applications, but the absorption varies quite noticeably with temperature and frequency (Figure 6-7). Absorption in liquid water at 5 GHz is about 150 Nepers/m or 13 dB/cm: there is no point in inventing a submersible 802.11-equipped portable digital assistant! It can also be expected that humans, being mostly composed of water, will act as strong absorbers.

The data in Figures 6.6 and 6.7 should also put to rest the common misconception that the microwave oven frequency of 2.45 GHz represents a resonant absorption of the water molecule. The fundamental rotational frequencies of free water are at several hundred gigahertz (the 22-GHz absorption represents a transition between two neighboring highly excited rotational states). There is no special resonant process at 2.45 GHz; this frequency was chosen for its availability, though in this frequency range cooking uniformity is enhanced by the decrease in absorption with increasing temperature (the hot exterior permits penetration into the cool center of the item being heated).

2.4. Reflection by Metals

In metals, the conductivity is so high that the change in amplitude of the transmitted signal takes place over a distance much shorter than a wavelength. This distance, known

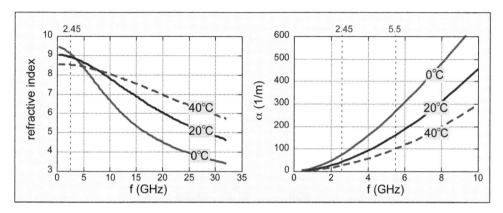

Figure 6-7: Refractive Index and Absorption Coefficient for Liquid Water as a Function of Frequency and Temperature (Based on Data for Water From Martin Chaplin, http://www.lsbu.ac.uk/water/microwave.html)

as the *skin depth* δ, is typically a few microns for common metals at microwave frequencies (Table 6-1). If the thickness of the metal is much larger than the skin depth (which will certainly be the case except for very thin foils and coatings), essentially no signal will be transmitted into the bulk of the metal.

Because the amplitude change at the metal surface is so abrupt (recall that a wavelength is 6 or 12 cm, thus on the order of 10,000 skin depths), on the scale of a wavelength we can essentially assume that the potential goes to zero instantly at the metal surface. An impinging plane wave does not provide such a result, but the addition of a reflected wave of opposite sign does:

$$\underbrace{A_{inc}e^{i(\omega t - kx)}}_{\text{incident wave}} + \underbrace{A_{ref}e^{i(\omega t + kx)}}_{\text{reflected wave}} = 0 \ @(x = 0) \text{ IF } A_{ref} = -A_{inc} \qquad [6.6]$$

surface

Thus, a bulk metal surface reflects almost all the impinging energy, with a sign change in the reflected wave (Figure 6-8). The angle of reflection, measured with respect to the

Table 6-1: Skin Depth for Common Metals at ISM and Unlicensed National Information Infrastructure Frequencies

Metal	δ (2.4 GHz) (μm)	δ (5.5 GHz) (μm)
Copper	1.3	0.9
Aluminum	1.6	1.1
Steel	3.2	2.1
Titanium	6.6	4.4
Mn-steel	8.5	5.7

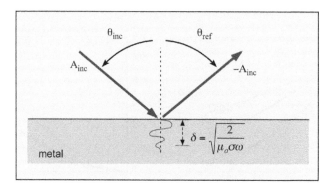

Figure 6-8: Reflection From a Metal Surface

normal to the surface, is equal to the angle of incidence. (This equality is necessary for the incident and reflected waves to remain in phase laterally over the surface.)

Pure reflection of this nature is valid when the lateral extent of the metal surface greatly exceeds a wavelength. For the sort of finite metal obstacles encountered in indoor environments, whose size is on the order of some tens of wavelengths, the reflected beam is broadened and will generally interfere with the incident beam (if present) to create a complex potential distribution in the reflected direction. Analysis of the effect of finite-sized obstacles on propagating waves—*diffraction*—requires a more detailed examination of the effects of the induced surface currents, which we shall put off until section 2.7, in which we also formulate a more appropriate criterion for when an obstacle must be treated as of finite size.

2.5. Reflection and Refraction by Dielectrics

Typical dielectric materials create both reflected and refracted waves (the latter also being subject to significant absorption for many materials of practical interest). The amount of incident radiation that is reflected and refracted is a function of the refractive indices of the media, the angle of incidence, and the polarization of the radiation.

Let us first consider normal incidence (Figure 6-9). In this case, all polarizations must be equivalent by symmetry (unless the medium is itself asymmetric) and by symmetry the transmitted radiation must also be along the normal.

We can derive the reflected and transmitted potential by requiring that the potential and its derivative be continuous at the interface. The transmitted, incident, and reflected waves are

$$\vec{A}_{trans} = \vec{A}_{tr0}e^{i(\omega t - n_2 kr)}$$
$$\vec{A}_{inc} = \vec{A}_{in0}e^{i(\omega t - n_1 kr)} \qquad [6.7]$$
$$\vec{A}_{ref} = \vec{A}_{rf0}e^{i(\omega t + n_1 kr)}$$

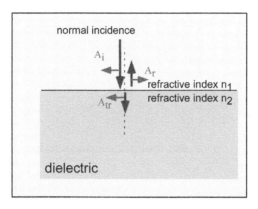

Figure 6-9: Reflection and Refraction of Radiation Normally Incident on a Dielectric Interface

Continuity of the potential at $r = 0$ requires

$$\vec{A}_{in0} + \vec{A}_{rf0} = \vec{A}_{tr0} \qquad [6.8]$$

Taking the derivative in r brings down a factor of $in_i k$. Continuity of the derivative at $r = 0$ is thus

$$-in_1 k\vec{A}_{in0} + in_1 k\vec{A}_{rf0} = -in_2 k\vec{A}_{tr0} \qquad [6.9]$$

We now have two equations and so can solve for the reflected and transmitted amplitudes in terms of the incident amplitude. Defining R as the ratio of the reflected to the incident wave, we find

$$R = \frac{\left(1 - \dfrac{n_2}{n_1}\right)}{\left(1 + \dfrac{n_2}{n_1}\right)} \qquad T = \frac{2}{\left(1 + \dfrac{n_2}{n_1}\right)} \qquad [6.10]$$

The behavior of the reflection coefficient with varying refractive index ratios is shown in Figure 6-10. As one might expect, there is no reflection when the indices are equal. For large disparities, a large reflection results, with the sign being dependent on the relative location of the large and small index regions.

Because realistic values of the ratio are on the order of 2:1, about a third of the impinging radiation is reflected at an interface for normal incidence. Because the power is the square of the amplitude, this means that at normal incidence only about 10% to 30% of the power is lost due to reflection at a single interface. For a finite thickness obstacle such as a wall, reflections will occur at both interfaces; their relative phase must be accounted for in determining the overall reflection. In the case where the reflections add in phase, the total reflection coefficient can be quite large even when the individual interfaces reflect modestly. This can amount to about 4 dB of reflection loss for a typical

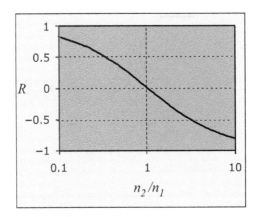

Figure 6-10: Reflection Coefficient at Normal Incidence

refractive index of 2.5 corresponding to, for example, a thin slab of plywood—small compared with the overall link loss budget of around 40 dB. Because it is the amplitudes and not the powers that add, a wall composed of a number of reflecting slabs could produce quite significant reflection losses if the thicknesses were arranged so as to have all the reflections in phase.

The procedure for finding reflection and transmission coefficients at arbitrary angles of incidence is similar but much more laborious, because amplitudes and derivatives must be resolved into components parallel to and perpendicular to the surface. In this case one must make a distinction between polarizations: by convention, we speak of the potential as either being parallel or perpendicular to the *plane of incidence* formed by the incident, transmitted, and reflected rays (Figure 6-11). The reader should note that the plane of *incidence* is perpendicular to the plane of the *interface*, a distinction that may lead to some confusion if not carefully attended to.

The angle of reflection is equal to the angle of incidence, as in the case of metals (Figure 6-8). However, the angle of transmission is not: as shown in Figure 6-8, the sines of the angles vary in inverse proportion to the refractive indices—*Snell's law* of refraction. This result arises simply because the boundary conditions can only be met if the refracted and incident waves have the same phase relationship in the plane of the interface and is thus true for any polarization.

The resulting reflection coefficients are shown in Figure 6-12 as a function of incident angle for a range of values of refractive index. Here it is assumed that the radiation is incident from air ($n_1 = 1$) to a region of higher dielectric constant. The amplitude of the reflection coefficients decreases monotonically to $R = -1$ as the angle nears 90 degrees: at glancing angles *all the incident radiation is reflected*. However, for **A** in the plane of incidence, the reflection coefficient is initially positive, so to become negative, R must cross the axis (have a value of zero) at some angle: *Brewster's angle*. As a consequence,

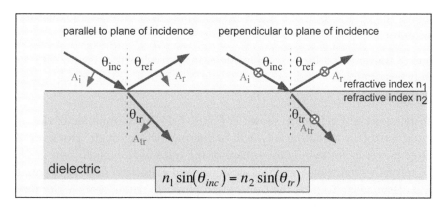

Figure 6-11: Reflection and Refraction at a Dielectric Interface

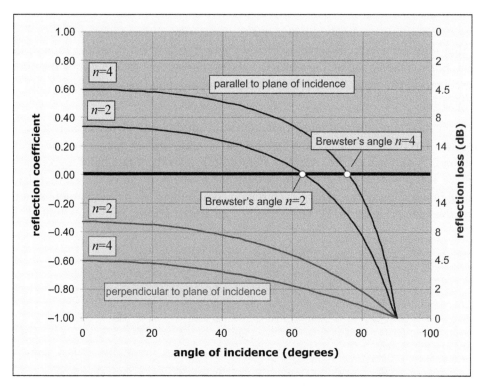

Figure 6-12: Reflection Coefficients at a Dielectric Interface ($n_1 = 1$)

reflection is quite small over a wide range of angles for polarization in the plane of incidence (e.g., a vertical antenna bouncing off a floor). Reflection is on the average more significant for **A** perpendicular to the plane of incidence (a vertical antenna bouncing off a wall). Because of this discrepancy, the polarization of a ray is likely to change each time it is reflected or transmitted through a dielectric, so that in a complex environment the polarization of a received signal may bear little resemblance to that of the transmitter.

The angle of transmission for varying incident angles is shown in Figure 6-13. For reasonably large refractive indices, the transmitted beam is nearly normal even at glancing angles.

The transmission coefficients are shown in Figure 6-14. The amplitude of the transmitted wave falls monotonically with increasing incident angle. However, the transmitted power is not merely the square of the transmission coefficient in this case: the refractive index and angle of incidence and refraction must be accounted for. Rather than trying to treat the transmitted power in detail, it is simpler merely to assume that any power that was not reflected (Figure 6-12) must have been transmitted.

2.6. Refraction in Continuous Media

Long outdoor links generally use highly directional antennas and narrow beams; they are thus quite vulnerable to anything that deflects the direction of propagation of the beam. For completeness, we introduce the necessary concepts for estimating the effects of gradients in the refractive index of a continuous medium (the atmosphere) on propagation.

Consider radiation propagating in a medium with varying refractive index. If the variation in refractive index is small over a wavelength, the ray approximation (section 2.1) can be used. We are then interested in how the direction of a given ray changes as it proceeds along its path through the medium. It can be shown that the change in direction of the path at each point is determined by the local gradient in the refractive index (Figure 6-15): The resulting equation is known as the equation of

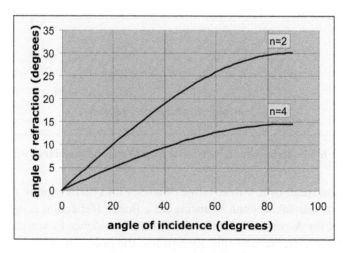

Figure 6-13: Angle of Refraction at a Dielectric Interface ($n_1 = 1$)

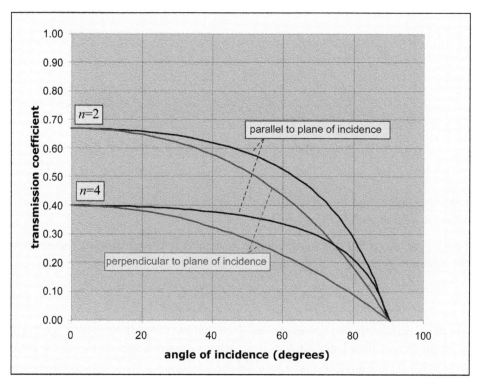

Figure 6-14: Transmission Coefficients at a Dielectric Interface ($n_1 = 1$)

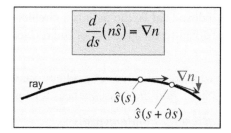

Figure 6-15: Ray Propagation With Varying Refractive Index

geometric optics. (The mysterious form of this equation can be elucidated by expanding the derivative using the chain rule; all it is saying is that the change in direction of the path is determined by that part of the gradient in n that is perpendicular to the path.)

The net result is that the ray is deflected toward regions of higher refractive index. To see how this equation is used, let's look at a relevant example: a ray initially propagating in a horizontal direction (i.e., parallel to the ground) in a region with a vertical gradient

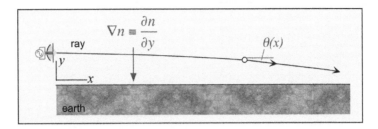

Figure 6-16: Propagation of a Horizontal Ray in a Vertical Gradient

in refractive index (due to a gradient in temperature, humidity, or density). The situation is summarized in Figure 6-16.

For each point along the ray, we can define an angle θ to the horizontal. The geometric optics equation becomes

$$\frac{\partial}{\partial s}[n(\cos\theta,\ \sin\theta)] = \left(0,\ \frac{\partial n}{\partial y}\right)$$ [6.11]

where [6.11] represents two equations, one for x and one for y. If we assume the angle is very small, then $\cos\theta \approx 1$ and $\sin\theta \approx dy/dx$, where y is the height of the ray. It is then easy to verify by substitution that a solution is

$$y \approx \frac{Gx^2}{2n_o} \Rightarrow \text{radius of curvature R} = \frac{n_o}{G}$$ [6.12]

where n_o is the refractive index at the beginning of the path and $G = dn/dy$ is the gradient of the refractive index. That is, the path is a circular arc, with a radius of curvature inversely proportional to the gradient of the refractive index, normalized to the index itself.

In Chapter 8, we examine what sort of values of gradient occur, how they are related to the behavior of the atmosphere, and the consequences for long radio links.

2.7. Diffraction and Scattering: Fundamentals

So far all our discussions have assumed laterally uniform plane waves, producing textbook results, which are very useful approximations for treating the propagation of visible light in ordinary objects. To apply the results, we merely trace rays and determine their direction and amplitude based on reflection, refraction, or absorption events (Figure 6-17).

Unfortunately for this simple picture (but in some cases fortunately for link functionality), WLANs based on microwave radios operate in a very different world, filled with objects whose lateral size is comparable with the wavelength of the radiation. The presence of an object along the path of a ray doesn't guarantee zero intensity (a shadow) in that direction; the absence of an object along a ray doesn't guarantee

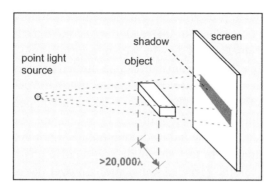

Figure 6-17: An Object Casts a Geometric Shadow When All Dimensions Are Much Greater Than a Wavelength

unimpeded propagation (Figure 6-18). The shadow edges become diffuse, intensity within the shadow varies, and energy may be scattered into high angle rays.

At first glance, such results appear challenging to an intuition based on living in an optical world. However, we find that in essence they are the obvious consequences of the ideas we already introduced: a potential impinging on a conductive object must necessarily induce currents in that object, and those currents in turn must radiate according to equation [5.1]. The program we follow thus consists of figuring out what the induced currents are and then calculating the resulting additional radiated potential: the *scattered* potential. The scattered potential is rather similar to the radiated pattern of an antenna shaped like the obstacle and varies with angle in a smooth and reasonably simple fashion. However, the total potential is the sum of the incident and scattered potentials, and their relative phase varies rapidly with position when the path lengths are long (as is the case for a typical WLAN link). It is this rapid variation in relative phase that gives rise to the complex spatial variation of the diffraction pattern.

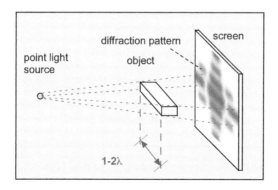

Figure 6-18: Objects Comparable With a Wavelength Cast Diffracted Shadows

A key assumption of the approach is that the incident potential continues merrily right through thick metallic obstacles. For the reader accustomed to thinking of metal surfaces as being ideal shields, this may be a counterintuitive notion. However, it is perfectly possible to treat both circumstances in a unified fashion (Figure 6-19). The induced currents in the metal surface create a potential that cancels the incident potential very close to the current sheet, where the surface appears to be infinite. The very same currents induce a potential at distant locations, but here the sheet no longer appears of infinite extent, and the induced potential need not and does not exactly cancel the incident potential. In fact, depending on relative phase, the two may add to create an enhancement in the received signal strength over the unobstructed signal, even within a geometric shadow.

Having exhausted the section's budget for philosophical digressions, let us proceed to the gritty task of calculation. We must first set some conventions for describing the geometry of interest; these are depicted in Figure 6-20. A source at average distance R_{inc} illuminates an obstacle (here a disk) at an incident angle θ_{inc}, defined counterclockwise with respect to the outward normal in the source direction. A receiver is placed at an observation point R_{obs} from the obstacle, inclined at angle θ_{obs}, here defined counterclockwise with respect to an oppositely directed normal vector. (This curious choice is made so that in the original direction of propagation $\theta_{obs} = \theta_{inc}$.) The incident potential itself follows the direct path to the observer of length R_{dir}.

We now need to calculate how the local induced current at the metal surface is related to the incident potential. If the current succeeds in screening the incident potential, the potential is zero in the metal just behind the induced currents. As long as the current sheet is very much thinner than a wavelength (which it certainly is at a few gigahertz; see Table 6-1), we can integrate through the layer to produce a relationship between the

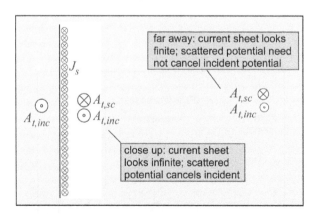

Figure 6-19: Currents That Screen Incident Potential Near the Surface May Not Do So at Large Distances (Dot, Potential Directed Into the Plane of the Paper; Cross, Potential Directed Out of the Plane of the Paper)

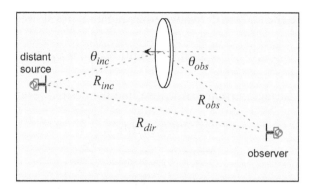

Figure 6-20: Geometry for Diffraction Calculations

current density and the derivative of the potential at the surface, just beyond the current sheet, following the nomenclature of Figure 6-21:

$$\frac{\partial A}{\partial n} = -\mu_o J_s \qquad [6.13]$$

Note that J_s is a current per unit area here rather than a current per unit volume.

The potential on the left-hand side of the current sheet is the sum of the incident potential and the reflected potential (see section 2.4). To obtain the normal derivative of the incident wave, assuming **A** to be perpendicular to the plane of incidence for simplicity, we take the derivative in the direction of propagation and multiply by $-\cos\theta_{inc}$. The negative sign arises because the incident wave is pointed in the opposite direction to the normal; the cosine is simply the projection of the derivative onto the direction perpendicular to the surface. A similar calculation is performed for the reflected wave; recall from section 2.4 that the angle of reflection is the same as that of incidence but the direction of propagation is opposite. Noting that only the transverse component of the potential A_t can couple, we obtain

$$\frac{\partial}{\partial \hat{n}} A \bigg|_{surface} = A_{t,inc}\left(-ik\frac{e^{-ikR_{inc}}}{R_{inc}}(-\cos(\theta)) - (-ik)\frac{e^{-ikR_{incl}}}{R_{inc}}(\cos(\theta))\right) \qquad [6.14]$$

$$= 2ikA_{t,inc}\cos(\theta)\frac{e^{-ikR_{inc}}}{R_{inc}}$$

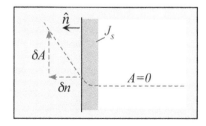

Figure 6-21: Geometry for Defining the Surface Derivative

Setting this equal to the surface derivative from equation [6.13], we get

$$\frac{\partial}{\partial \hat{n}} A \bigg|_{surface} = 2ikA_{t,\,inc} \cos(\theta) \frac{e^{-ikR_{inc}}}{R_{inc}} = -\mu_o J_s \qquad [6.15]$$

and solving for the surface current density, we find the desired result:

$$J_s = \underbrace{\frac{-2ikA_{t,\,inc}}{\mu_o}}_{(V/R)} \underbrace{\cos(\theta)}_{geometry} \underbrace{\frac{e^{-ikR_{inc}}}{R_{inc}}}_{propagation} \qquad [6.16]$$

The expression for the current is composed of three pieces. We can rewrite the first chunk using $k = \omega/c$ to get something of the form $(i\omega A/\mu_o c)$. Remember that $i\omega A$ is the characteristic voltage for a potential (equation [5.2]) and $\mu_o c$ is the impedance of free space: this piece is just of the form of Ohm's law, $(V/R) = I$. The second piece, the geometry factor, is often just set equal to 1, as we are usually interested in incident angles close to 0 (most common objects are fairly convex and thus always have a near-normal face toward the observer). The final piece is simply the phase and amplitude of the incident potential at the obstacle, given that the amplitude was $A_{t,\,inc}$ at the source.

Now that we have an expression for the induced current at the surface, we can calculate the scattered potential in the usual fashion by integrating over the current using equation [5.1], as depicted schematically in Figure 6-22. For simplicity we assume that the obstacle, source, and observer are all in a plane perpendicular to the potentials and currents, so that the scattered potential is purely transverse; the general case adds a more complex geometry factor but doesn't qualitatively change the result.

Inserting the expression for the surface current density from equation [6.16], we obtain

$$A_{sc,\,obs} = \frac{-ikA_{t,\,inc}}{2\pi} \int \cos(\theta_{inc}) \frac{e^{-ikr_{inc}}}{r_{inc}} \frac{e^{-ikr_{obs}}}{r_{obs}} ds \qquad [6.17]$$

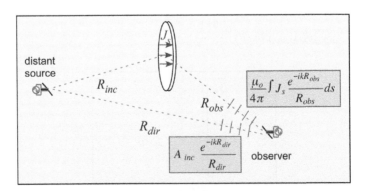

Figure 6-22: Scattered and Direct Potentials at the Observer

Note that here we have written *r* to denote the local distance from the source or observer to a point on the obstacle, rather than the average distance *R*, so that we can integrate over the surface of the obstacle. Just as in the case of an antenna problem, the changes in distance are very tiny relative to the average, and so will have little effect on the $(1/r)$ term but if comparable with a wavelength will still induce large variation in the exponential.

The form of equation [6.17] has a simple interpretation: the source radiates to an intermediate point on the disk and induces a current proportional to the resulting phase and amplitude. This current then radiates its own potential, with relative phase and amplitude proportional to e^{-ikr}/r as usual. We integrate over the obstacle to obtain the resulting scattered field; the potential complexities arise from the variations in phase across the obstacle surface.

Because we know that a current element radiates equally in all azimuthal directions, we will not be surprised to see that the scattered field is present well outside of the geometric shadow. However, for finite inclinations to the normal direction, it is easy to see that the geometric shadow and geometric reflection directions are likely to play special roles. The largest scattered field will occur in directions where the phase of contributions from various locations on the surface is about the same: the directions of stationary phase. Figure 6-23 shows two incident rays to neighboring points on the surface of an obstacle, separated by some small distance δ. The distance from the source to the lower point is smaller by δ sin θ than the distance to the upper point. However, in the direction of the geometric shadow, this discrepancy is exactly compensated by the excess distance from the lower point to the distant observer. The same thing happens in the direction of the geometric reflection. The strongest scattered fields are to be expected in these directions. Their interaction with the incident fields determines the total signal detected at distant points.

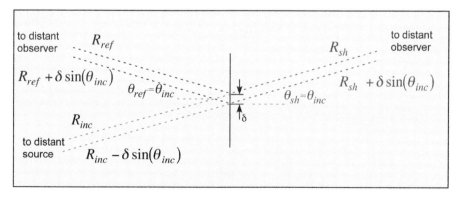

Figure 6-23: Shadow and Reflection Directions Are Directions of Stationary Phase With Respect to Small Displacements on a Surface

2.8. Diffraction and Scattering: Examples

To see how this all works out, let's analyze a particular example. The astute reader may have already have noticed that Figures 6.20 and 6.22 depict a disk-shaped obstacle; in a work of fiction, this would have constituted an example of foreshadowing, though here the author was merely trying to reduce the effort of creating the illustrations. Diffraction by a disk illuminated on-axis is a famous early example of diffraction theory, and therefore the correct answer is available to verify our calculations, so let us attempt it.

The geometry is shown in Figure 6-24. A disk of arbitrary diameter d (not necessarily small compared with a wavelength) is illuminated on axis by a source at on-axis distance R_{inc}, with the distance to a particular point on the disk denoted r_{inc}, both presumed much larger than d and λ. A similarly distant observer, also on-axis, receives the incident and scattered potentials. What intensity does the observer detect?

Because the source and observer are distant, we can impose the *paraxial approximation*: the relative variation in r across the disk surface is miniscule and can be ignored as far as the $(1/r)$ terms are concerned. Furthermore, as we are on-axis, the cosine is simply $= 1$. Equation [6.17] becomes

$$A_{sc,obs} \approx \frac{-ikA_{t,inc}}{2\pi R_{inc}R_{obs}} \int e^{-ikr_{inc}} e^{-ikr_{obs}} ds \qquad [6.18]$$

If we can figure out how to express the distances as a function of position on the disk surface, we should be able to perform the integral. The path length to any point at a given distance ρ from the center of the disk can be obtained by the theorem of Pythagoras (Figure 6-25), but in this form is inconvenient for analytic integration. Because the diameter is much smaller than the distance to either source or observer, the square root can be approximated by a Taylor expansion to obtain a path length that is quadratic in the displacement

$$r_{inc} = \sqrt{R_{inc}{}^2 + \rho^2} = R_{inc}\sqrt{1 + \underbrace{\left[\frac{\rho}{R_{inc}}\right]^2}_{small}} \approx R_{inc}\left(1 + \frac{1}{2}\left[\frac{\rho}{R_{inc}}\right]^2\right) = R_{inc} + \frac{1}{2}\left[\frac{\rho^2}{R_{inc}}\right]$$

$$[6.19]$$

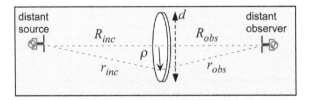

Figure 6-24: Diffraction by a Disk Illuminated On-Axis by a Distant Source

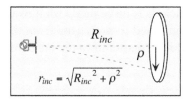

Figure 6-25: Path Length _r_ to a Point on the Disk Surface

An identical development applies for the observer distance r_{obs}. In this context such a quadratic expansion of the path length is known as *Fresnel diffraction*. We can now substitute for the two path lengths in equation [6.18] in terms of the single variable ρ:

$$A_{sc,obs} \approx \frac{-ikA_{t,inc}}{2\pi} \frac{e^{-ikR_{inc}} e^{-ikR_{obs}}}{R_{inc}R_{obs}} \int e^{-\frac{ik}{2}\left[\frac{1}{R_{inc}} + \frac{1}{R_{obs}}\right]\rho^2} 2\pi\rho \, d\rho \qquad [6.20]$$

The integral is now of the form of an exact differential (i.e., the derivative of ρ^2 is $\rho d\rho$) so we introduce a new variable $u = \rho^2$. It is also convenient to name the quantity in brackets:

$$\frac{1}{R_{av}} = \frac{1}{R_{inc}} + \frac{1}{R_{obs}} \qquad [6.21]$$

With these substitutions we obtain

$$A_{sc,obs} = \frac{-ikA_{t,inc}}{2} \frac{e^{ikR_{inc}} e^{-ikR_{obs}}}{R_{inc}R_{obs}} \int_{0}^{d^2/4} e^{\left[\frac{-ik}{2R_{av}}\right]u} du \qquad [6.22]$$

The integral is now simply that of an exponential. Recall that the integral of such an exponential is merely the exponential divided by the constant part of its argument. Evaluating it we obtain

$$A_{sc,obs} = A_{t,inc} \frac{e^{-ikR_{inc}} e^{-ikR_{obs}}}{R_{inc}R_{obs}} R_{av} \left(e^{-\frac{ik}{2R_{av}}\frac{d^2}{4}} - 1\right) \qquad [6.23]$$

It is possible to simplify this still-cantankerous expression to obtain a more elegant (and perhaps surprising) result. First we substitute for the average distance, obtaining

$$\frac{R_{av}}{R_{inc}R_{obs}} = \frac{1}{R_{inc}R_{obs}} \frac{R_{inc}R_{obs}}{R_{inc} + R_{obs}} = \frac{1}{R_{inc} + R_{obs}} \qquad [6.24]$$

Plugging this back into equation [6.23], we obtain

$$A_{sc,obs} = A_{t,inc} \frac{e^{-ik(R_{inc}+R_{obs})}}{\underbrace{R_{inc} + R_{obs}}} \left(e^{\frac{-ikd^2}{8R_{av}}} - 1\right) \qquad [6.25]$$

$$= A_{no\,disk} \left(e^{\frac{-ikd^2}{8R_{av}}} - 1\right)$$

The factor in front of the brackets is just exactly what would have happened to the incident potential if it had continued to propagate from the source to the observer, ignoring the disk. We now add the scattered to the incident to obtain the total potential at the observer:

$$A_{total} = A_{no\ disk}\left(e^{-\frac{ikd^2}{8R_{av}}} - 1\right) + A_{no\ disk} = A_{no\ disk}e^{\frac{-ikd^2}{8R_{av}}} \qquad [6.26]$$

The effect of this presumably opaque metal disk stuck right on the direct path between the transmitter and receiver is to multiply the potential that would have been obtained in the absence of the disk by a phase factor (which hardly matters because we generally don't know the absolute phase anyway). The intensity is the same as it would have been in the absence of the disk:

$$I_{total} = |A_{total}|^2 = I_{no\ disk} \qquad [6.27]$$

The disk *fails to cast a shadow on axis*! This counterintuitive result is known as *Poisson's bright spot*. It is a tip-off that real shadows are not just black and white. The bright spot is very puzzling if one tries to understand it in terms of an impinging potential that is screened by the metal disk to produce zero potential within the metal and thereafter, but it is a very sensible result if one simply adds up currents. The induced currents produce a scattered potential that is a function of the diameter d and average distance R_{av}. The scattered potential runs in a circle around a characteristic value of the incident potential, which is a very sensible result if these same currents must produce that characteristic value when the point of observation is moved very close to the disk. The sum of this scattered potential and the incident potential produce a result that runs around the origin as the size or distance are changed but with no change in amplitude (Figure 6-26). The scattered potential varies from 0 to twice the incident potential.

It is interesting to note that, having figured out what a disk does, we can immediately obtain the result for the related problem of a hole in an effectively infinite screen (Figure 6-27).

The trick we use is known as *Babinet's principle*: The sum of a disk and a hole the size of the disk is an intact infinite screen, which certainly blocks all impinging radiation. Therefore, the sum of the scattered potentials of disk and hole must be the negative of the incident potential, or equivalently the sum of all the potentials must be zero:

$$A_{no\ screen} + A_{hole} + A_{disk} = 0 \qquad [6.28]$$

Because we know the incident (no-screen) potential and the scattered potential from the disk, we can subtract to obtain the scattered potential from a screen with a hole:

$$A_{hole} = -A_{no\ screen}\left(e^{-\frac{ikd^2}{8R_{av}}}\right) \qquad [6.29]$$

The total potential is obtained by adding back in the incident (no-screen) potential; after a bit of algebra to convert the exponentials, we get

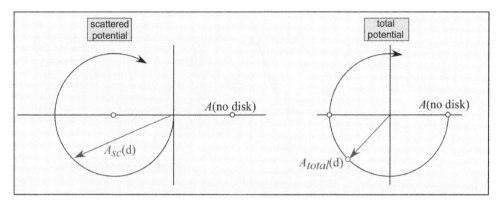

Figure 6-26: Scattered and Incident Potentials Combine to Produce Not Much Change

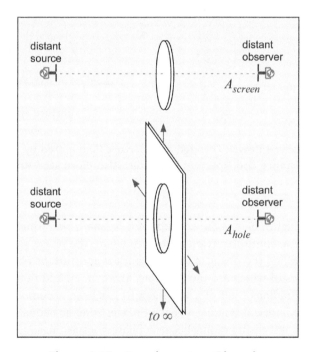

Figure 6-27: Complementary Obstacles

$$A_{total,\ hole} = A_{no\ screen}e^{\frac{ikd^2}{16R_{av}}}\left(2i\sin\left(\frac{kd^2}{16R_{av}}\right)\right)$$ [6.30]

The resulting variations in $|\mathbf{A}|^2$ as a function of distance and aperture size for 2.4-GHz illumination are shown in Figure 6-28. It should not be surprising that the power varies from 0 to four times the incident power: the scattered potential is the same size as the incident potential, but in this case interference may be either constructive or destructive, as illustrated in the left side of Figure 6-26.

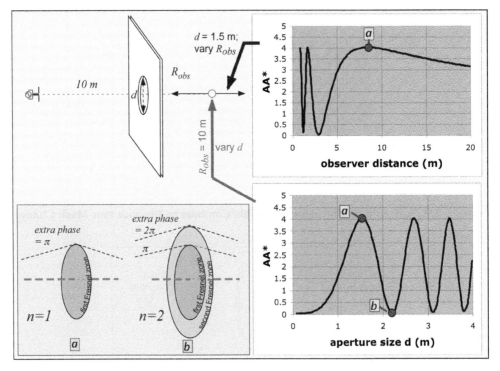

Figure 6-28: Received Power Through an Aperture Illuminated by 2.4 GHz Radiation, vs. Distance and Aperture Diameter; Also Illustrated Are the First Two Fresnel Zones

Also illustrated in Figure 6-28 is the concept of a *Fresnel zone*, which is often useful in examining normal-incident diffraction problems, where the phase varies quadratically with displacement along the obstacle. The cylindrical aperture or obstacle is divided into zones, the edges of which are the locations at which an additional path length corresponding to π radians (half of a wavelength) has been accumulated. In the particular case of cylindrical symmetry, each Fresnel zone has equal area: the zones grow narrower with increasing radius because phase goes as the square of the distance, but area increases as the square of radius. Thus, each Fresnel zone contributes the same amount to the integral but with an alternating sign. An odd number of Fresnel zones creates a maximum in the potential and an even number of zones creates a minimum, as shown in Figure 6-28. In the cylindrical geometry, the object size in Fresnel zones is

$$n \approx \frac{1}{\lambda} \left[\frac{[d/2]^2}{R_{av}} \right] \qquad [6.31]$$

A WLAN receiver will not necessarily sit exactly on the axis of an obstacle. For practical purposes we should like to know how an obstacle such as the disk affects received power for more general locations of the receiver. Unfortunately, the general problem is not readily solved analytically because the simple quadratic expansion for phase is no longer valid and we must resort to numerical simulation. A numerical estimate of the magnitude of the scattered potential, normalized to the incident potential at each location, from a 1-m disk illuminated on axis by a source 10 m away at 2.4 GHz is shown in Figure 6-29. The scattered potential is quite complex but can be generally described as comparable with the incident potential within the geometric shadow and falling rapidly toward zero as the point of observation moves outside of the shadow. Along the axis, the scattered potential shows maxima and minima as we might expect from the left side of Figure 6-26.

A numerical evaluation of the total potential behind the disk is depicted in Figure 6-30. Variations in the total potential are notably more drastic and spatially abrupt than those in the scattered potential; this is because they result from the small difference between the incident and scattered potential in regions where the two are nearly equal. Figure 6-30 also provides a graph of relative signal power in a cross-section at 5 m from the disk. Poisson's bright spot is clearly visible as a central region of little or no attenuation. Surrounding this are narrow shadowed regions in which the signal is attenuated by as much as 20 dB, though these regions of deep shadow are only 10–20 cm across. Through most of the shadowed region, the signal is attenuated by around 10 dB. Outside the geometric shadow significant fluctuations still occur in the received power. As we move away from the disk, Poisson's bright area grows, as does the shadowed region, roughly in proportion to the geometric shadow. The depth of the

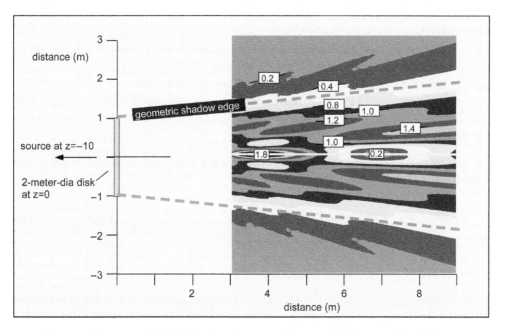

Figure 6-29: Numerical Estimate of Scattered Potential vs. Location for a 2-m-Diameter Disk Illuminated by a 2.4-GHz Source at 10 m Distance

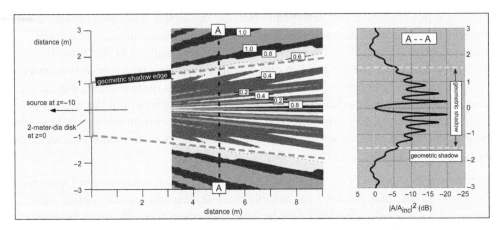

Figure 6-30: Numerical Evaluation of Total Potential to the Right of a 2-m-Diameter Disk Illuminated by 2.4 GHz Radiation at 10 m Distance

deepest shadowed regions falls slowly with increasing distance; numerical calculations suggest that a 15-dB shadowed region is still present 30 m from the disk. When the observation point is far from the disk, the scattered potential is essentially the radiation pattern of the disk acting as an antenna, with power density falling as $(1/r^2)$. When r is large compared with the separation between the source and the obstacle, the potential from both falls at about the same rate, so the relative magnitude of the scattered potential compared to that direct from the source changes very slowly at long distances.

A second class of obstacles that is worth examining is the straight edges of semi-infinite planes. Indoors one might encounter this sort of geometry at the end of a large metal (or thick concrete) wall or partition; outdoors, we will find that the roofs of buildings, and for long links the curved edge of the earth, can be seen approximately as straight edges diffracting the impinging signal. A representative configuration is shown in Figure 6-31. Here for simplicity we consider the case where the source is far from the edge and thus $R_{av} \Rightarrow R_{obs}$.

For normal incidence, we can again expand the path length quadratically in the lateral displacement (equation [6.32]).

$$r_{obs} = \sqrt{R_{obs}^2 + \rho^2} \approx \underbrace{R_{obs}\left(1 + \frac{1}{2}\left[\frac{\rho}{R_{obs}}\right]^2\right)}_{\text{if } \dfrac{\rho^2}{R_{obs}} \ll 1} = R_{obs} + \frac{1}{2}\left[\frac{x^2}{R_{obs}}\right] + \frac{1}{2}\left[\frac{y^2}{R_{obs}}\right] \qquad [6.32]$$

In the quadratic approximation the distance is the sum of the path length due to the x and y displacements, so the integral over the surface can be split into an x-integral and a y-integral. Inserting the definition of distance into the scattering integral (noting that the incident potential is simply a plane wave of constant phase on the obstacle surface), we get

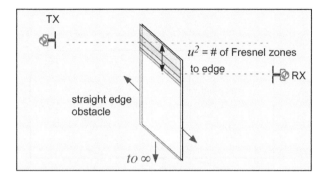

Figure 6-31: Diffraction by the Edge of a Semi-infinite Obstacle

$$A_{sc,obs} \approx \frac{-ikA_{t,inc}}{2\pi} \frac{e^{-ikR_{obs}}}{R_{obs}} \int e^{-\frac{ik}{2}\left[\frac{1}{R_{obs}}\right]y^2} dy \int e^{-\frac{ik}{2}\left[\frac{1}{R_{obs}}\right]x^2} dx \qquad [6.33]$$

where the x-integral extends to infinite in both directions, but the y-integral terminates at some finite distance above or below the observer. We can write the integrals a bit more elegantly by substituting a normalized distance in terms of the square root of the number of Fresnel zones:

$$u = \sqrt{\frac{2}{R_{obs}\lambda}}(x \text{ or } y) \qquad [6.34]$$

resulting in

$$A_{sc,obs} \approx \frac{-ikA_{t,inc}}{2\pi} \frac{e^{-ikR_{obs}}}{R_{obs}} \left[\frac{R_{obs}\lambda}{2}\right] \int_{-\infty}^{u_o} e^{-\frac{i\pi}{2}u^2} du \int_{-\infty}^{\infty} e^{-\frac{i\pi}{2}u^2} du$$

$$= -\frac{iA_{t,inc}}{2} e^{-ikR_{obs}} \int_{-\infty}^{u_o} e^{-\frac{i\pi}{2}u^2} du \int_{-\infty}^{\infty} e^{-\frac{i\pi}{2}u^2} du \qquad [6.35]$$

The form of the potential appears fairly simple: it is proportional to the incident potential multiplied by a phase factor corresponding to propagation from the obstacle to the receiver. However, there remains a pair of rather mysterious integrals (in the boxes). This sort of integral appears in any plane diffraction problem in the Fresnel approximation, and its properties have been elucidated; when plotted in the complex plane, the values with increasing u form a pretty curve known as the *Cornu spiral*, shown in Figure 6-32. (Note that the Cornu integrals are conventionally defined using positive arguments in the exponentials, as shown in Figure 6-32. Thus, for our purposes we need to take the complex conjugate of the spiral: graphically, one reflects it through the real (x) axis.)

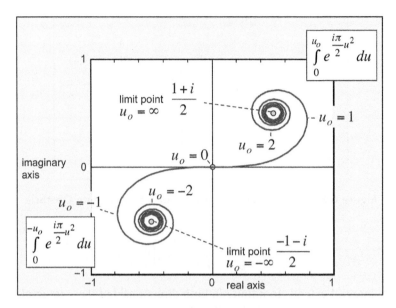

Figure 6-32: The Cornu Spiral

The form of the spiral is worth a bit of examination. Recall that the normalized variable u is just a length across the surface. If we start in the center of the graph ($u = 0$) and proceed, for example, to the right, the spiral gives us the complex value of the integral from 0 to whatever ending value of u, u_o, we choose. For small values of u_o, corresponding to the region of little phase shift along the surface, the integral is a smooth curve with a quadratically increasing imaginary part. However, for $u_o > 1$ the curve qualitatively changes its behavior, becoming a spiral around a limit point $(1 + i)/2$. What this says physically is that the scattered field from the region near the axis of the object (less than one Fresnel zone) essentially adds in phase but more distant portions change phase so rapidly with increasing distance that their contribution to the scattered potential spirals rapidly around in the complex plane (and doesn't add up to much).

The second boxed integral in equation [6.35] is simply the distance from one limit point of the spiral (at $-\infty$) to the other (at $+\infty$): taking the complex conjugate, we obtain $(1 - i)$. Thus, we can rewrite the scattered potential as

$$A_{sc,obs} \approx (1 - i) \frac{-iA_{t,inc}}{2} e^{-ikR_{obs}} \int_{-\infty}^{u_o} e^{-\frac{i\pi}{2}u^2} du$$

$$= -(i + 1) \frac{A_{t,inc}}{2} e^{-ikR_{obs}} \int_{-\infty}^{u_o} e^{-\frac{i\pi}{2}u^2} du$$

[6.36]

For a particular value of u_o, the scattered potential can be found graphically by tracing the Cornu spiral from the limit point at $-(1 + i)/2$ (bottom left) to the location on the curve corresponding to the value of u_o and extracting the integral as the difference vector. For example, at the geometric shadow edge $u_o = 0$, the integral is the negative of the limit point, or $(1 + i)/2$. Taking complex conjugates, we get

$$A_{sc,obs} = -\frac{A_{t,inc}}{2} e^{-ikR_{obs}} \left(\frac{1-i}{2}\right)(i+1) = -\frac{A_{t,inc}}{2} e^{-ikR_{obs}}$$

$$= -\frac{1}{2} A_{no\ screen}$$

[6.37]

When we add the incident potential to the scattered potential, we get just 1/2 of the value that would have been obtained in the absence of the screen. This is precisely the result we could have obtained from Babinet's principle, because the scattered potential of two identical screens, each subtending half the plane (so that the whole is an infinite screen) must be exactly $(-A_{inc})$ so that the total is 0.

The general result for finite displacements is shown in Figure 6-33. The received power fluctuates somewhat when the observer is above the edge of the screen; as the observer moves into the geometrically shadowed region, the signal falls rapidly to a level 20–30 dB below the unobstructed case and thereafter drifts slowly toward 0. To achieve the same signal power as an unobstructed receiver, the observer ought to be at a normalized height of about 0.6 above the geometric shadow edge, that is, rather more than half of a Fresnel zone needs to be exposed.

Because Figure 6-33 is shown in terms of the normalized distance u, the graph may be used for any real configuration once the appropriate scaling factor is obtained from equation [6.34]. As we noted, this graph is useful to keep in mind when examining diffraction around large objects such as walls, roof edges, or mountains.

Realistic indoor obstacles are likely to be of finite size and are more likely to be rectangular than circular in cross-section. Let us first take a qualitative look at how such an object forms a shadow. In Figure 6-34 we show a series of images of a square obstacle "viewed" from various locations, where in each case the obstacle surface is colored according to the difference between the direct path length from source to observer and the path length to each point on the obstacle surface, measured in half-wavelengths, that is, these are a sort of generalized Fresnel zones. (They are not identical to Fresnel zones, because we are measuring exact path lengths, not necessarily using the quadratic approximation to distance.) Each image is accompanied by a numerical estimate of the scattered amplitude at that point (normalized to the incident amplitude at the same location). At short distances, viewed from the axis, the obstacle appears to subtend many Fresnel zones. Because each full zone contributes the same amount to the integral, the scattered potential would vary rapidly as distance changes (running around the circles on the left half of Figure 6-28). However, the corners mess up the perfect cancellation of successive Fresnel zones that round obstacle would have; we can

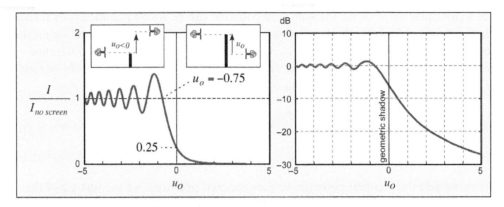

Figure 6-33: Potential and Signal Power Detected Near the Edge of a Straight Obstacle Illuminated by a Plane Wave

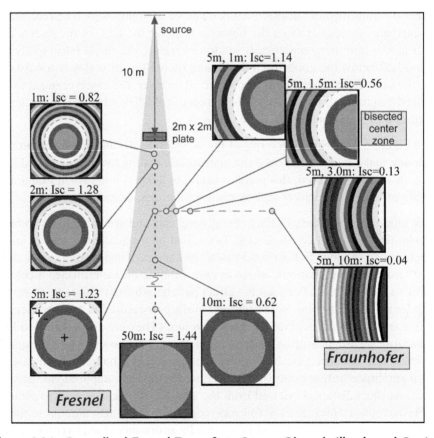

Figure 6-34: Generalized Fresnel Zones for a Square Obstacle Illuminated On-Axis

anticipate that the extremes of scattered potential (0 and twice the incident potential) will not appear in the case of a rectangular obstacle and indeed the largest amplitude is about 1.5 and the smallest around 0.6 on axis.

At long distances, the path length becomes dominated by the source-obstacle distances; at 50 m, the first Fresnel zone just touches the edges of the obstacle, and we can anticipate about the same appearance at 100 or 500 m. As we move laterally toward the shadow edge, the first Fresnel zone gets clipped by the edge of the obstacle; at the geometric shadow edge, only half of the first Fresnel zone is visible. As we move outside the geometric shadow, we see that the phase changes very rapidly across the surface of the obstacle, so we would expect that the contributions of the zones, which roughly alternate in sign, would add up to very little, explaining how the scattered potential rapidly falls outside the geometric shadow. Well outside the shadow edge, the shape of the zones becomes more linear, and their width becomes roughly constant. In this region, the path length is mostly due to the angle of view: as a function of the coordinate x across the obstacle, the excess path length is about $x \sin(\varphi)$, where φ is the angle of inclination and is thus linear across the surface of the obstacle. This regime, where the phase is linear rather than quadratic in position, is known as *Fraunhofer diffraction*.

To obtain a more quantitative understanding of how signals might vary within the shadowed region, we need to take a closer look at the Cornu spiral introduced in Figure 6-32. Consider for simplicity a tall obstacle of finite width. Here we must introduce the average distance R_{av} as in equation [6.21], with a corresponding modification of the normalized distance u (equation [6.34]), but otherwise the math is pretty much unchanged. The integral of the scattered potential will again split in two, as in equation [6.35]. We'll let the y integral proceed to effectively infinite limits, so that it simply multiplies the scattered potential by a constant value.

What happens to the integral in x when we move our point of observation within the shadowed region? The limits of the integral now correspond to the edges of the obstacle, measured relative to the intersection of the direct path to the observer. In Figure 6-35 we show examples for the on-axis and shadow-edge cases, for a normalized width of 1. The value of the integral is the difference between the endpoints of the dotted region, which corresponds to the part of the spiral between the limits of integration. At the center of the shadow, the integral proceeds from $-1/2$ to $+1/2$. At the shadow edge, the integral proceeds from -1 to 0; at intermediate positions the limits vary linearly. A nice way to imagine the situation is to picture the dotted region as a tube of fixed length sliding along the spiral, with its endpoints defining the ends of the arrow denoting the integral value.

We can see that when the normalized width of the obstacle is less than 1, the magnitude of the scattered potential (the length of the arrow) will always be significantly less than the limiting value of v2 (from one limit point to the next), and thus the scattered potential can never cancel the incident potential: this is the region of weak shadowing. We can also see that, even for a large obstacle, when we get near the geometric shadow

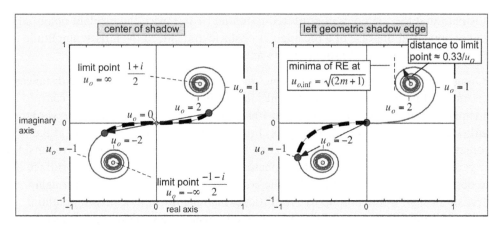

Figure 6-35: Traveling Along the Cornu Spiral on Axis or at the Shadow Edge, for a Normalized Obstacle Width of 1

edge (so that one of the limits u_{ol} or u_{or} is < 1/2 or thereabouts), the length of the arrow will also be smaller than v2, so the periphery of the geometrically shadowed region is always weakly shadowed.

For larger obstacles, our imaginary tube will spin around the limit point following the spiral. For $u > 1$ the distance from the spiral to the limit point can be accurately approximated as $0.33/u$. The minima and maxima of the real part (i.e., the left and right edges of the spirals) occur at the square roots of odd integers and thus become very closely spaced for large values, that is, the spiral runs around the limit points very rapidly for large u. Thus, for large normalized obstacles near the center of the shadowed region, the ends of the tube can be visualized as spinning rapidly around their respective limit points as we move from the shadow center to shadow edge. If we denote the distance from the limit point as r_l for the limit corresponding to the left edge of the obstacle and r_r for that corresponding to the right edge, the largest value the magnitude of the scattered potential can take is roughly ($\sqrt 2 + r_r + r_l$) and the smallest is ($\sqrt 2 - r_r - r_l$), each producing a total potential of normalized magnitude $(r_r + r_l)/\sqrt 2$ when the incident potential is subtracted. The exact scattered potential will vary rapidly with position as the two spirals twist, and deep shadows will occur when the magnitude of the scattered potential happens to be exactly equal to 1 (the depth depending on the precise inclination of the difference vector at that point). Deep shadowed regions are necessarily narrow as the change in phase rapidly moves the two ends of the tube in opposite directions. Thus, substituting for the distance r, we can estimate a sort of characteristic shadow depth at any position as

$$\langle S \rangle_{dB} \approx 20 \log \left(\frac{0.33}{2} \left[\frac{1}{u_{o,l}} + \frac{1}{u_{o,r}} \right] \right)$$ [6.38]

An example of the procedure is illustrated in Figure 6-36 for a 2-m-wide tall obstacle, illuminated on axis at 2.45 GHz from 10 m away. Within the geometric shadow, we construct the contours of the region at which one of the limits $u_o = 1$. (These appear as nearly straight lines in the figure but are in fact subtly curved.) Halfway from these contours to the geometric shadow edge is the region of weak shadowing (typically < 5 dB deep). As we move away from the obstacle, the effective distance R_{av} increases and thus the normalized width of the obstacle decreases, but when the observer becomes very distant, R_{av} changes slowly, because it can never be larger than the source distance. Thus, the normalized width converges to a limit (here about $2u_{o,ctr} = 2.6$) at large distances.

Within the region of strong shadowing, the typical value of the total potential can be roughly estimated using [6.38]. In Figure 6-36 the estimate (shown in Table 6-1 and in the bulleted circles in the shadow) is performed with limits corresponding to a position half-way between the axis and the contour of $u = 1$. (Note that the Fresnel approximation reaches its limits at an observer distance of 1 m from the plate; closer than this, the $(1/r)$ dependence of the scattered potential cannot be removed from the integral, and numerical methods must be used. We don't really need to know the details here: the shadow will be deep so close to the obstacle and only reflections from other objects in a room will save our link budget.)

Our estimates of the average shadow depth are not very different from those one might have obtained in the case of a disk. For example, with reference to Figure 6-30, the slice at 5 m could reasonably be interpreted as showing an average shadow depth of around 14–15 dB away from the Poisson bright spot in the center. It seems plausible to assert that shadows resulting from indoor obstacles of modest size (bookshelves, desks, columns, etc.) will have a typical depth of 10–20 dB, varying more or less as depicted in Figure 6-36, with narrow regions of much stronger (and weaker) shadowing within them. For purposes of estimating link budgets, it is reasonable to treat the typical shadow depth as providing the average signal power and the variations in shadow depth as a source of *fading*, just like any other multipath propagation. The signal strength variations due to fading can be mitigated in a number of ways, discussed in section 3 below.

Let us examine one more case that might be encountered in the real world: a tall narrow rectangular obstacle such as a building column (Figure 6-37), illuminated *off-axis*. We will approximate the obstacle as two-dimensional. Because the obstacle is narrow and we assume the transmitter and receiver are distant relative to the width, we can remove the $(1/r)$ terms and the cosine from the integral (although we can no longer assume the cosine term is equal to 1). The height can be treated as if infinite if it is much greater than about 10 Fresnel zones, which will be the case for typical tall indoor obstacles, thus contributing a factor of $(1 - i)$ as before. However, for finite values of the incident angle φ, the phase shift horizontally across the obstacle will be dominated by a roughly linear contribution from the projection of the surface distance onto the line from the

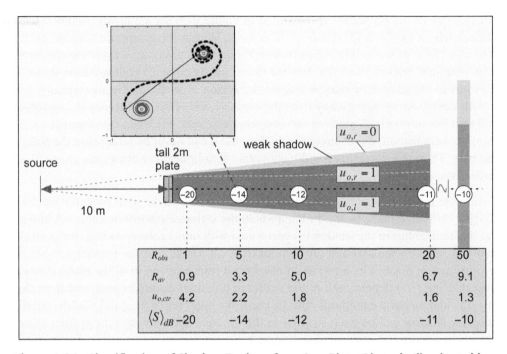

Figure 6-36: Classification of Shadow Regions for a 2-m Plate Obstacle Illuminated by a Source 10 m Distant; $u_{o,\,ctr}$ Is the Normalized Distance to Either End of the Spiral Viewed From the Center of the Shadow (i.e., Half the Normalized Width of the Obstacle); < S > Is the Estimated Average Depth of Shadow Half-Way Between the Center and the u = 1 Contours, as Shown in the Inset for an Observer 5 m Distant

transmitter rather than the (small) quadratic term of the Fresnel approximation: as we noted above, this is the regime of *Fraunhofer* diffraction.

The resulting formula for the scattered potential is rather more complex than those we have examined above. First, the diffracted potential varies as something like the (3/2) power of the inverse distance instead of (1/r). Roughly speaking, this dependence arises from illuminating an object that then radiates isotropically in a plane: the illumination goes down as (1/r) and the planar scattering contributes a factor of $\sqrt{(1/r)}$. The second novelty is a term that looks like (sin x/x). This form generally arises any time the phase of a radiator varies linearly across its surface: it is commonly encountered in some antenna problems such as the extended line source, which were a bit beyond the scope of our discussion in Chapter 5. The full result is

$$A_{sc} = A_{t,\,inc}(i-1)\underbrace{\frac{e^{-ik(R_{inc}+R_{obs})}\cos(\varphi_{inc})}{\sqrt{2\lambda R_{inc}R_{obs}(R_{inc}+R_{obs})}}}_{\dfrac{e^{-ikr}}{r\sqrt{r}}}\underbrace{\frac{w\sin[\frac{kw}{2}(\sin(\varphi_{inc})-\sin(\varphi_{obs}))]}{[\frac{kw}{2}(\sin(\varphi_{inc})-\sin(\varphi_{obs}))]}}_{\dfrac{\sin(x)}{x}} \qquad [6.39]$$

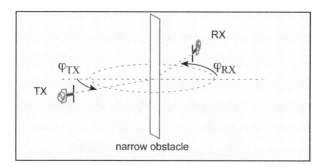

Figure 6-37: Tall Narrow Obstacle Illuminated and Viewed at Varying Azimuthal Angles

For a narrow obstacle (w a few wavelengths), this is less frightening than it looks. In Figure 6-38 we show the scattered potential, normalized to the value of the incident potential at the observer, and the total. The scattered potential is mostly composed of a smooth lobe in the shadow direction and another lobe in the specular reflection direction, with a few little bumps elsewhere. The two lobes are actually of the same magnitude, but the reflected lobe is smaller relative to the incident potential because the receiver is closer to the source on that side. (The receiver is presumed to be isotropic in the plane—e.g., a vertically oriented dipole.) The scattered potential is fairly simple and slowly varying; the received intensity is complex mostly because of the complex phase relationship between the incident and the scattered waves.

Generally speaking, the obstacle casts a shadow, with a partial Poisson bright spot in the middle; the shadow is much wider than the geometric shadow but mostly not very deep (the narrow minima are about 10 dB down). The reflection is broad and fluctuates rapidly with receiver position because of the rapid change in relative phase: the direct path to the receiver at an angle of 150 degrees is much shorter than the reflection path from the obstacle.

To summarize this very long discussion, obstacles of finite size relative to a wavelength do not create the abrupt shadows we are accustomed to in the optical world. Rather, the obstacles scatter weakly outside the geometric shadow (and strongly in the reflected direction if present). The scattered potential is similar in magnitude to the incident potential within much of the shadow (except on axis for a disk) and nearly opposite in direction. The exact magnitude of the scattered potential varies from point to point because of the changes in the relative phase of various parts of the obstacle, and these variations are magnified when the incident potential is added because the difference is small, so that narrow regions of deep shadowing occur within a background shadow with attenuations of typically 10–20 dB relative to the unobstructed power. For obstacles many Fresnel zones across, in directions of strong scattering, the Fresnel approximation (quadratic phase variation across the obstacle) is appropriate, whereas when the obstacle is small relative to a Fresnel zone, the linear Fraunhofer approximation may be used.

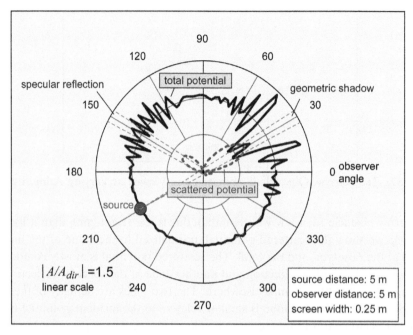

Figure 6-38: Scattered Potential From a Tall Narrow Obstacle Illuminated at an Angle (Normalized to Direct Potential at the Observer) and Total Potential

3. Multipath Propagation: Fading and Delay

We've examined a host of mechanisms that conspire to subtract power from the direct path between the source and receiver while creating alternative indirect paths. As the receiver is placed farther and farther from the transmitter, the signal from the direct path becomes weaker, both due to simple $(1/r)$ scaling and the effects of absorption and diffraction from partitions, walls, and other obstacles. The signal received from alternative paths, such as reflections from walls, floors, or ceilings, becomes a larger part of the total. Because the relative phase of the various contributions can vary considerably with small changes in the positions of transmitter, receiver, and reflector, the total received amplitude becomes a sensitive function of position: This effect is known as *fading*. In addition, the time delay associated with the longer paths can be significant on the scale of bit times, leading to multiple time-delayed copies of the intended signal. We have alluded to this situation briefly in Chapter 2, section 3.2. *Multipath delay* is particularly important when higher order modulations are used, because large signal-to-noise ratios (S/N) are needed for reliable demodulation. Recall, for example, that 64 quadrature-amplitude-modulation requires a (S/N) of about 25 dB; delayed signals that are much smaller than the direct path may still cause problems.

Let's look at a very simple example to see how even in an unobstructed environment fading can still be very significant. Consider an open room with a concrete floor and a

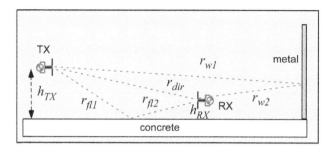

Figure 6-39: Open Room With Direct Beam, Floor Reflection, and Wall Reflection

reflecting back wall (Figure 6-39). How does the received signal vary as we move the receiver along the length of the room? We consider only three rays: direct propagation from the transmitter to the receiver, a reflection from the floor to the receiver, and a reflection from the metal wall at the back of the room. Even this elaboration introduces two additional angles, four distances, and two reflection coefficients to be found. On average, the reflected beams represent extra power, so the received power should fall more slowly than $(1/r^2)$. However, the actual signal at a particular point may be small if the beams interfere destructively.

For vertical polarization, the floor reflection could be large at short distances but will show a minimum value at a distance corresponding to Brewster's angle (Figure 6-12). The path followed by the floor reflection is not too different in length from the direct path, so the relative phase will vary slowly: the floor reflection will cause gradual changes in signal strength with position.

On the other hand, the path to the back wall grows shorter as about the same rate that the direct path grows longer. The relative phase of these two beams fluctuates rapidly and would be expected to lead to abrupt changes in signal power with position. A variation in relative phase of 4π will result for a displacement of a wavelength (about 12 cm at 2.4 GHz or 6 cm at 5 GHz). To accurately predict signal strength, we must be much more precise than this, requiring that all path lengths be known to around 1 cm accuracy—obviously not practical in most situations. We cannot hope to predict the received signal versus position even in this simple environment; the best we can do is to estimate the range of signal strength expected as we move around in some general region of the room.

The simulated result for a particular geometry, assuming vertical polarization, is depicted in Figure 6-40, where *LOS* denotes the direct (line-of-sight) path from the transmitter to the receiver. The floor reflection is significant within about 5 m of the transmitter but becomes small near Brewster's angle. The wall reflection becomes comparable with the LOS signal as the receiver gets close to the wall, and its rapid phase changes lead to short-range fading on a scale of a few centimeters, as shown in the inset. The peak power falls rather more slowly than the 20 dB/decade that we'd expect

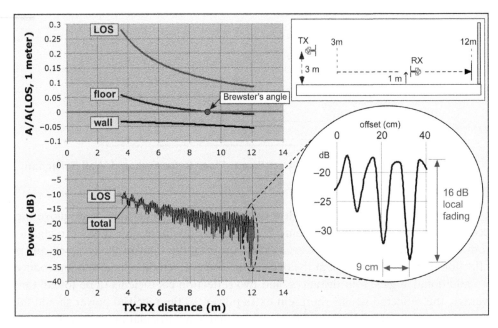

Figure 6-40: Simulated Amplitude of Potential From Direct and Indirect Paths, and Resulting Signal Power, vs. RX Position; f = 2.45 GHz, Floor n = 2.1, Wall 14 m From TX

for open space, but wild fluctuations occur locally on a scale of a few centimeters. Note that just as in the case of diffraction, slowly varying amplitudes of the constituents lead to large fluctuations of the resulting signal due to rapid changes in relative phase.

Real environments are much more complex. In addition to the likelihood of scattering and absorption in the direct path from cubes, cabinets, and partition walls, we must expect reflections from the side walls and ceiling constituents as well as time-dependent reflection and absorption from people. Local fades of 10–20 dB in a single-frequency signal must be expected. This is a lot! Recall from Chapter 2 that at 60 m the ideal free-space signal strength would be about −50 dB from a milliwatt (dBm), and we ought to allow a couple of wall absorptions and reflections in an indoor environment. We'll be fortunate to get an average received power of −70 dBm indoors: still enough for good high-rate reception but not enough to tolerate 15 dB of additional local fading without falling back to slow 1 megabit per second transmission or losing the link altogether. Fading is also time dependent and unpredictable in detail. To make a robust radio, we need to have some means of mitigating the effects of this sort of local fluctuation.

Fading mitigation techniques in radio design fall into three general categories. The most common is the use of diversity antennas: two antennas spaced far enough apart so that if one antenna is in a deep fade, the other isn't. A second approach is to spread the spectrum of the transmitted signal enough so that it includes frequencies that are not captured by the fade. The third is to use more directional antennas that can exclude some of the paths that are leading to destructive interference. The first two approaches are used in almost all WLAN technologies, and the third is becoming more common. Let's take a closer look at each.

3.1. Fading Mitigation With Diversity Antennas

Fading results from the near-perfect cancellation of two or more propagation paths from the transmitter to the receiver. Such cancellation depends on the exact phase relationship between the paths; it is unlikely to persist if we move the antenna, unless by wild coincidence the beams are exactly symmetrical with respect to the direction of motion. Because it is inconvenient for a user to optimize the position of their laptop on a packet-by-packet basis, it is more common to use two antennas, known as *diversity* antennas. In Figure 6-41 we show the situation where two incoming beams perfectly cancel at antenna 1, leading to a deep fade. The path lengths to antenna 2 for these beams (presumed to come from distance sources) differ by δr_1 and δr_2 from the original path lengths, leading to a corresponding shift in relative phase. In all likelihood, the shift will destroy the cancellation and lead to an improved signal on antenna 2.

How far apart do we need to place the two antennas? Recall that in the vector picture at the right of Figure 6-41, one wavelength corresponds to a 360-degree rotation of a signal vector about the axis. Displacements of much less than a quarter of a wavelength will accomplish little; large displacements of many wavelengths just spin the vectors randomly a few extra times without adding any value. The smallest beneficial offset is about a half-wavelength: 6 cm for ISM or 2.5 cm for Unlicensed National Information Infrastructure. This spacing is not achievable in a PC-card form factor (roughly 5.5 cm wide) at ISM frequencies. Access points often have antennas mounted on opposite sides of the package and thus spaced at rather more than a wavelength apart.

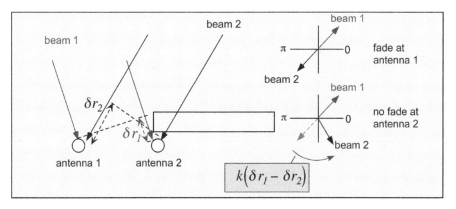

Figure 6-41: Beams 1 and 2 See Different Path Lengths to Antenna 1 and Antenna 2

It is in principle possible to combine the signals of the two antennas, achieving thereby a slight improvement in performance, but most WLAN systems simply provide switching and use the antenna with the best signal. In receive mode, this is very simple to implement by testing both antennas during a packet preamble. Transmit mode diversity is much more subtle, requiring some sort of measure of link error rate such as a count of missing acknowledgments. Furthermore, in general it is not necessary to implement diversity on both sides of a link: it is only the total path length that matters for cancellation, and the path length can be adjusted at either end. Therefore, WLAN radios generally use only receive-side diversity, implemented using an RF switch (Figure 6-42); this block diagram should be familiar from Chapter 4.

3.2. *Fading Mitigation With Spread-Spectrum Modulation*

Another way we might disrupt the conditions for cancellation is to change the frequency, thereby changing the phase accumulated along the various paths. In Figure 6-43 we show a simple example of a direct and reflected beam (presumed of equal amplitude due to some absorption or scattering of the direct beam). If we make a

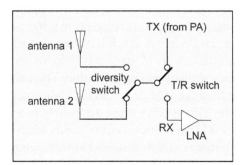

Figure 6-42: Switched-Diversity Block Diagram

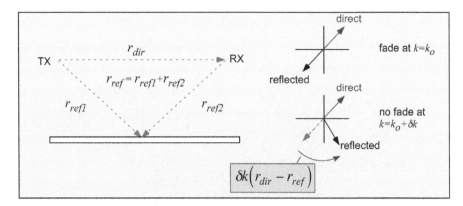

Figure 6-43: Conditions for Cancellation Are Disrupted by a Change in *k*

change in frequency and thus propagation constant k, the relative phase between two paths will be shifted by difference in length between them multiplied by the change δk. An antenna that is in fade at one frequency is generally not in fade at nearby frequencies.

To obtain some estimate of the sort of bandwidth needed to take advantage of this effect, we need to calculate how changes in frequency affect the propagation constant:

$$k = \frac{\omega}{c} = \frac{2\pi f}{c} \rightarrow \delta k = \frac{2\pi}{c}\delta f \qquad [6.40]$$

If the product of the change in k and the path length difference is less than about π/8 (45 degrees), we won't have accomplished much; thus, we can establish a relationship between the change in frequency and the corresponding path length difference below which the antenna will still be in fade:

$$\delta k \cdot \delta r = \frac{2\pi}{c}\delta f \cdot \delta r < \frac{\pi}{8} \rightarrow \delta r < \frac{c}{16}\frac{1}{\delta f} \qquad [6.41]$$

For the 16-MHz bandwidth used in 802.11 WLAN protocols, a path difference of more than about 1.2 m is sufficient to ensure that at least part of the signal will escape cancellation. Because fading is mostly a concern for relatively longer links with minimum path lengths of tens of meters, such nearly perfect agreement is unlikely, except for special cases such as low-angle reflections along a long corridor.

The manner in which a spread-spectrum signal is created also has some influence on its robustness to multipath fading. An 802.11 classic signal has 11 chips per symbol. Multiple paths with delays that differ by a time comparable with the chip time can be separately demodulated by convolving the input signal with Barker sequences (see Figure 3-7) offset in time: This approach, known as a *rake receiver* because of the resemblance of the multiple parallel correlators to the fingers of a rake, is widely used in the related field of code-division multiple access cellular telephony. However, for the higher rate modulations used in the 802.11b protocol, the symbol rate is equal to the chip rate and rake techniques are not helpful. A narrowband fade in such a signal corresponds to a "notch" in which transmission is anomalously low; in the time domain, such a notch leads to intersymbol interference, which is combated to some extent by the coding used (complementary code keying or packet binary convolutional coding; see Chapter 3, section 3.5).

In an orthogonal frequency-division multiplexing (OFDM) signal, a narrowband fade corresponds to degraded (S/N) on one or more subcarriers. From equation [6.41] we can see that for plausible path length differences of 5 to 10 m fades are likely to be 1- to 2-MHz wide, with the worst impact in the middle, thus knocking out at most two to three 802.11a/g subcarriers. Because the information in the OFDM signal is interleaved uniformly across all the subcarriers, this level of disruption is generally dealt with successfully by the convolutional codes. It is this simultaneous resistance to narrowband fading as well as longer multipath delays that accounts for the popularity of OFDM-like approaches in high-data rate wireless networking.

3.3. Fading Mitigation With Directional Antennas

Fading often arises from interference between multiple paths. Generally speaking, such multiple paths have distinct angles of arrival at the receiver. If the receiver were to use a directional antenna, one of the paths would be rejected, eliminating the fade. Such an approach is also likely to reduce or eliminate large multipath delays and consequent intersymbol interference. In a strictly fixed environment, a directional antenna solution can be implemented using a fixed antenna, but most real indoor and outdoor environments have time-dependent propagation characteristics because of the movement of people and objects, and many applications of wireless links are predicated on mobility of the transmitter or receiver or both. Thus, the use of a directional antenna can only be contemplated if some sort of adaptation is provided to allow an optimal configuration to be found for each packet to be transmitted.

There are several approaches to achieving the requisite adaptability. The simplest is the use of *sectorized* antennas. Sector antennas are typically used at the access point, each connected to its own radio transmitter and receiver. A number of sector antennas are placed at a central location. Each antenna covers a portion of the azimuth, like a slice of pie. For example, let us imagine that the 360 degrees of azimuth is divided into four segments, each primarily served by one antenna. Further, assume a few decibels (say 3) of gain from directing the radiation of the antenna primarily in the horizontal plane, as is usually appropriate in indoor installations. Recalling that the beam solid angle is inversely proportional to the directivity, we estimate about 6 dB sector gain and thus 9 dBi (dB relative to an isotropic antenna) directivity for each antenna. Each antenna will primarily see only the direct or nearly direct beams of the clients within its sector. High-angle multipath is minimized. Note that sectorized antennas cannot address collinear fading and don't help much with diffraction or shadowing from line-of-sight obstacles, save in the increase in signal resulting from higher directive gain.

In a sector configuration, the antennas do not need to be adaptive but the software should be: in a sort of higher order diversity, the different sectors must *hand off* clients so that the best antenna is used to serve each client. Hand offs enable the system to adapt to moving clients and changing propagation conditions. Very sophisticated hand-off capabilities are provided in cellular telephony standards, but 802.11 has historically left exchanges between access points outside of the standard, leading to incompatible proprietary solutions.

A more subtle implementation of sectorization involves the use of a phased array with a phase-shifting network such as the Butler phase shifter discussed in Chapter 5 (section 6.11). In this case, each input port of the matrix provides a different beam direction. By attaching a radio to each port, multiple locations can be served with directive multipath-rejecting beams from a single array antenna. Hand off between directions must again be provided to adapt to changing conditions and enable clients to move from one location to another while remaining associated.

A different approach is to used a switched adaptive array, in which the system can search through a finite number of states, each corresponding to a differing beam direction, to find the best one for the conditions pertaining during a given packet (Figure 6-44). Because the approach is switched at the RF level, a single radio can be used. If the number of possible states is small, each can be tested during the packet preamble and the best state for that packet selected. Such an approach is simple and inexpensive to implement using a microstrip antenna array, but to obtain good directivity the antenna must be physically large and is thus inappropriate for a portable client. Typically, the state used for reception is reused to transmit to the same client. One disadvantage of such an arrangement is that broadcast packets from the access point must be sent several times (once for each antenna direction) to reach all clients.

A similar but simpler alternative is to use two to three coplanar patches and a small state space, providing a modest 2–3 dB of additional directive gain. Such an antenna is compatible with PC-card or portable digital assistant form factors and by minimizing power consumption may provide significant performance advantages from this modest gain improvement.

Finally, a fully adaptive array can be used (Figure 6-45). Here by adjusting the phase and amplitude shifts between successive array elements, a desired beam can be

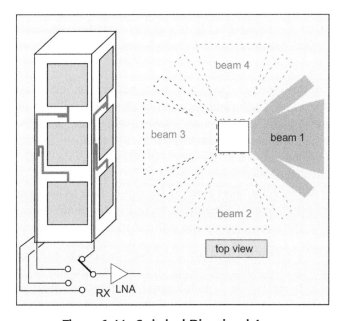

Figure 6-44: Switched Directional Array

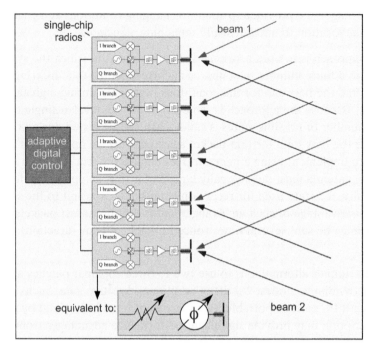

Figure 6-45: Digital Adaptive Array for Rejecting Multipath (or Other Interference)

accepted and an undesired one (such as a multipath fade but also including an interferer such as a microwave oven or portable phone) can be rejected (Figure 6-46). The exemplary adaptive array shown has one analog front end for each antenna but may use only one digital signal stream, created by the phased combination of the antenna signals.

Adaptive arrays of this type are capable of doing better than just selecting the best single direction for a link. Each radio link can be characterized by a distribution of signal amplitude with phase and direction, known as the *spatial signature* of that source. With the appropriate optimization, an adaptive array can *invert* this signature, thus combining all the beams from a single source in phase to arrive at an increased total power over that available even from the main beam. Such a procedure is analogous to the use of a rake receiver but takes place at RF rather than at baseband after digitization of the signal. Spatial signature inversion is thus much more powerful than a rake receiver but is correspondingly more difficult to implement.

Let us consider a simple example of a source whose main power is in two beams arriving from slightly differing directions, impinging on a pair of antennas (Figure 6-47). The beams arriving at antenna 2 differ in path length from those at antenna 1 by δr_1 and δr_2, which are just the product of the separation of the antennas s and the sine of the angles of arrival. The antenna voltages are thus

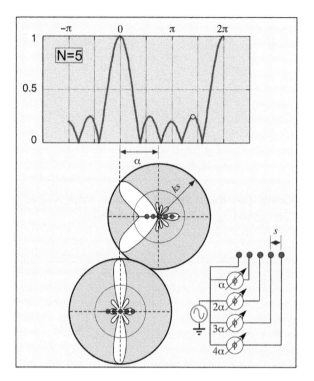

Figure 6-46: Array-Factor Construction Shows That Varying Phase Offset of an Adaptive Array Corresponds to Beam Steering

$$V_1 = b_1 + b_2$$
$$V_2 = b_1 e^{-iks \sin(\theta_1)} + b_2 e^{-iks \sin(\theta_2)}$$

[6.42]

Adding the signals from the two antennas will give differing results depending on the phase offset in the array. An example of this procedure for a pair of incoming beams that nearly cancel at antenna 1 is shown in Figure 6-48. By choosing the optimal phase offset before combining voltages, almost twice as much power is obtained as would have resulted from simply choosing antenna 2 (i.e., from simple diversity switching).

The power of this technique grows as more array elements are used, because more individual beams can be resolved. An eight-element array can provide 8- to 15-dB improvements in fade resistance over a single antenna (see Okamoto's book in section 6). However, the complexity of implementation also increases: for an 8-element array, seven relative phase shifts and seven relative gains must be determined for each client. Extraction *de novo* of the spatial signature for each client for every packet appears likely to involve excessive computational complexity. Field tests have shown that the spatial signature of stationary sources in an indoor environment may change only modestly (a few tenths of a decibel in amplitude and less than 5% in angle) over a

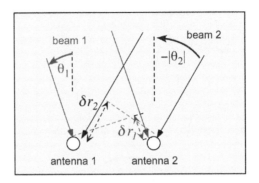

Figure 6-47: Simple Two-Beam Spatial Signature With Angles of Arrival and Path Length Corresponding to Relative Phase Shifts

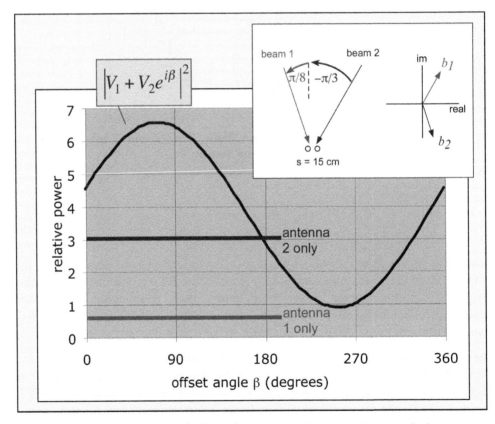

Figure 6-48: Optimal Phasing of Antenna Voltages to Constructively Combine Multiple Beams

period of several hours (Okamoto, Figures 6-5 and 6-6). However, a moving client endures significant changes in spatial signature over displacements of 3 or 4 cm, so dealing with high-velocity clients requires frequent updates and is complex to administer.

A further elaboration is to place an array of antennas in both the transmit and receive locations. This approach is known as *space-time coding* or *multiple in multiple out* (MIMO) transmission, though a number of other acronyms (BLAST, STREAM) have also been used. By controlling the transmitted signatures as well as the received signatures, powerful additional capabilities can be obtained, particularly in the case where significant reflections at large angles (typical of an indoor environment) are present. The closely spaced receive antennas could not by themselves resolve the signals from the closely spaced transmit antennas, but the combination of the direct and reflected signals acts to synthesize an effective separation between the antennas much larger than the physical separation. Thus, nearly orthogonal signatures can be constructed to simultaneously transmit to multiple users at the same bandwidth or to transmit distinct data sets on the same bandwidth but with differing spatial signatures, increasing capacity over the Shannon limit for a single beam (equation [2.11]). MIMO systems have been shown to reduce signal-strength variations over time by 15 dB or more in outdoor links versus single-antenna systems. MIMO techniques have been standardized in metropolitan-area technologies such as IEEE (Institute of Electrical and Electronic Engineers) 802.16. Such approaches are also under consideration for LANs in the new 802.11 task group n, but standards-compatible solutions are not yet available.

4. Statistical Modeling of Propagation

We have mentioned that propagation loss in real environments is time dependent: Just because I measure a signal strength of − 70 dBm at 2:32 p.m. on Thursday, I cannot guarantee that the signal will be the same from minute to minute or day to day. To say something useful about signal amplitude over time, we need to treat the signal as a random variable and try to understand the distribution of signal voltage or power, given that we know something about the average. Modeling of the distribution of arrival times is also necessary for examining the performance of modulations and implementations; we touch briefly on this issue as well.

The approach typically used in this area is to attempt to fit measured data to an analytic probability distribution, thereby extracting a handful of parameters that characterize the likelihood of a given signal amplitude and that can be conveniently summarized. A large number of analytic formulae are used by different researchers in the field, with no simple consensus on which to apply in what circumstance. Therefore, we limit our examination to two very common distributions, the *Rayleigh distribution* and the *Rician distribution*, which correspond, respectively, to the case of many equal beams or to one large beam with many equal small beams. In each case, the equal beams are assumed to

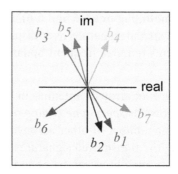

Figure 6-49: Equal-Amplitude Randomly Phased Beams

be present with random phase distributions, which is quite plausible in view of the discrepancy between typical path lengths of meters and wavelengths of a few centimeters. The former case might correspond to a link having a number of reflected or diffracted beams between the transmitter and receiver but no direct path. Thus, these distributions seem to provide physically plausible representations for the common conditions of a non-LOS (Rayleigh) or LOS (Rician) link, perhaps accounting for their popularity.

In the Rayleigh distribution, we assume that the antenna voltage is the result of many different signals, whose amplitude is equal but whose phase is randomly distributed over all possible values from 0 to 2π (Figure 6-49).

The voltage is the sum of the individual voltages:

$$V_{total} = v\left[e^{i\varphi_1} + e^{i\varphi_2} + e^{i\varphi_3} + \ldots + e^{i\varphi_N}\right] \qquad [6.43]$$

where N is the number of beams and v the voltage of each. The real and imaginary (or in-phase and quadrature) parts of the signal thus vary randomly from 0 to $\pm Nv$. When N is reasonably large (in practice, greater than about $N = 6$), the real and imaginary voltages are normally distributed with a standard deviation σ proportional to the number of beams:

$$V_{total} = \sqrt{v_i^2 + v_q^2}; \; \Pr(v_{i,q}) = \frac{1}{2\pi\sigma^2}e^{-(v_{i,q}/6)^2}$$
$$\sigma \approx vN/2\pi = V_{max}/2\pi \qquad [6.44]$$

The probability distribution of the total voltage is then the Rayleigh distribution:

$$\Pr(V)dV = \frac{V}{\sigma^2}e^{-\left(V/\sqrt{2\sigma}\right)^2}dV \qquad [6.45]$$

This function is plotted in Figure 6-50. The most probable value of the voltage is σ, which is about (1/6) of the maximum possible voltage: this would be the voltage corresponding to the average power obtained in a number of repeated measurements.

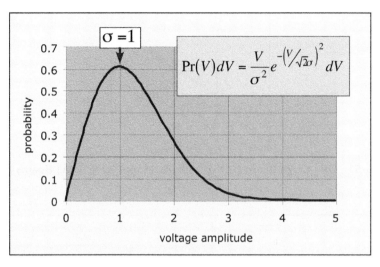

$$\Pr(V)dV = \frac{V}{\sigma^2}e^{-\left(V/\sqrt{2}\sigma\right)^2}dV$$

Figure 6-50: Rayleigh Distribution

The voltage will be between 0.3σ and 2.4σ about 90% of the time. Thus, the likely range of power around the average is about 17 dB wide at 90% probability: around 7 dB higher and 10 dB lower than the average power. If we were confident that the measured power in a given location obeyed the Rayleigh distribution, to ensure 95% availability of a link we would require that the average power be about 10 dB above the minimum threshold for obtaining the desired data rate (e.g., −85 dBm for a typical WLAN link). This 10-dB increase is the *fade margin* imposed on the average power to ensure that the link is almost always available at the desired performance level.

In the case where an LOS path exists between the transmitter and receiver, one might expect that the received power would be dominated by one beam propagated along that path (in the absence of a strong floor or ceiling reflection). The smaller randomly phased beams give rise to some residual variation in the received signal strength. The situation is depicted in Figure 6-51.

In these LOS conditions, the received signal voltage is described by a Rician distribution:

$$\Pr(V)dV = \frac{V}{\sigma^2}e^{-\left(\frac{v_m^2+V^2}{2\sigma^2}\right)}I_o\left(\frac{v_m V}{\sigma^2}\right)dV \qquad [6.46]$$

Here v_m is the amplitude of the dominant (LOS) beam. This distribution is plotted in Figure 6-52 for the case of the main-beam voltage three times larger than the expected value of the minor beams; the modified Bessel function I_o, shown in the inset, behaves as something like a generalized exponential, rising rapidly with increasing argument. The product of the two produces a most probable voltage approximately equal to v_m,

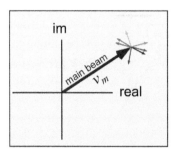

Figure 6-51: One Large Beam (LOS) With Random-Phase Small Beams (NLOS)

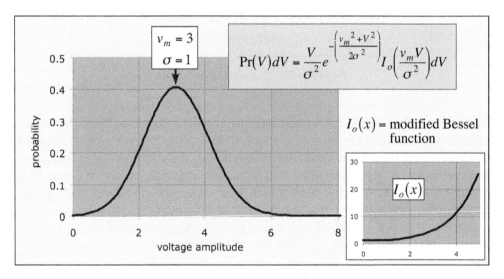

Figure 6-52: Rician Distribution

with a distribution of voltages about 2σ wide. When v_m is small compared with σ (i.e., there isn't really a dominant beam), the Bessel function is equal to 1 and the Rician distribution becomes the Rayleigh distribution.

The Rician distribution does not provide as convenient a guideline for fade margin as the Rayleigh distribution, because in addition to the average received power, one must also know the portion of that power attributable to the main beam. However, the Rayleigh distribution will always represent a worst-case limit, so using the 10-dB fade margin derived from the Rayleigh distribution provides a conservative approach that should ensure adequate signal power in both cases.

We have mentioned multipath delay a number of times but have not examined it in detail. How much delay ought one to expect? The basis for approaching this problem is the speed of light: because 1 m meter adds 3 nsec of delay to a path, the size of characteristic features in the environment sets the expected scale for the variation in delay. Typical indoor environments have scales of 3 m to perhaps 50–60 m for a large building, so we might expect delays associated with multiply-reflected paths to extend from a few tens of nanoseconds to perhaps 100–200 nsec with respect to the direct path. Some representative scenarios and data for indoor multipath are further discussed in Chapter 7, section 5.

Distributions of delay may be readily characterized by, for example, a root-mean-square (RMS) excess delay or a maximum value of delay for some threshold of amplitude, where the excess delay is that over the delay associated with the shortest available path. For WLAN purposes, we can use the rule of thumb that significant energy at >100 nsec excess delay will necessitate an OFDM protocol to achieve high data rates, whereas RMS delays of 200 nsec are likely to be tolerable if a low-rate 802.11 classic link is adequate for the application under consideration.

5. Capsule Summary: Chapter 6

When lateral variation over a wavelength is small, propagation can be studied based on tracing the paths of rays. Key events in the life of a ray are absorption as it passes through large blocks of material, reflection from dielectric surfaces, and diffraction from finite-sized obstacles. Diffraction, scattering, and reflection from obstacles of size generally comparable with a wavelength can be understood in terms of the currents induced in the obstacle by the impinging radiation; the overall diffracted potential near the obstacle has strong local variations in amplitude because of the rapid change in relative phase of the incident and scattered potentials. The narrowness of the resulting deep shadow regions relative to the geometric shadow helps to make wireless links possible under obstructed conditions.

The complexity of real environments causes time-and space-dependent amplitude variations (fading) that cannot be predicted or controlled in detail. These variations are mitigated by building spread-spectrum radios, providing diversity antennas, or using more complex directional arrays. The statistics of received signal amplitude can be used to provide rules of thumb for how much margin to allow over the minimum allowed power from the reported radio sensitivity to ensure a reliable link.

6. Further Reading

Waves and Wave Propagation

Physics of Waves, W. Elmore and M. Heald, Dover, 1985: *An elegant and readable examination of waves in all sorts of circumstances, mechanical, acoustic, and electromagnetic. Includes a thorough treatment of diffraction according to the classic approach employing Huyghens' principle.*

Adaptive Arrays for WLAN

Smart Antenna Systems and Wireless LANs, Garret Okamoto, Kluwer, 1999: *A detailed description of an experimental smart-antenna system, including both simulations and indoor and outdoor experimental results.*

Statistical Propagation Modeling

"The Indoor Radio Propagation Channel," H. Hashemi, Proc. IEEE, vol. 81, p. 943 (1993) and references therein: *This is a useful tutorial/survey summarizing the many approaches that have been taken to characterize amplitude and delay distributions, including a vast number of further citations into the literature.*

Chapter 7

Indoor Networks

1. Behind Closed Doors

Indoor environments are defined by walls, floors, ceilings, and, where applicable, windows and doors. Inside these constraining structures we find incidental obstacles such as the human occupants. The structural features and interior contents determine the propagation characteristics of indoor environments. Because major features don't change rapidly within a building, it makes some sense to give thought to their effects on propagation, but after thinking is done we shall certainly need to measure. Finally, buildings are occupied by people, often users of electronic equipment that can interfere with wireless local area networks (WLANs).

In view of these observations, we pursue an understanding of the RF side of indoor networks by first examining how buildings are built and the implications for propagation at microwave frequencies. After examining some surveys of signal strength for various sorts of facilities, we consider the properties of common sources of interference and conclude with some examples of tools to assist in setting up indoor coverage networks.

2. How Buildings Are Built (with W. Charles Perry, P.E.)

2.1. Some Construction Basics

We divide buildings into residential and commercial construction. Within commercial construction, we consider low-, mid-, and high-rise and very large structures.

Buildings are designed to do three things: stand up, serve the needs of their occupants, and burn slowly if at all. All other features of a building are subservient to these key requirements.

The first obligation of a building is to remain standing. This requirement is fulfilled in rather different fashion in residential or low-rise construction and in larger buildings. Residential buildings, and commercial buildings up to a few floors, rely on some of the building walls to carry part of the load of the roof and upper floors. Walls that bear such structural loads are known as *load-bearing* or *shear walls* (from the requirement to tolerate significant shear stresses without collapse); other walls serve merely to divide the interior space into convenient regions and are commonly known as *partition walls*. Shear walls are an integral part of a building structure, not subject to modification without risk to the structural integrity of the building. Partition walls are generally lightly built and lightly removed or moved. The exterior walls of a building are almost always shear walls; some interior walls may also be load bearing. (Note that interior columns may also be provided for additional support without excessive consumption of space.) The construction materials and techniques used in shear walls are different from those used for partition walls, particularly in commercial construction; their microwave properties are also somewhat different (see section 3).

Mid-rise and high-rise construction are based on load-absorbing steel or reinforced-concrete frames. In a frame-based design, the exterior walls bear no significant loads and are not required to provide structural support but merely to protect the interior from the weather and provide privacy to the occupants. Thus, a wide variety of materials may be used for the exterior walls of such buildings, constrained by material cost, construction cost, appearance, and maintenance requirements.

Commercial construction practices are generally similar the United States, western and central Europe, and much of modern Asia, with certain minor exceptions related primarily to the cost of local assembly labor relative to local materials. Residential construction practices differ rather more noticeably across the world, because they are more subject to the vagaries of culture and personal preference.

2.2. Low-Rise Commercial Construction (One to Three Stories)

Low-rise commercial buildings typically share the load of supporting the structure between exterior walls, interior shear walls, and interior columns. A representative structure is shown in Figure 7-1. The exterior walls were traditionally hand-laid brick or concrete masonry units (known as cinder blocks to former college students) with steel reinforcing bars. This approach is still used on small buildings or where assembly labor is inexpensive and easily available. In most modern construction, exterior walls are made of reinforced concrete panels, which are fabricated on site and lifted into place: this approach is known as *tilt-up* construction.

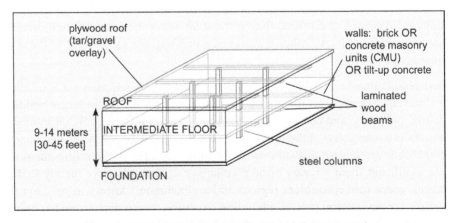

Figure 7-1: Typical Low-Rise Commercial Construction

A typical tilt-up panel is shown in Figure 7-2. Panels are fabricated by populating a wooden mold with reinforcing bars and then pouring concrete; the panels incorporate such major building features as openings for windows and doors and attachment points to facilitate connection to the cranes that lift them into place after cure. A single panel will usually form one complete exterior wall of a generally rectangular building.

The lowest floor of the building is often laid directly on top of a concrete slab foundation. Beams that provide strength for the intermediate floors are themselves supported by ledges built into the tilt-up panels and by saddles attaching them to interior support columns (Figure 7-3).

The floor beams for small buildings are generally wood or wood laminates, nominally 4 × 12 inches (10 × 30 cm) in cross-section, though slightly smaller in actual dimensions. The intermediate floors are typically constructed of panels that rest on the floor beams (Figure 7-4). Each panel is built of plywood (thin layers of wood laminated together with misoriented grain for improved strength), with intermediate nominal 2 × 4-inch wood support members (*joists*), hanging from metal joist hangars, between the main floor beams providing local support. The plywood floor may optionally be covered with a thin coating of lightweight concrete, typically reinforced with a coarse wire grid.

Intermediate floors form the ceiling of the story below them but rarely is this fact visible. A *false ceiling*, often constructed of pressed cellulose (i.e., paper) about 2–3 cm thick, is generally hung from metal T-bar supports attached to the floor beams to provide an

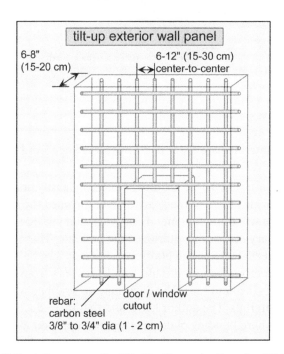

Figure 7-2: A Segment of a Tilt-Up Concrete Exterior Wall Panel

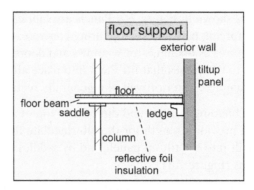

Figure 7-3: Intermediate Floors Are Supported at the External Walls and Internal Columns

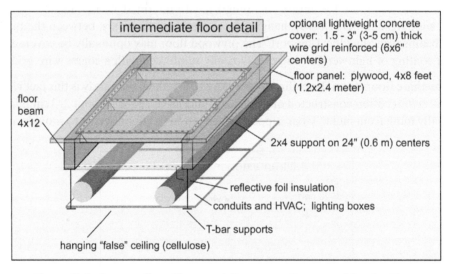

Figure 7-4: Intermediate Floor and Structures Supported by the Floor

aesthetically pleasing interior. These cellulose panels are easily popped out of their support frames to provide access to the space above. Between the false ceiling and the intermediate floor one will often find metal ductwork that provides heating, ventilation, and air conditioning services for the interior of the building. These ducts are typically constructed of thin sheet steel, though plastic ducting may also be encountered. Both round and rectangular cross-sections are used, and sizes vary from around 10 inches (25 cm) to 30–40 inches (around 1 m). The ducts may be wrapped with fiberglass or other thermal insulation materials. The author has found that these ducts are not always particularly close to where building plans say they are supposed to be; if in doubt, it is a good practice to pop out the ceiling panels and look. Fluorescent lighting panels, typically sheet metal boxes 0.5–2 m wide and 1–3 m long, also hang from the floor beams.

An alternative approach to construction of intermediate floors, with important consequences for microwave propagation, uses a thin corrugated steel deck covered with poured concrete to form the floor surface (Figure 7-5). Open-web steel joists may be used in this case instead of wooden joists, though the beams are still often wood laminates. As we discuss in section 3, conventional plywood/concrete floors cause a modest microwave loss (5–10 dB), whereas a corrugated steel floor in a low-rise building is a very effective shield, such that propagation between floors can be considered negligible. Thus, the type of floor used in a building is an important factor in determining whether neighboring floors represent common or distinct network coverage areas.

Interior shear (load-bearing) walls, if present, are typically constructed in the same fashion as exterior walls, using tilt-up concrete or hand-laid reinforced masonry. Interior partition walls are generally formed using gypsum (calcium sulfate dihydrate, $CaSO_4 \cdot 2H_2O$) wall boards assembled onto sheet-metal studs (Figure 7-6). Gypsum is popular because it is inexpensive, not readily flammable, and releases moisture upon heating. The gypsum is generally laminated with paper covering, and the wall board is usually painted or covered with patterned paper for aesthetic reasons. Studs are a few centimeters across and laid at 0.4-m (16-inch) intervals, so that they provide some

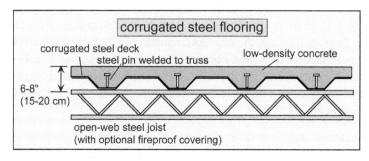

Figure 7-5: Corrugated Steel Intermediate Floor Construction

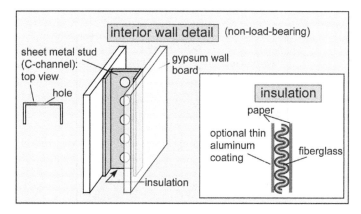

Figure 7-6: Interior Wall Construction, Commercial Buildings

scattering but little impediment to propagation at ISM or Unlicensed National Information Infrastructure (UNII) bands. Interior walls may be insulated with fiberglass or similar material, typically wrapped in a paper cover. The paper can be coated with a layer of vacuum-deposited aluminum or aluminum foil; roof insulation is always aluminized, but the use of aluminized insulation for exterior walls is less frequent and is optional for interior walls. The thickness of the aluminum appears to vary from about $6\,\mu m$ (deposited film) to as much as $50\,\mu m$ (foil).

Roof construction is similar to intermediate floor construction (Figure 7-7). Wooden-laminate beams and joists support plywood roof panels, on which is laid a waterproof covering. The latter is typically composed of a layer of protective felt covered with tar-and-gravel-treated paper in commercial construction, though conventional shingled roofs using asphalt, tile, or ceramic-coated steel shingles may also be encountered. Conduits and ducting for heating, venting, and air conditioning may be suspended from the roof in the manner used for lower floors. Roofs are almost always thermally insulated, and aluminized insulation is generally used.

2.3. Mid-Rise and High-Rise Commercial Construction

Buildings taller than three floors rarely rely on load-bearing external or internal walls. Instead, structural integrity is provided by a frame composed of welded or bolted steel girders (Figure 7-8). In some structures, reinforced concrete columns are used instead of or in addition to steel framing. A *shear tower* made of reinforced concrete is often present at or near the center of the building, providing key services such as plumbing access, electrical services, and the elevator shafts and stairways that allow user access. The shear tower provides a relatively fireproof and mechanically robust escape path in the case of emergency.

Because the external walls are no longer load bearing, a wide variety of materials may be used in their construction. Interior construction techniques and finishings are generally similar to those used on low-rise buildings; intermediate floor construction is also similar to that described in section 2.2.

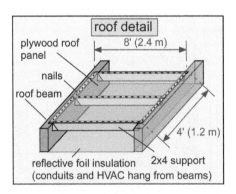

Figure 7-7: Roof Construction, Low-Rise Commercial Buildings

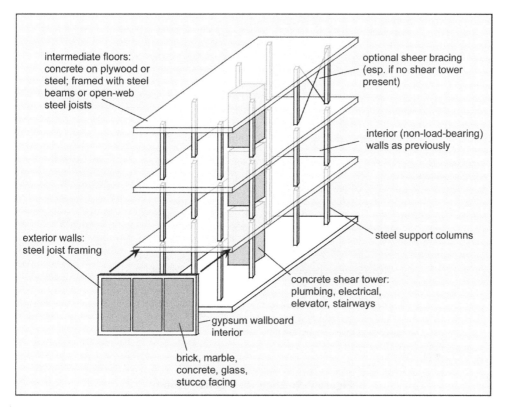

intermediate floors:
concrete on plywood or
steel; framed with steel
beams or open-web
steel joists

optional sheer bracing
(esp. if no shear tower
present)

interior (non-load-bearing)
walls as previously

exterior walls:
steel joist framing

steel support columns

concrete shear tower:
plumbing, electrical,
elevator, stairways

gypsum wallboard
interior

brick, marble,
concrete, glass,
stucco facing

Figure 7-8: Steel-Framed Mid-Rise Construction (3–15 Stories)

Very tall buildings—skyscrapers—use elaborations of the steel framing approach used in mid-rise construction (Figure 7-9). Very large steel beams greater than 1 m in extent are required at the base of large structures, tapering to a more modest size at the top. Beams in large buildings are always provided with fire protection, either using gypsum or concrete sheathing. Concrete columns, formed from high-density concrete poured around a core of steel reinforcing bars wrapped in a steel confinement structure, are used instead of steel I-beams in some structures. Interior partitions and intermediate floor construction are similar to those used in low- and mid-rise buildings.

Exterior walls of tall buildings are often dominated by glass window area. The glass is often coated (*glazed*) to control insolation and consequent heating of the building interior. A wide variety of films is used. Commonly encountered materials are zinc oxide (ZnO, an insulator), tin oxide (SnO_2, a semiconductor with typical conductivity of around 3–10 $\mu\Omega$-m or 300–1000 $\mu\Omega$-cm), Ag, Ti, silicon nitride (Si_3N_4, an insulator), porous silicon dioxide (SiO_2, also an electrical insulator), titanium nitride (TiN, a fair conductor similar to tin oxide), stainless steel, and titanium dioxide (TiO_2, a good insulator). An example coating structure uses 40-nm (400 Å) layers of tin oxide

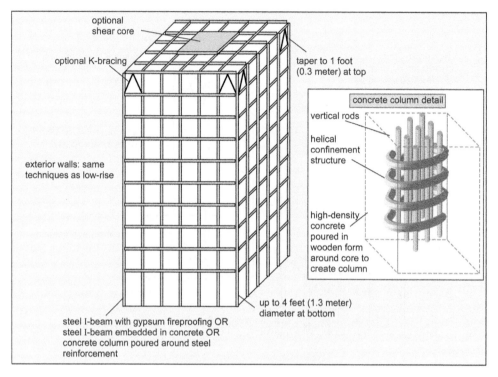

Figure 7-9: High-Rise Construction

separated by 10-nm layers of silver; thus, the whole structure is less than 140 nm (0.14 μm) thick, much less than a skin depth even for a good conductor. It is reasonable to expect that such windows are modestly reflective at microwave frequencies, but such thin layers should not result in strong absorption within the layer. The author is aware of anecdotal reports of glass coatings specifically designed to provide high absorption at microwave frequencies, purportedly used in airports and other facilities to impede cellular telephony coverage (and thus increase airport revenues through repeater charges), but has not been able to verify these reports to date.

2.4. Residential Construction

Residential construction practices around the world are more disparate than commercial construction, being more strongly guided by culture, history, and traditional preferences. The discussion here focuses on practices common in the United States.

In the United States, small single-family homes and small apartment buildings have traditionally been constructed using wooden frames for structural strength. Wood is also used for studs and rafters to support interior and exterior walls and roofing. Exterior walls are reinforced with plywood facing to improve shear strength of the framing. In seismically active regions like California, additional shear bracing of various kinds is used to ensure structural integrity in the event of an earthquake. Exterior wall materials are chosen for decorative value and robustness and vary widely. Wood paneling, stucco (about which we have a bit more to say in a moment), brick, and aluminum siding are all common. Brick may be laid unreinforced in seismically stable areas.

Interior walls are typically gypsum (*dry wall* or *wall board*) over wooden studs. Services (electrical wiring, plumbing, vents) are provided in the interwall spaces. Ceilings are generally constructed of plaster laid over gypsum wall board. Roofs are plywood laid on wooden beams and rafters. Roofs are generally constructed of shingles (individual tiles overlaid so that the point of attachment of the shingle is shielded from rain water). Shingle materials include asphalt mixed with gravel, treated wood, and metal-backed ceramic. Tar-and-gravel roofs are rarely used in residential construction (though they are encountered in "Eichler" homes common in the southern Peninsula region of the San Francisco Bay Area, where the author lives).

New U.S. construction is moving rapidly to sheet-metal studs and beams for interior and exterior walls, because the cost has fallen to below that of typical wood framing. The studs and beams are configured as C-shaped sheet metal channels as shown in Figure 7-6. Exterior walls are stiffened with metal straps or sheets. The total reinforced area is small because of the excellent tensile strength of metal beams. Metal plates are used at the windows to distribute lateral stresses. The exterior walls are covered with polycarbonate insulation followed by decorative coatings. Interior walls continue to use gypsum wall board to ensure fire resistance. Roof construction practices resemble those used in low-rise commercial construction, with plywood laid over sheet-metal rafters.

Plaster exterior wall construction has evolved somewhat over the last century (Figure 7-10). In the early part of the twentieth century, stucco exteriors were formed

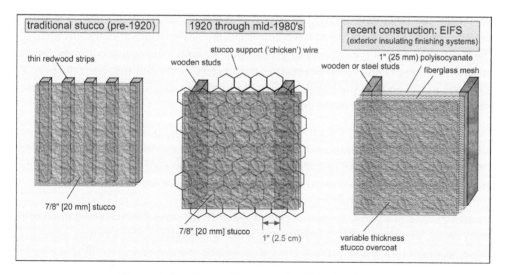

Figure 7-10: Stucco Through the (Modern) Ages

by laying plaster over thin strips of redwood. From about 1920 through the 1980s, it was more common to place a layer of hexagonal stucco support wire as reinforcing material for the plaster layer. In recent construction, *exterior insulating finishing systems* are generally used. These consist of a polyisocyanate insulating layer on which is laid a fiberglass mesh, forming a bed for the plaster layer. Wire support may still be used for stucco layers of thickness greater than 1 cm (3/8 inch).

2.5. An International Flavor

Significant differences in construction practices (and construction quality, but that's another story) exist around the world, conditioned mainly by the local costs of labor and materials. In Russia and Poland, exterior walls are masonry and fill, with masonry interior walls common, because of low local labor cost and excellent thermal insulation properties of masonry. Lightweight concrete walls are becoming popular as well. Intermediate floors and ceilings are often a composite of fired clay tiles, which are shaped to provide conduits for services and ventilation, and reinforced concrete.

Western European construction uses brick, masonry, and concrete in internal walls to a much greater extent than is typical in the United States, both in commercial and residential construction. Concrete masonry units are also very popular in Mexican construction.

Taiwanese commercial construction is very similar to U.S. practice. Japanese commercial construction uses reinforced concrete extensively for its robustness under seismic stress. Residential construction varies widely. Japanese homes are generally wood framed, often with heavy tiled roofs. Hong Kong folks generally live in large apartment buildings constructed along commercial lines.

2.6. Very Large Structures

Structures that must have a large unsupported interior span, such as auditoriums, airport terminals, theaters, and convention centers, face special structural problems. The basic challenge is the support of a roof span without intervening columns. Small structures can use *cantilevered* beams (the term refers to the fact that both the position and slope of the ends of the beam are fixed to minimize sagging), but large structures exceed the shear strength of available materials.

Various solutions for large-span roofs exist. Ancient engineers used masonry arches and domes to construct cathedrals and public buildings; these rounded structures effectively convert gravitational forces to compressive stresses in the roof and walls, which stone and masonry are well fit to resist, but at the cost of heavy expensive roofs. Modern solutions often use long-span *truss* structures, composed of triangles of steel wires or beams arranged in a plane, so that stresses applied to the truss are converted into tensile stress on the members of the constituent triangles. A similar approach uses a *space frame*, in which tetrahedra formed from steel beams are assembled to fill space with minimal-length structures. Again, shearing is not possible without tensile stress on the members of the truss. Another approach used for very large structures is to suspend a fabric roof from a network of support cables in the manner of a suspension bridge.

Tensile stresses from the weight of the roof must be transferred to the walls. Walls are formed of reinforced concrete or steel columns. In a technique known as *infill* construction, the space between the columns is taken up with non–load-bearing, decorative, weather-resistant materials such as brick, concrete masonry blocks, or glass blocks. Tilt-up construction is not practical for large structures, nor are wood beams used in modern practice.

3. *Microwave Properties of Building Materials*

From our examination of construction practices, it is apparent that the microwave properties of several common materials are needed to understand likely propagation results in indoor environments. We should like to know about the behavior of (at least) concrete, wood, gypsum, glass, and masonry.

A number of individual studies of particular materials are available in the literature, but the most complete single examination of building material properties at microwave frequencies the author has been able to obtain was performed by Stone and coworkers at the U.S. National Institute of Standards and Technology's Gaithersburg laboratories. Stone and colleagues measured attenuation after transmission through varying thicknesses of concrete with and without reinforcement, brick on concrete, concrete masonry units and brick walls, wet and dry soft lumber, plywood, drywall, and glass, all with a fixed test setup. Thus, the data set forms a valuable reference for relative attenuation of a wide variety of materials, measured in a consistent fashion.

There are some important limitations of the data that should be mentioned. No attempt was made to measure or correct for reflection from the samples, although the data were time gated to remove any reflections from the ambient around the test setup. Because the measurements were made in normal incidence, reflection coefficients of most materials are less than about 0.5 (see Figure 6-10), so the correction is of the order of 2 dB or less for a single interface. A complete correction would involve accounting for multiple reflections from both interfaces, but in the cases where this might be relevant, absorption is also high so that the correction is not of great significance. This correction is thus of modest import except in the case of samples with very little attenuation, where the distinction between absorption and reflection is of modest import in any case. Data were taken using separate systems in the 0.5- to 2-GHz and 3- to 8-GHz range; the low-frequency data appear consistent and sensible, but the higher frequency data show a complex frequency dependence and are quantitatively inconsistent with the lower frequency data. Thus, here we examine only the 0.5- to 2-GHz results, from which we must optimistically extrapolate to the frequency ranges of interest for WLAN applications.

The low-frequency results of Stone et al. are summarized in Figure 7-11. We can generally conclude that drywall provides minimal attenuation (on the order of 1 dB at normal incidence) and that dry plywood also has very little absorption or reflection. Wet plywood incurs an additional dB of attenuation (and/or reflection). Brick and soft lumber in typical thicknesses involve attenuations of 5–10 dB, dependent on thickness. Again, wet lumber incurs a decibel or two of additional absorption over dry lumber.

Concrete, by itself or in combination with brick, appears to represent a very strong attenuator, reducing the incoming signal by as much as 35 dB at 2 GHz for a 20-cm-thick layer. The presence or absence of reinforcement ("rebar") has a modest effect on concrete attenuation.

Results are also shown for grids of reinforcing bars in air, using 7- and 14-cm spacings. Recall that a wavelength at 1 GHz is about 30 cm and the rod diameter is about 19 mm, so $7 - 2 = 5$ cm holes are less than a quarter of a wave at 1 GHz. We see that half-wave grids provide little impediment to propagation, but quarter-wave grids are more significant. The reflection/scattering of the 7-cm grid extrapolated to 2.4 GHz (0.4 λ openings) is still 3 dB larger than that of the 14-cm grid at 1 GHz (also 0.4 λ openings), showing that the size of the steel rods is not negligible for the smaller grid. Nevertheless, we can conclude that conductive grids with openings significantly larger than half a

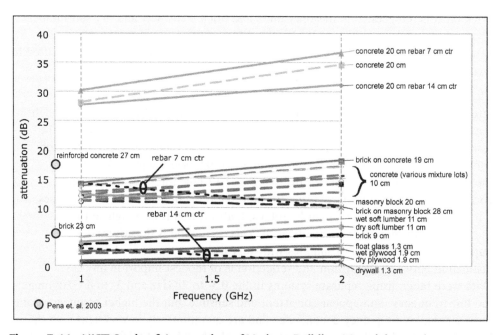

Figure 7-11: NIST Study of Attenuation of Various Building Materials at Microwave Frequencies; "Rebar" Denotes the Presence of Reinforcing Steel Bars Within Concrete or the Properties of Grids of Reinforcing Bars in Air (After Stone, NISTIR 6055)

wavelength at the frequency of interest introduce only a few decibels of reflection and scattering. Note that the effect of rebar spacing within concrete is much less than that which would have been expected if the bare grid attenuation was added to that of unreinforced concrete; the large refractive index of the concrete means the wavelength within the medium is less than that in air, thus making the openings larger in terms of local wavelengths.

Also shown on the graph are measurements of the properties of brick and concrete obtained by Peña and coworkers in a separate study reported in 2003. These measurements were performed at roughly 900 MHz. The researchers obtained and modeled angle-dependent data, with full correction for reflections, to extract the real and imaginary dielectric constants and thus reflection and absorption behavior of the materials. Their result for a 23-cm brick wall (about 6 dB absorption) is generally consistent with that obtained for a 9-cm brick wall by Stone and coworkers. However, their results for a reinforced concrete wall show much lower (albeit still significant) absorption than that reported by Stone for similar thickness.

Recommended attenuation values for transmission through various materials provided by Wireless Valley, Inc., a commercial organization involved in both propagation modeling and installation consulting for indoor networks, are summarized in Table 7-1. These recommendations are generally consistent with the measurements of Stone et al. for glass, wall board (recalling that one wall penetration encounters two layers of drywall), wood, and brick, although Peña et al.'s results for brick suggest rather lower absorption. Reifsneider's recommendations for concrete walls are generally consistent with the data of Peña et al. (recalling that a 27-cm wall is unusually thick for U.S. construction) but not with that of Stone et al. In the course of some related work, I measured the absorption of a typical tilt-up reinforced concrete wall 20 cm thick at only 5–6 dB at 5.3 GHz.

Thus, we can see that multiple sources report more or less consistent results for common materials, but large discrepancies occur in reported behavior of concrete walls. What's going on? To understand why different authors report such differing behavior

Table 7-1: Recommended Values of Attenuation for Various Interior Structures

Obstacle	*900 MHz*	*1.8 GHz*	*2.4 GHz*
Interior wall (drywall)	2	2.5	3
Brick, concrete, masonry block wall	13	14	15
Cubicle wall	1	1.5	2
Wooden door	2	2.5	3
Glass window	2	2.5	3
Glass window, insolation "doping"	10	10	10

From Indoor Networks Training Course, Reifsneider, September 2003.

for this material in particular, we must undertake a brief examination of the nature of this ubiquitous modern building material.

Concrete, more formally *Portland concrete cement* (sometimes abbreviated as PCC), is a mixture of Portland cement and *aggregate*. The aggregate is crushed stone or other nonreactive material and plays an important role in the physical properties of the material but is probably not particularly microwave active. The cement is a powder consisting of tricalcium silicate $3CaO \bullet SiO_2$, and dicalcium silicate $2CaO \bullet SiO_2$, with smaller amounts of tricalcium aluminate $3CaO \bullet Al_2O_3$ and tetracalcium aluminoferrite $4CaO \bullet Al_2O_3 \bullet Fe_2O_3$. When mixed with water, the tricalcium silicate forms a hydrated calcium silica gel, $3CaO \bullet 2SiO_2 \bullet 3H_2O$, and releases Ca^+ and OH^- ions $(Ca(OH)_2$, which is dissociated in aqueous solution). It is the hydrated gel that provides mechanical strength to the mixture after curing. The reactions that form the gel are complex, and the structure of the result is not well understood. It is well known that in the related process of formation of gels from pure silica, the nearly reversible hydrolysis of water by silica bridge bonds,

$$\equiv Si - O - Si \equiv +H_2O \Leftrightarrow \equiv Si - OH + HO - Si \equiv$$

(where $\equiv Si - O$ denotes a silicon atom attached to three other oxygen atoms, presumably within the bulk of the gel or particle), plays an important role in the formation of gel particles and in the redissolution and merging of particles to form an extended gel. It seems plausible to infer that an analogous set of nearly reversible reactions support the creation of a connected if porous gel from the initial powders during the curing of concrete. It is also well known from research into the interaction of water and silicon dioxide that water in the presence of silica can exist in three forms, distinguishable by their infrared absorption: free water molecules, hydrogen-bonded water, and silanol groups (Figure 7-12). The key relevance is that silanols, although still polar to a similar extent to water molecules, are not free to rotate in response to an electromagnetic potential. The vibrational frequencies of the attaching bonds are greatly in excess of microwave frequencies. Thus, silanol will not contribute to

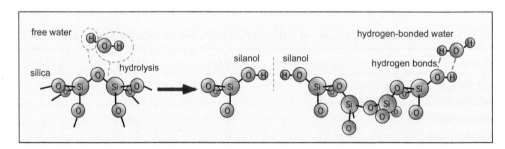

Figure 7-12: Interaction of Water With Silica: Silanol Groups Are Produced by Hydrolysis of Bridge Bonds and May Thereafter Form Hydrogen Bonds to Each Other or to Nearby Water Molecules

microwave absorption. Hydrogen-bonded water molecules are also relatively unable to move, attached as they are to the fairly rigid silica lattice, and will contribute little absorption at microwave frequencies. Thus, microwave absorption in concrete cement seems likely to be dominated by the residual free water molecules and should slowly fall as the concrete cures and water is taken up into bonded sites.

This supposition appears to be borne out by the microwave absorption data at 300 MHz obtained by Pokkuluri, shown in Figure 7-13. If we use Stone et al.'s data as a rough guide for the frequency dependence of absorption in concrete, we infer a change of about 20%/GHz, so that we should expect an absorption of about 5 dB/20 cm after 3 years of cure at 2 GHz, in qualitative agreement with the data of Dobkin and Peña (and perhaps also that of Wireless Valley, because they do not provide a reference thickness).

More to the point, we find that for samples cured some tens of days, we would expect an attenuation roughly two times higher than observed in samples cured for years, approximately accounting for the discrepancy between the results of Stone et al. and those of other workers. Stone et al.'s data include various mixtures of concrete incorporating differing amounts of the constituents and show significant (30%) variations in absorption resulting. The corresponding variation in fully cured concrete might be expected to be at least as large and perhaps larger, because absorption

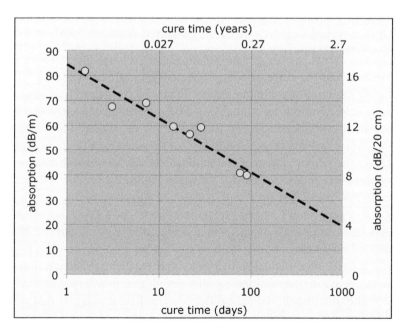

Figure 7-13: Microwave Absorption at 300 MHz in Concrete as a Function of Cure Time (Based on Data From Pokkuluri Thesis, Virginia Polytechnic Institute, 1998)

depends on the residual water available after hydration and thus on details of porosity and stoichiometry, likely to be influenced by initial composition and curing conditions. This supposition is supported by the data of Kharkovsky and coworkers taken on 15-cm-thick samples at 8–12 GHz. They found that two mixtures with water-to-cement ratios of 0.4 and 0.7 had absorption of 16 and 28 dB, respectively, after 6 months of cure. Kharkovsky et al.'s data also show slow continuous changes in absorption at long times: they found about 2 dB/month at 5–6 months of cure, roughly consistent with Pokkuluri's data at much lower frequency. Thus, we can conclude that a typical 20-cm-thick concrete wall is likely to represent about 5–10 dB of absorption at 2.4 GHz and perhaps 2–3 dB more at 5 GHz, with the exact value significantly dependent on composition and curing conditions.

The author has been able to find only one brief reference to plaster walls (Ali-Rantala et al.) that suggested the absorption of plaster is slightly less than that of concrete. Referring to Figure 7-10, one can estimate that 2 cm of stucco would represent 0.5–1 dB of absorption, comparable with that of similar layers of plywood or gypsum. The use of wire reinforcement of typical dimensions, noted in Figure 7-10, appears likely to represent little impediment to propagation at 5–6 GHz; the present author has verified that a layer of wire composed of 2-inch (5 cm) hexagons attenuates a 5.3-GHz signal by only about 1.5 dB, and 1-inch (2.5 cm) hexagons impose 3.5 dB attenuation. At 2.4 GHz, 2-inch hexagonal screen represented an almost-insignificant 1.4-dB loss, but 1-inch hexagonal screen caused a decrease of 7.1 dB in signal strength from open space. Because stucco walls are typically very thin, the dielectric effects on wavelength would be expected to be modest, so one might propose that a residential wall using stucco reinforced by the typical support wire can represent as much as 7–8 dB of transmission loss in the 2.4-GHz ISM band.

Refractive indices and the corresponding reflectivity at normal incidence for various building materials are summarized in Table 7-2. It is clear that most building materials have refractive indices around 2–3. The reflection coefficient of a real slab of material will vary depending on its thickness if the material absorption is small (e.g., plywood or glass or gypsum in typical thicknesses) due to multiple reflections from the two interfaces. Estimated reflectivity versus slab thickness over the range of refractive index expected is shown in Figure 7-14. For $n = 2$ appropriate for those materials most often used in thin slabs (wood, glass, and gypsum), the reflectivity is at a maximum $\Gamma \sim 0.6$ at typical thicknesses of 1.3 to 1.9 cm (0.5–0.75 inches) versus the normal reflectivity from a single interface of about 0.4. Note, however, that even this maximum reflection coefficient only corresponds to a loss in the transmitted signal of about 2 dB ($10 \log (1 - 0.6^2)$). Thus, reflection plays a modest role in decreasing transmitted signal strength except at glancing angles of incidence. On the other hand, the high reflection coefficients suggest that reflected signals will be common in indoor environments, leading to multipath delay and fading as discussed in Chapter 6.

Table 7-2: Refractive Index and Reflection at Normal Incidence (From a Single Interface) for Various Building Materials

Material	n	Γ (Normal Incidence)	Source; Remarks
Brick	2	0.33	Pena et al. op. cit.[1]; Landron et al. IEEE Trans Ant Prop 44 p. 341, 1996
Concrete	2.5	0.43	Pena et al. op. cit.[1]; similar to CRC Handbook value for $CaSO_4$-$2H_2O$ of 2.3
Glass	2.5	0.43	CRC Handbook values for Corning 0080, 0120 (soda-lime glass, soda-lead glass)
Coated glass		0.7[2]	Landron et al. op. cit.
Limestone	2.7	0.46	Landron et al. op. cit.
Gypsum	2.2	0.37	Tarng & Liu IEEE Trans Vehic Tech 48, no. 3, 1999
Wood	2.2	0.37	Tarng & Liu op. cit.

[1]Pena data allows a range of $n = 1.7-2.2$ for brick and 2.3–2.7 for concrete.
[2]Landron data measured on exterior (coated) side, presumably dominated by coating.

Some empirical data for attenuation resulting from transmission between floors are shown in Figure 7-15. Seidel et al. does not describe the means of construction of the floors studied, but by comparison with the data obtained by the present author, it is reasonable to infer that building 1 used concrete-on-wood or wood floors. The fact that loss does not increase linearly with the number of floors is ascribed to diffraction and/or

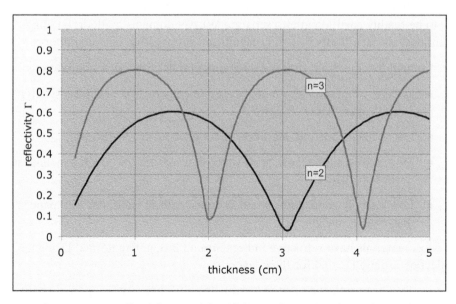

Figure 7-14: Reflectivity vs. Slab Thickness for Refractive Index $n = 2$ and 3; Slab of Refractive Index n Surrounded by Air ($n = 1$)

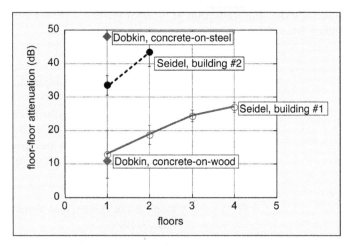

Figure 7-15: Attenuation Between Floors of a Building (Data From Seidel, IEEE Vehicular Tech Conf., 1992, p. 814, at 915 MHz; Unpublished Data by the Present Author at 2.4 GHz)

reflection from neighboring buildings, which are not simply related to the number of floors separating transmitter and receiver. As one might expect, a corrugated steel floor of the type depicted in Figure 7-5 acts as an effective shield, leading to negligible direct propagation between floors. Seidel and coworkers also reported higher losses (on the order of 35 dB for one floor) for a second building, suggesting that they were also working with corrugated steel flooring in some cases. Buildings with steel floors are likely to be readily partitioned into separate WLAN domains by floor; buildings with conventional wooden floors will allow significant propagation between floors.

We can summarize the discussion of building materials with a few observations:

1. Absorption of materials commonly used in construction at thicknesses typically used can be classified as follows:

    -   Lossy: thick concrete, masonry, or solid wood; 10–20 dB/wall

    -   Modest loss: single-layer brick walls, thin masonry, wooden paneling, stucco walls; 5–10 dB/wall

    -   Low loss: plywood, glass, wooden doors, cubicle walls, dry wall panels: less than 5 dB/obstacle

2. Most materials have refractive indices around 2–3, giving normal incidence reflection around $\Gamma = 0.4$ for thick layers.

    -   Thin layer reflection depends strongly on thickness, varying from about 0.1 to 0.6; however, even at the peak this represents modest transmission loss of a couple of decibels.

- Reflection loss will become significant for vertically polarized radiation on walls at angles of incidence more than 70 degrees.

- Reflections from floors and ceilings are subject to Brewster's angle for vertically polarized antennas.

Thus, the received power in most cases will be dominated by the loss and reflection along the direct path between the transmitter and receiver. A first estimate of signal strength (in the absence of large metal obstacles) can be obtained by counting walls along the direct ray. Glancing-angle wall reflections will subtract significant power from the direct ray and add to average power along long narrow corridors at the cost of increased fading.

4. *Realistic Metal Obstacles*

In Chapter 6 we discussed in some detail how the scattered potential from ideal thin metal obstacles interacts with the incident radiation to create a shadowed region of varying depth. In this section we examine how realistic metallic obstacles affect propagation and compare the results with theoretical predictions.

Measurements were performed at 5.216 to 5.228 GHz in an open room with a concrete floor using a microstrip array antenna with approximately 20 dB power gain relative to an isotropic antenna to minimize the effects of reflections from ceiling features, walls, and floor on the results. Two obstacles were tested: a single 19-inch rack 0.65 m wide, 0.56 m deep, and 2 m high and a triple 19-inch rack 1.8 m wide, 0.85 m deep, and 1.6 m high. (These racks are very commonly used to hold server computers and other electronic gear in commercial and industrial facilities and provide representative examples of typically sized metallic obstacles.) The racks were unpopulated and thus consist of frames with sheet-metal front doors perforated by ventilation slots and a mostly open back side with support members crossing the access area at the top and center height. The test arrangement is depicted schematically in Figure 7-16.

The signal strength at four frequencies was averaged and compared with that obtained in the unobstructed case to produce a measure of shadow depth in decibels for each

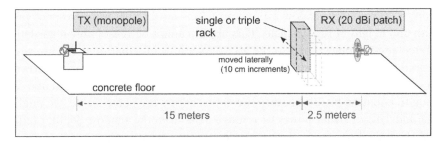

Figure 7-16: Test Range for Diffraction/Shadowing From Realistic Metallic Obstacles

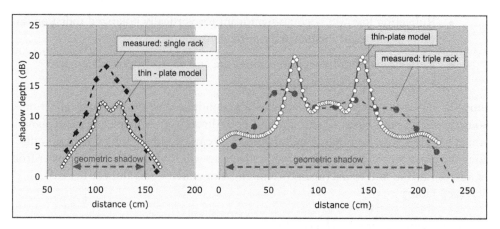

Figure 7-17: Shadow Depth vs. Obstacle Position for Single and Triple Racks; Average Over Geometric Shadow: 11.0 dB (Single Rack) and 10.1 dB (Triple Rack)

position of the obstacle. The results are summarized in Figure 7-17 and compared with estimates of signal strength obtained by numerical integration of the scattering equations for ideal flat plates of the same lateral dimensions.

It is apparent that the qualitative features of the shadows cast by the racks are well represented by the thin-plate models. The narrow deeply shadowed regions expected from the flat-plate model of the triple rack are not observed in the data; it is not clear if this is an artifact of the limited lateral resolution of the measurements.

In normalized terms (equation [6.34]) the single rack is $u_o = 2.4$ wide at 5 GHz for the given configuration ($R_{av} = 2.1$ m) and can be reasonably regarded as tall. Equation [6.38] predicts a shadow depth half-way from the center line of 8 dB, in reasonable agreement with the observed shadow averaged over the geometrically shadowed region. Note, however, that the shadow depth in the central region is 5–7 dB deeper than expected for a flat plate and almost 10 dB deeper than the estimate of equation [6.38], presumably due to the complex actual geometry and finite depth of the rack.

The triple rack is about 7.4 normalized units wide by 6.6 normalized units high. Equation [6.38] provides shadow estimates of 18.5 and 17.5 dB using, respectively, the width and height of the obstacle. This object cannot be reasonably regarded as a tall rectangular object, and so the approximations of equation [6.38] are suspect. A plausible approximation is to add the received power from two ideal shadowing objects (one very wide and one very tall), with each typical shadow calculated according to equation [6.38]; because the two shadows are comparable, we might as well just add 3 dB of signal power or remove 3 dB from the shadow depth, predicting a typical shadow of about 15 dB, slightly deeper than what is observed half-way from the edge.

We can summarize by saying that the observed shadows of complex objects of realistic dimensions are on the order of 10–20 dB deep, as we had predicted based on simple idealized models. The detailed behavior of obstacles of finite depth is not accurately predicted by simple flat-plate models, but the differences are comparable with fade margins of ±5 dB that must be assigned to all estimates of field strength in any case. In practice, the effective shadow depth of obstacles on average will be reduced by the added power due to reflected rays following indirect paths within the indoor environment.

5. Real Indoor Propagation

We've now reviewed the sort of materials and obstacles we might encounter in an indoor environment and what their likely effects on a microwave signal might be. What does all this imply for operating a radio in an indoor environment? We first look at a simplified example of indoor propagation to get some idea of what to expect and then compare the results of the conceptual exercise with real data obtained in real buildings.

In Figure 7-18, the indoor propagation cartoon first introduced in Figure 6-2 reappears, this time with some exemplary propagation paths (rays) added. Along each path we've added a simplistic loss budget, accounting for 3 dB (reflection and absorption) in passing through interior partition walls, 5- to 7-dB loss upon reflection depending on angle, and 15 dB from thick concrete exterior walls. At the end of the process, each ray has accumulated a certain path loss due to distance traveled (not shown in the figure)

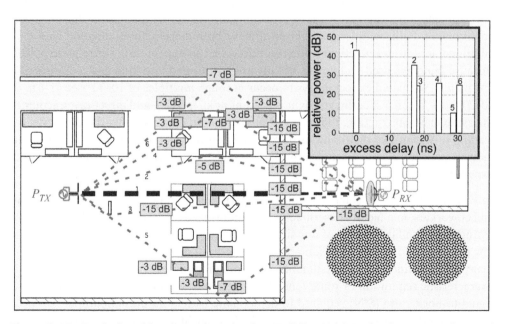

Figure 7-18: Typical Multipath Propagation in a Building With Reflection, Scattering, and Absorption Losses; Power (Arbitrary Units) vs. Delay Is Shown in the Inset

and a certain additional delay relative to the direct ray; the result can be plotted on a graph of power versus excess delay, assuming a typical room size, as is done in the inset. (Note that the delay here is much too imprecise to be used to specify the relative phase of the various ray paths, so we can add powers to get an average power, but we can't specify an actual power with any confidence.) Even though this example appears somewhat complex, it is hardly realistic: only a few reflected rays and one scattered ray are included, and we have ignored multiple reflections and reflections from the floor and ceiling or obstacles therein.

From the inset graph, we can see that the direct ray is the largest contributor to signal power, even though the configuration is hardly line-of-sight (LOS). This is because the large concrete shear wall attenuates the direct and indirect rays equally. The lower power rays will contribute both to fading, as their relative phase varies, and intersymbol interference, the latter being determined by their delay relative to a symbol time. Here, because we have included only a few rays in a small pair of rooms, the maximum delay of 30 nsec would be unobjectionable for 802.11b (recall from Chapter 3 that the chip time is 90 nsec and the highest modulation used is quaternary phase-shift keying): the delayed rays would muddy the symbol transitions but have little effect in the middle of the symbol where sampling is typically done. However, we can see that higher symbol rates and higher modulations might start to encounter significant intersymbol interference even in this small propagation example.

From this simple example, we tentatively conclude that in a real environment we expect to see a number of rays of decreasing relative power, with characteristic delays set by the path length between the various reflecting and scattering features in the environment. Paths with longer delays will have both more path attenuation and additional loss from reflections and absorptions and so appear at lower relative power. Because buildings used by humans usually are partitioned into human-sized areas of a few meters, we would expect delays to occur in rough multiples of 10–20 nsec (3–6 m or 10–20 feet). When a clear line of sight is present or all indirect and direct rays encounter the same main obstruction, the received power will be dominated by the direct ray, leading to Rician-like distributions of received power (Chapter 6, section 4); when an indirect path suffers the least absorption, we might see many rays of equal power and expect Raleigh-like behavior.

In Figure 7-19 we show some examples of real indoor propagation data, from Hashemi and coworkers, in a facility described as a medium-sized office building. The qualitative features are similar to those of our simple example, though we see many more rays at low power and long delay, presumably resulting from multiple reflections. There are many fairly distinct peaks, presumably corresponding to discrete ray paths, with characteristic separation of some tens of nanoseconds. Figure 7-19 also depicts the signal-to-noise ratio (S/N) required to demodulate 64 quadrature-amplitude-modulation (QAM) symbols (see Table 2-2). In clear line of sight, the power is obviously dominated by the direct ray, and a 64QAM symbol would see a multipath

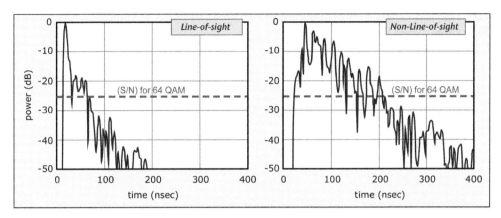

Figure 7-19: Indoor Power-vs.-Delay Data; Medium-Sized Office Building, 5 m TX-RX (From "The Indoor Radio Channel," H. Hashemi, Proc. IEEE 81, p. 943 (1993), Data by David Tholl of TR Labs); based on original image ©1993 IEEE

delay of about 60 nsec. The non-LOS measurement shows about 18 dB attenuation of the direct path relative to the best indirect path; in this case, because of attenuation of the direct ray, many more indirect rays are comparable in power with the largest ray, and a 64QAM symbol would need to deal with delays up to about 200 nsec. We can also see why Rician-like behavior would seem reasonable in this case for the LOS received power, where the received power is dominated by the direct ray with three or four small rays about 20 dB down, and Rayleigh-like behavior would be reasonable for the non-LOS case where about 11 roughly equal rays (counting down to 20 dB relative attenuation or an amplitude of 0.1) make contributions to the received power. The LOS result has a beam at 15-nsec delay about 18 dB larger than the largest delayed beams, corresponding to a Rician v_m of about 10σ. A gradual decrease in beam power extends out to about 200 nsec, after which very little power is observed: One might guess that this corresponds to the size of the building in which the measurements were taken, with longer delay times resulting only from multiple interior reflections or objects beyond the building walls and thus significantly attenuated.

To provide some idea of how delay varies over several different facilities, Table 7-3 shows summarized data at 1900 MHz from several different buildings, measured by Seidel and coworkers. They found characteristic delays of around 100 nsec, similar to the average behavior seen in Figure 7-19. The longest RMS delays, as large as 1500 nsec, were observed when transmitter and receiver were on differing floors; the authors suggested that the long delays are the result of reflections from neighboring buildings.

The distinction between LOS and non-LOS conditions is not as clear-cut as it may appear. It is apparent from the discussion of section 3 that an interior partition wall made of

Table 7-3: RMS Delay Spread and Variation in Received Power at Numerous Locations in Three Large Office Buildings

Building	Median RMS Delay Spread (nsec)	Maximum RMS Delay Spread (nsec)	No. of Locations
1	94	440	91
2	77	1470	83
3	88	507	61

From Seidel et al. (Op. Cit.).

gypsum wall boards has little effect on signal strength, so that an LOS path may exist between neighboring rooms for microwaves even though people cannot see through the relevant walls. On the other hand, as the path length increases, the direct ray becomes less dominant and more variation in signal strength and effective delay may be expected, even when an unobstructed path exists between the transmitter and receiver. This effect is demonstrated in Figure 7-20, where received signal strength is shown for measurements performed in a large nominally open area, with a 5-m-wide unobstructed path from transmitter to receiver in all cases. The average received power shows deviations of as much as 8 dB from the free-space prediction for the direct ray. At long distances, the average power appears to be increasing over the free-space prediction due to added power from the floor reflection. The minimum signal as a function of frequency is often much lower than the average power, as shown in detail in the inset at right; locations with lower average power have more variations in power with frequency, suggesting that in those locations near cancellation of the direct ray by reflections from nearby obstacles enables rays with much longer path lengths to significantly contribute to the received power. For example, at 9.15 m, variations of as much as 18 dB are seen in signal power versus frequency. Note that the lateral resolution of the measurements is inadequate to show small-scale fading that may result from, for example, counterpropagating reflected rays from the walls of the room.

The distribution of average signal strength that results from these complex interactions with the indoor environment is shown in Figure 7-21 for a typical small wood-framed residence and in Figure 7-22 for a moderate-sized two-floor tilt-up office building with concrete-over-plywood floors. These data were obtained using an uncalibrated radio card, so relative signal levels may be meaningful but absolute signal power should not be taken too seriously. In the residence data were taken at 1- to 2-m intervals and in the larger building at roughly 5-m intervals; the contours are a smooth interpolation of the measured data points. The reported values at each location average a number of packets, with the measurement position varied over a few tens of centimeters.

Turning first to Figure 7-21, in the region where only one internal wall interposes itself between transmitter and receiver (the horizontal dotted arrow), the signal strength is

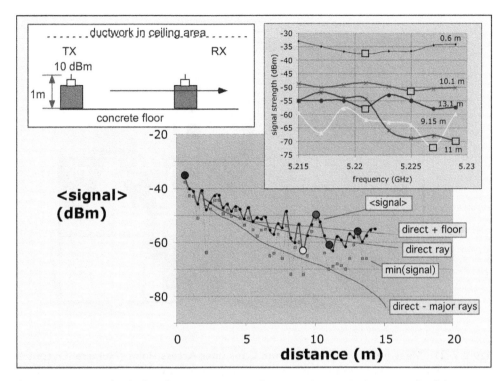

Figure 7-20: Received Signal Power Averaged Over 5.215–5.228 GHz and Minimum Signal Power Over Frequency vs. Direct Path Distance Compared With Free-Space Propagation of Direct Ray, Sum of Direct and Floor Reflection Power, and Worst-Case Power Where Major Rays (Floor, Walls, Ceiling) Subtract From Direct Ray; Insets Depict Measurement Configuration and Power vs. Frequency at Several Path Lengths (Indicated by Enlarged Symbols in Signal Power Plot)

quite close to that expected in free space. In directions where multiple internal and external stucco walls lie on the direct path (the inclined dashed arrow to bottom right), the detected signal strength is as much as 40 dB below the free space value. If we account for the fact that in this direction the direct path may involve as many as three stucco walls (8 dB each) and four internal partition walls, each involving two layers of dry wall ($8 \times 2 = 16$ dB loss, for a total of about 40 dB excess loss), simple wall counting leads to results in reasonable agreement with observed signal strength. The deep shadow at −91 dB from a milliwatt (dBm) is plausibly accounted for by a large metal obstacle—a central-heating furnace and associated ductwork. A first approximation to the distribution of signal strength in real complex environments is obtained just by counting losses along the direct path, allowing for diffraction.

Figure 7-22 shows similar data for a larger commercial building. On the upper floor, where the access point is located (at about 2 m height above the floor), signal strength through much of the measured area is remarkably close to that expected for free space

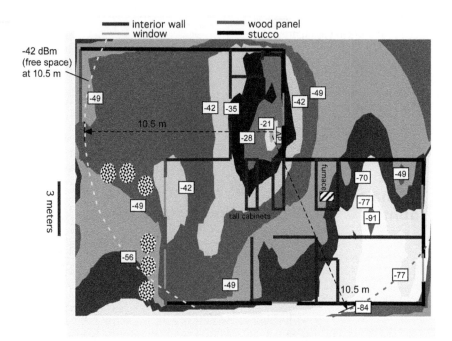

Figure 7-21: Measured Signal Power From Consumer Access Point (Nominal Output ≈ 14 dBm, Marked AP) in Wood-Framed Residence; Boxed Numbers Denote Value at Contour Edge and Dashed Lines Show Circle of Radius 10.5 m From Access Point, at Which Estimated Power for Free Space Propagation Is −42 dBm

propagation (to within the limitations of this simplistic measurement). We can infer that the cloth-and-pressboard cubicles that occupy most of the upper floor constitute little impediment to propagation at 2.4 GHz, despite the sheet-metal shelves they are equipped with. The region at upper right of the upper floor shows reduced signal strength due in part to measurement points located within rooms contained in partition walls (appearing as regional reduction in signal strength on this low-lateral-resolution contour plot) and measurement points at which the direct ray passes through external walls.

The lower floor, which is also occupied by cubicles on the right and only some exposed conduit on the left, is also close to free space once one corrects for a floor loss of about 9 dB (measured separately). There is some suggestion of enhanced signal strength near the lower wall, perhaps due to reflections from the partition walls on the upper floor or the concrete external wall. Signal strength is reduced at the upper right where again interior partition walls and exterior walls interpose themselves onto the direct ray from the access point to the measurement point. The interior concrete shear wall at top left of the lower floor leads to modest additional attenuation. Thus, we can generally conclude that attenuation and obstacles along the direct ray once again provide

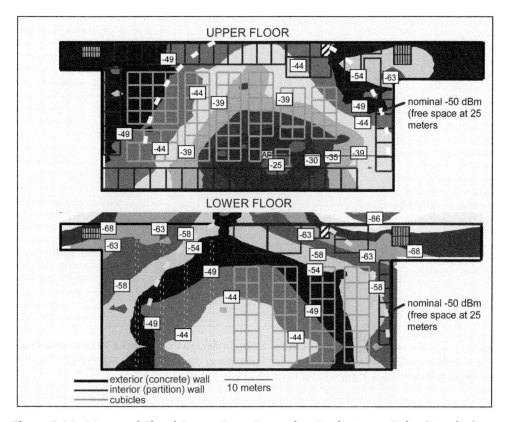

Figure 7-22: Measured Signal Power From Enterprise-Grade Access Point (Nominal Output ≈100 dBm, Marked AP) in Tilt-Up Two-Floor Commercial Building; Boxed Numbers Denote Value at Contour Edge and Dashed Lines Show Circle of Radius 25 m From Access Point, at Which Estimated Power for Free Space Propagation Is −50 dBm

decent guidance to signal strength. A caution to this sanguine conclusion is the deep minimum (−86 dBm) at top right, which is not in line with the elevator shaft (striped box) or any other known metallic obstacle and is a bit of a mystery. (Complete access to all the rooms and interior wall spaces was not available.) Incomplete knowledge of building features, a likely circumstance in many cases, provides a limit on one's ability to model the results: measured data are always needed to complement theory.

Figure 7-23 depicts survey data for a much larger facility (a convention center). Convention halls are typically very large open spaces with concrete walls and floors and ceiling height of more than 10 m. There are few significant obstacles to propagation in most cases within the open area, because display booths are typically formed mostly of plastic panels with plastic or metal rods for mechanical support, and in any case do not exceed a few meters in height. Note, however, that convention halls may be filled at

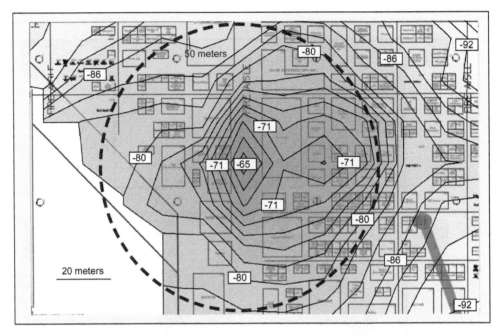

Figure 7-23: Survey of Signal Strength for a Single Access Point on the Trade Show Floor of a Convention Center; Estimated Signal Strength in dBm at Contour Edge Depicted by White Boxes (Image Courtesy of Jussi Kiviniemi, Ekahau Inc.)

floor level with mobile microwave absorbers—human beings! For good coverage, an access point ought to be placed several meters above floor level to allow the direct rays to clients to pass through only a few people or obstacles.

In this case, the exact location of the access point being monitored is not known and may be on a separate floor from that surveyed, so overly detailed examination of the data is inappropriate. However, it is clear that a region some tens of meters on a side is covered at a very reasonable signal strength of -75 dBm, and signal strength over much of the surveyed area exceeds typical low-rate sensitivity for 802.11 radios and will provide some sort of coverage. Indoor regions of more than a hundred meters in extent can be covered with a single access point.

Before we leave the subject of indoor propagation, it is worthwhile to add a few brief remarks on a subject we have heretofore rigorously ignored: *propagation exponents*. In free space we know that the received power will decrease as $(1/r^2)$, where r is the distance between transmitter and receiver. It would certainly be convenient if we could model indoor propagation using the Friis equation in exactly the same fashion, just using a different value for the exponent of r—that is, it would be nice to say indoors that received power is proportional to $(1/r^n)$ where n is not necessarily 2. A great deal of work in the literature has been devoted to deriving values of n that fit a particular data set. However, wide variations in the value are seen between different researchers and different data sets: values as small as 1.8 and as large as 5 have appeared in the literature.

It is the opinion of this author that such work resembles the old joke about looking for your wedding ring under the lamp post not because you lost it there but because the light is better. It would be nice if indoor data fit a modified exponential model with a substantially invariant value of n, but it doesn't. It's quite possible to see why: the dependence of signal on distance is inextricably tied up with what obstacles are encountered as the receiver moves away from the transmitter and thus uniquely dependent on the physical arrangement and construction of a given facility. Trying to force fit such particularized data sets into a modified distance dependence is like trying to evaluate the probability of checkmate in a chess position based on measuring the distance from your queen to your opponent's king with a ruler: the measurement is perfectly valid and quite irrelevant. Indoor propagation cannot be understood in the absence of knowledge about what is inside the doors. Although it might be possible to specify useful path loss exponents for specific types of room configurations (e.g., drywall partitions spaced 3 m apart with normal incidence), it seems to me much more sensible to start with free space and count walls and other obstacles that subtract from signal strength than to introduce a functional form that has no physical basis.

6. How Much Is Enough?

Now that we have examined how much signal power one might find available in various sorts of indoor environments, it behooves us to ask how much do we need? The basic answer is that the amount of signal power needed depends inversely on the data rate that is desired. Recall from Chapter 2 that the (S/N) needed increases as the number of bits per symbol increase and that the absolute noise level is proportional to the bandwidth and thus increases if we increase the rate at which symbols are transmitted. The lower limit on noise is the thermal noise entering the system, modified by the excess noise of the receiver as determined by its noise figure. For bandwidths on the order of 10 MHz, as used in WLAN systems, the thermal noise is around −104 dBm, and a typical receiver noise figure of 5–8 dB (see Table 4-9) means that the effective noise floor is about −98 dBm for the radio chip. Adding a few decibels to account for board and switch losses, we might expect a noise floor of −95 dBm. To demodulate binary phase-shift keying reliably, we need a S/N of about 9 dB (see Table 2-2), but recall that for the lowest data rate of 802.11 classic (1 megabit per second [Mbps]), the receiver also benefits from around 10 dB of processing gain, because the true bandwidth of the signal is an order of magnitude less than the received bandwidth. Thus, it seems possible to achieve sensitivities in the mid-90s for 1 Mbps. To demodulate 64QAM, used in the highest rate modes of 802.11a/g, requires an (S/N) of around 26 dB and does not benefit from any spreading gain, so ignoring coding gain we might expect sensitivities of $(-95 + 26) \approx -70$ dBm at the highest rate of 54 Mbps.

Table 7-3 provides published sensitivities for a selection of commercial 802.11 radios, collected through early 2004. The results are in rather good agreement with the primitive estimates of the preceding paragraph. At low rates, the best sensitivity is -94 dBm and -91 dBm is a typical number, whereas at 54 Mbps sensitivity of -65 dBm is typical. Figure 7-24 shows actual measured data rate versus estimated received signal power for a number of commercial 802.11 radios. (The data rates reported in Figure 7-24 are those delivered from one TCP client to another across the wireless link. TCP, *transmission control protocol*, is the protocol used in Internet data transfers and delivers packets across a data link using *Internet protocol*, which in turn communicates with the Ethernet drivers that ship bits into the 802.11 clients. Thus, the reported rate includes delays due to overhead within the 802.11 packets and the Ethernet transport mechanism and reflects the true rate at which data can be moved across the link, very roughly 60% of the nominal peak rate for a given 802.11 transport mode.) The signal strength here is estimated from single-frequency measurements at corresponding positions and should be presumed accurate to no better than ±3 dB. The data rates behave more or less as one would expect from the sensitivities: orthogonal frequency-division multiplexing data rates begin to fall for signal power less than -65 dBm or so as slower modulations must be used, and rates extrapolate to zero at received powers in the mid-90s.

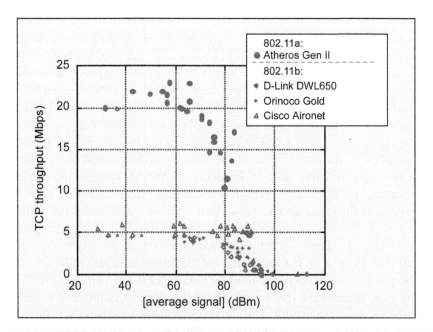

Figure 7-24: TCP Throughput vs. Signal Power (Estimated From CW Measurements) for Several Commercial 802.11 Radios (Slightly Edited Version of Data Published in "Correlating Link Loss With Data Throughput," Dobkin, High Frequency Electronics, January 2003, p. 22, by Permission of WJ Communications)

Reported data and measurements support the rule of thumb advanced by Jesse Frankel of AirMagnet that a received signal strength of about -75 dBm should be adequate to provide high throughput from most commercial 802.11b equipment, with sufficient margin (5–10 dB) to accommodate typical local fading and random variations in signal power. More aggressive guidelines can be used if control over the client population is available. By reference to Table 7-4 one finds that the reported sensitivities vary by as much as 7 dB, so the use of only high-quality clients and access points should allow perhaps another 5 dB of margin. However, system administrators wishing to follow such a path should recall that users are typically enamored of the client they have and would much prefer having it work adequately at their preferred locations than switching to a different client (which after all may create hardware and/or software problems for them). Using these guidelines, we can infer from Figures 7-21 to 7-23 that a single access point with typical output power around 14–16 dBm can be reasonably expected to cover a small residence and that a 100-mW (20 dBm) access point can cover a 25-m radius in an open industrial facility or perhaps 50 m in a large open room. Coverage is strongly affected by obstacles encountered by the direct ray, as noted in the previous section. In the sort of open-area cubicle environment often encountered in modern commercial facilities, the access point should be located high enough from the floor to avoid most obstacles and partitions for optimal coverage.

Note also that providing ample power at the corners of an open room with exterior walls using a high-powered centrally located access point will also probably result in

Table 7-4: Published Sensitivity for Various Commercial 802.11 Radios

Radio	Sensitivity at 54 Mbps (dBm)	Sensitivity at 11 Mbps (dBm)	Sensitivity at 1 Mbps (dBm)
Orinoco AP/Gold NIC card		−82	−94
Cisco Aironet 350		−85	−94
D-Link DWL900 AP		−79	−89
D-Link DWL650 NIC		−84	−90
D-Link DCF650 cf		−80	−88
D-Link DWLAG650	−73	−91	−94
Proxim 8550 AP		−83	−91
Surf'n'sip EL2511		−87	−95
Bewan USB11		−80	−88
Bewan 54G	−65	−80	
Trendware TEW403	−65	−80	
Senao 2511		−83	−91
Eazix EZWFDU01		−85	−93
Eazix EZWFM5-02	−65	−80	−87
Summary	−67 ± 4	−83 ± 3	−91 ± 3

significant leakage to the outside world (e.g., see Figure 7-21): if security is an important issue, the use of multiple low-power access points, preferably with inward-pointing directional antennas, is preferred. Such inward-looking access points mounted at the corners of a building, combined with a few centrally located access points, can achieve excellent high-rate coverage in the interior of a large building while keeping exterior leakage very small. Radiating cable can be used to provide supplementary coverage in corridors near exterior walls and for partitioned offices whose walls will cumulatively block coverage from the corner-mounted access points. (Even better security, as well as esthetic and environmental benefits, result from providing extensive evergreen foliage in the areas surrounding the building in question, so that would-be eavesdroppers in distant locations have no clear line of sight. We discuss the absorption due to foliage in more detail in Chapter 8.)

7. Indoor Interferers

7.1. Microwave Ovens

WLANs and wireless personal area networks mostly operate in the unlicensed bands around 2.4 and 5 GHz. In the United States, radios operating in these unlicensed bands are obligated under Federal Communications Commission (FCC) regulations to accept any interference they encounter, but the users don't have to be happy about the results. What are the likely sources of such interference?

High on everyone's list of suspects is the humble microwave oven. Though we spent some time in Chapter 6 debunking the myth that 2.4 GHz plays some unique role in the physics of liquid water, it is remarkably unlikely that microwave ovens will start to use other frequencies, given the huge installed base, vast experience, and the cost-driven nature of consumer markets. Therefore, WLAN users who choose to operate at 2.4 GHz must live with them when they are present. Note that in the United States the FCC does not regulate microwave ovens; they are, however, subject to Food and Drug Administration restrictions that require less than $1\,mW/cm^2$ of equivalent radiated power be detected at 5 cm distance from any location on the oven surface.

Averaged spectra and time-dependent spectra for a representative microwave oven interferer are shown in Figure 7-25. Emission peaks at around 2440 MHz, but it is apparent that significant emission is present across about half of the ISM band. Emission is seen to be sporadic in time. The likely explanation is that the oven RF power is actually on for roughly half of each AC cycle of 16.7 msec (making for a simpler power supply). At a scan rate of 500 ms/scan for 500 MHz, the spectrum analyzer spends only 1 msec in each 1-MHz region, and thus on any given scan the analyzer may record peak RF output, reduced power, or nothing at all, even though the average oven peak power is constant from cycle to cycle.

Figure 7-26 shows measured power emitted from a number of standard commercial microwave ovens at distances from 1 to 5.5 m, corrected to a standard distance of 5 m

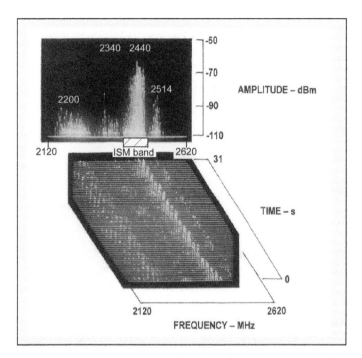

Figure 7-25: Emission Spectrum Due to a Microwave Oven Averaged Over Time (Top) and vs. Time (Bottom, 0.5-sec Scans) (From "Literature Search and Review of Radio Noise and Its Impact on Wireless Communications," Richard Adler, Wilbur Vincent, George Munsch, and David Middleton, US Naval Postgraduate School, For FCC TAC, 2002)

assuming free-space propagation. As can be seen in Figure 7-25, the emission spectrum of a microwave oven is complex and contains numerous peaks; the data in Figure 7-26 include only a few of the highest peaks for each oven. Note that different ovens do not have the same power spectrum, though all share the same general features of strong emission around 2.45 GHz. These data also show much stronger emission around 2.35 GHz than seen in Figure 7-25 for several ovens (though that is outside the nominal ISM band and hopefully of little import to 802.11 users).

Recall from Table 3-1 that Institute of Electrical and Electronic Engineers (IEEE) 802.11 WLANs divide the ISM band into 14 5-MHz channels centered from 2.412 to 2.484 GHz, though only channels 1, 6, and 11 are normally available as nonoverlapping channels in the United States. From Figure 7-26 it is clear that most ovens will have a modest effect on the low-numbered channels but will interfere significantly with channels 6 through 11. Fortunately, the sporadic nature of microwave oven interference suggests that only modest effects will be felt for short packets. This supposition was confirmed in the work of Chang and coworkers, who found that the same microwave oven interference that caused packet error rates of 30–40% in 400-byte packets for

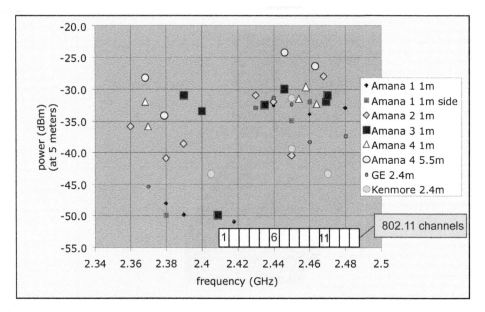

Figure 7-26: Measured Power at Selected Frequencies for Six Conventional Microwave Ovens; Measurements at 1–5.5 m, Using 2.4-GHz Quarter-Wave Monopole 1 m From Floor, Acquired With a Spectrum Analyzer in Peak-Hold Mode, Corrected to 5 m Assuming Free-Space Propagation

distances up to 20 m from the oven caused less than 5% packet errors in 100-, 200-, and 300-byte packets even when the receiver was only 4 m from the oven. Chang et al. describe their equipment as 802.11 direct-sequence spread spectrum (DSSS); if we assume they were using an 802.11 classic link at 1 Mbps, a 400-byte packet is roughly 3.3 msec (the exact value is unclear, because the authors have not recorded whether the 400-byte total covers only data or includes headers). It is difficult to reconcile the measured 50% duty cycle of oven radiated power (e.g., see Unawong et al.) with the sudden threshold for bad packets they found above 300 bytes, but we can still qualitatively conclude that short packets have a good chance of finding their way around oven interference pulses. Because much higher data rates use shorter packets, it is reasonable to expect that their resilience to oven interference will be good. For example, at 11 Mbps, a maximum-length 1500-byte packet will be only a bit more than 1 msec in duration; at least half of the packets will encounter no interference. Note that when interference of this type is limiting performance, backing off to lower rates may actually increase packet error rate. Most control interfaces provide a mechanism for the user to adjust the threshold for packet fragmentation (sometimes known as interference robustness), which is the proper response to problems with oven interference.

Chang and coworkers report high bit error rates out to about 22 m from the microwave oven for 400-byte packets, with the radius of disturbance shrinking to about 17 m when a concrete wall was interposed between the oven and the receiver. Unfortunately, they provide no information on transmitter position or received signal strength, so we cannot generalize their results. As a rule of thumb, access points should be positioned so that locations of frequent usage are significantly closer to the nearest access point than to the nearest microwave oven, and ovens should be sequestered behind partition walls (concrete if possible) to further reduce their interference potential. Because the peak received power of -30 dBm or so at 5 m is quite comparable with that of an access point at a similar distance, interference will likely be serious if the oven is closer to the user than the desired access point and may still represent a limitation on high-rate data even when relatively distant. Finally, in view of the typical usage pattern for most microwave ovens in industrial locations, users stuck near an oven should consider breaking for lunch at lunch time.

7.2. Other WLAN Devices

Multiple access points within a single building or single floor of a large building are likely to provide overlapping regions of coverage. If all the access points are under the control of a single organization, their channels can be arranged so as to reduce interference. In the U.S. ISM band, as noted previously, only three nonoverlapping channels exist (1, 6, and 11; see Figure 3-9): assuming good-quality radios, access points on distinct nonoverlapping channels should not interfere with each other. However, as previously noted, at least four channels are needed to provide a robust tiling of the plane with no adjacent interferers, and seven or more are preferred. The situation is worse if multiple-floor buildings with penetrable floors are present. To design a network that is truly interference free while providing complete coverage, it is desirable to operate in the UNII band where an adequate set of nonoverlapping channels is available (see Figure 3-15).

The potential achievability of noninterfering channel allocations is somewhat irrelevant when overlapping access points are not under the control of a single organization, as is likely to be the case in multitenant facilities, particularly when the tenants are separated only by interior partition walls. Fortunately, 802.11, like other Ethernet-style medium access controls (MACs), is robust to collisions and interference as long as loading is not too heavy. In general, both MACs will detect a packet failure and randomly back off for retransmission, resulting in a modest degradation in performance as long as the collisions are infrequent. However, Armour and coworkers showed that the HiperLAN MAC suffers more degradation in performance from an uncontrolled neighboring access point on the same frequency than does carrier-sense multiple access with collision avoidance (CSMA/CA), because HiperLAN does not expect unscheduled transmissions to overlap its assigned time slots.

Even if interference between neighboring access points is eliminated, the shared-medium nature of wireless links ensures the potential of cochannel interference between clients sharing the same access point. Interference between multiple clients is handled by the MAC layer of the protocol; in 802.11, the CSMA/CA MAC treats interference between two users as a dropped packet if it leads to a packet error and resends the packet after random backoff. In practice, this approach provides for up to five to seven simultaneous users, each receiving around 1 Mbps of throughput, in a single 802.11b cell. Because typical users are only occasionally transferring network data, up to 20 workstations or potential users can be accommodated in one access point. In HiperLAN and other centrally coordinated schemes, negligible interference between users in the same domain occurs because each is allocated transmission times.

As with microwave ovens, implementation of *request to send/clear to send* and packet fragmentation, described in Chapter 3, will increase robustness to interference from other clients and access points at the cost of reduced peak throughput. Another practical consequence of multiple access points at similar power levels is changes in association state of a client if it is set to associate with the strongest access point signal. Such "flipping" can be very irritating for the user but can be dealt with by requiring association to only the desired BSSID (see Chapter 3).

7.3. Bluetooth vs. Wi-Fi

Recall from Chapter 3 that Bluetooth devices transmit in 75 1-MHz channels, randomly selected 1600 times per second. Thus, in any given millisecond, collocated Wi-Fi or 802.11g and Bluetooth devices can potentially interfere. The extent of the interference is likely to depend both on the ratio of the desired signal (Wi-Fi or Bluetooth depending on whose viewpoint one takes) to the interferer (BT or Wi-Fi): the *signal-to-interference ratio* (S/I). The (S/I) is in turn determined by two path losses: that from the wanted transmitter to the receiver and that from the interferer to the victim receiver. Interference is a bigger problem when the victim receiver is at the edge of its range. Finally, because Bluetooth is a frequency-hopping protocol and Wi-Fi is not, the frequency separation between channels is another variable; when the separation greatly exceeds the width of the Wi-Fi channel, little interference is likely in either direction.

Model results, reviewed by Shelhammer, show that the interference effect on Bluetooth devices of an interfering 802.11b device is mostly confined to the Bluetooth slots within about 6 MHz of the Wi-Fi channel. Because there are 75 slots (in the United States) that can be used, this means that about $(11/75) = 15\%$ of slots are subject to interference. Bluetooth packets are generally (though not always) contained within a single frequency slot, so one can roughly say that the Bluetooth packet loss rate will not exceed 15% even when severe Wi-Fi interference is present. A 10–15% packet error rate has modest impact on data communications but will lead to noticeable degradation in voice quality. Proposed enhancements to the Bluetooth link standards, which provide simple retransmission capability, can nearly eliminate packet loss, but existing Bluetooth devices (of which tens of millions have been shipped) will not benefit from these improvements. Because of the robust Gaussian minimum-shift keying modulation used in Bluetooth (see Figure 3-24), even the central frequency slots with maximum overlap with Wi-Fi will achieve bit error rates of less than 10^{-3} if (S/I) > 5 dB; for a Bluetooth master and slave separated by 1 m, a Wi-Fi client device more than 4 m away from the victim receiver will have modest impact on the Bluetooth device. Note that if the Wi-Fi device in question is lightly used (as most are most of the time), effects will be proportionately smaller. These theoretical results were generally confirmed by the measurements of Karjalainen and coworkers. However, they found that much more severe effects could result if the WLAN and Bluetooth devices were placed together (as if on a single laptop device); in this case, the Bluetooth throughput could be degraded by as much as 80%, presumably due to the strong coupling of even the edges of the WLAN spectrum into the Bluetooth receiver.

Changing sides in the debate to become a Wi-Fi proponent, we find that the worst effects are again confined to Bluetooth slots within about 6 MHz of the Wi-Fi center frequency. At 1 Mbps, a Wi-Fi link can tolerate an (S/I) level of −3 dB (i.e., the Bluetooth signal is 3 dB *larger* than the Wi-Fi signal) before severe degradation of the bit error rate occurs because of the processing gain of the Barker code, even for the worst Bluetooth channel (which in this case is 1 MHz from the Wi-Fi center frequency). An 11-Mbps complementary code keying link, which has some coding gain but no processing gain, requires (S/I) > +4 dB for the worst Bluetooth channel to ensure unaffected performance (in this case the 0-MHz channel, because the complementary code keying spectrum is a bit different from the Barker spectrum). A Bluetooth client more than about 4 m away from a Wi-Fi device with good (S/N) will have little effect on the latter, though the potential for interference increases when the Wi-Fi device nears the end of its range. Note that just as in the microwave oven case, a Wi-Fi device operating in a high-rate mode will suffer *less* from the interferer than at 1 Mbps, because the shorter packets are less likely to encounter an interfering Bluetooth slot. An 11-Mbps link will only suffer packet loss of around 4% even in the presence of an adjacent Bluetooth device, whereas the same situation at 1 Mbps produces packet loss of more than 50%.

Soltanian and coworkers showed in modeling that adaptive notch filters could be used to remove Bluetooth interference from a direct-sequence 802.11 signal. Orthogonal frequency-division multiplexing packets, used in 802.11g, are potentially robust to narrow-channel interference, because it is in principle possible for the decoder to recognize and discard contaminated subcarriers. Doufexi and coworkers showed that with such adaptation, the impact of a Bluetooth interferer is almost negligible even at (S/I) of −11 dB and quite modest (S/N) ratios. However, because "g" radios must be capable of using direct-sequence preambles audible to older radios, problems with interference would persist. In any case, these advanced capabilities have not yet been incorporated into commercial radios as far as the author can ascertain (early 2004).

Wi-Fi–Bluetooth results are summarized in Table 7-5. The guard radius (the distance at which the interferer has little effect on the victim receiver) therein is only a guideline; for a specific physical arrangement, link loss for both transmitters and thus the (S/I) must be estimated.

Table 7-5: Wi-Fi and Bluetooth as Interferers

Interferer	Victim	Minimum (S/I) for Negligible Interference	Worst-Case Packet Loss	Guard Radius for Negligible Interference
Wi-Fi	Bluetooth	+5 dB	15%	4 m
Bluetooth	Wi-Fi 1 Mbps	− 3 dB	50%	2.5 m
Bluetooth	Wi-Fi 11 Mbps	+3 dB	4%	4 m

From Shelhammer, Figures 14-1–14-6.

7.4. Cordless Phones

Cordless phones are an extremely popular convenience (though the *in*convenience of a misplaced ringing cordless phone sometimes seems to exceed the benefits!) in homes and (somewhat less frequently) in office environments. Most purchasers appear quite comfortable with obtaining handsets and base stations from the same manufacturer, and there is little commercial pressure for interoperability, so cordless phones are generally unique proprietary designs subject only to compliance with regulatory requirements. As a consequence, both frequency-hopping and direct-sequence designs are used. Output power and signal bandwidth also vary noticeably, though the signals are quite narrowband from the view of an 802.11-type network, appropriate to the modest data rate (roughly 60 Kbps if uncompressed, as little as 10 Kbps if compression is used) needed for voice. A few representative examples are summarized in Table 7-6. The output power of most of the phones is comparable with or larger than typical consumer (30–50 mW) or enterprise (100 mW) access points.

The emissions spectra of portable phones as interferers are shown in Figure 7-27. The received power from these nearby interferers is as large as −30 dBm. The time-dependent results show that the interference is sporadic, so that even if the direct-sequence phone lies on top of the wanted Wi-Fi channel, data transfer will still occur during idle times, but peak data rates will be significantly reduced when the phones are active. Phones are also likely to be used more extensively than microwave ovens in most

Table 7-6: A Few Representative Cordless Phone Schemes

Vendor	Scheme	Power (mW)	Bandwidth (MHz)
Uniden	DSSS	10	2.8
V-Tech	FHSS	60	1
Sharp	DSSS	100	1.8
Siemens	FHSS	200	0.8

Source: FCC web site.

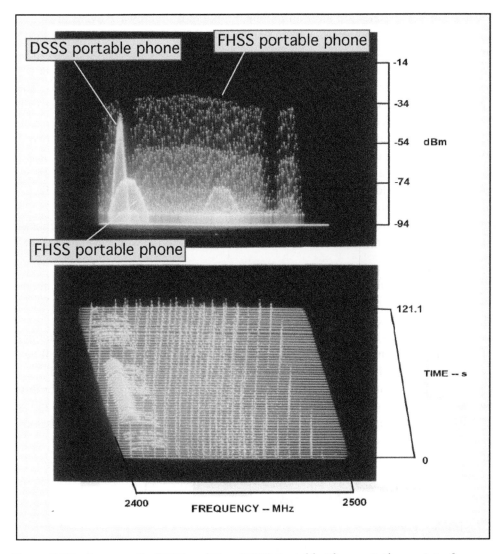

Figure 7-27: Spectra of a DSSS and Two FHSS Portable Phones Acting as Interferers: (Top) Averaged Power; (Bottom) Time-Dependent Power (Home Office, Davis, CA, 2002; Monopole Antenna ⇒ High Dynamic Range Preamplifier) (From *Literature Search and Review of Radio Noise and Its Impact on Wireless Communications*, Richard Adler, Wilbur Vincent, George Munsch, and David Middleton, U.S. Naval Postgraduate School, for FCC TAC, 2002)

environments (save perhaps fast-food restaurants), so they constitute an important source of interference for Wi-Fi. However, the narrowness of the DSSS signals in frequency (so that only a few Bluetooth hops would be affected) and the minimal overlap of two random frequency-hopping spread spectrum (FHSS) patterns imply that cordless phones will not be significant interferers for Bluetooth links.

Because cordless phones are after all mobile, unlike most microwave ovens, it is generally impractical to sequester them behind absorbing walls or relegate them to locations more distant from the relevant access point. FHSS phone interferers will resemble Bluetooth interferers, sporadically causing packet errors as they hop onto the Wi-Fi channel, with the effect being less serious for higher data rate packets. The higher power of the phones makes the guard radius much larger than that cited in Table 7-5 for Bluetooth (recall from Chapter 3 that Bluetooth stations typically transmit at 0 dBm). In simple environments with only one or two access points and one or two DSSS phones (assuming they are under control of one person or organization), it is possible to adjust the Wi-Fi channels and the phone channels to avoid overlap. Phone operating manuals don't always provide the channel frequencies, but these can be found (with some labor) in the FCC compliance reports. For more complex environments with numerous phones and access points, DSSS phones can be generally set around Wi-Fi channels 3 or 4 and 8 or 9 to minimize overlap with Wi-Fi devices at 1, 6, or 11.

In the case of a complex system in which the intermingling of numerous cordless phones and WLAN devices is inevitable, two alternative courses of action still remain. Wi-Fi–based portable phones, using *voice-over-Internet-protocol*, are becoming widely available (early 2004). Although these phones are Wi-Fi devices and thus will compete for the same medium as data, the CSMA/CA MAC ensures that coexistence will be more graceful than that of an uncoordinated proprietary interferer. The second approach is to relocate the data network to the 5-GHz band, where very few cordless phones operate today. The UNII band provides more bandwidth and channels and is not subject to interference from microwave ovens or Bluetooth devices. For mission-critical data applications in complex coverage networks subject to unavoidable interferers, 802.11a or dual-band networks should be considered.

8. Tools for Indoor Networks

8.1. Indoor Toolbox

Constructing and managing an indoor WLAN involves a number of tasks related to signals and propagation. Most networks use a number of fixed access points to provide service to a specific area. Planning the locations of the access points to ensure good coverage and minimize cost requires some thought be given to the unique propagation characteristics of the building or campus in question. A first estimate can be derived using building plans and the information presented previously in this chapter, but for complex installations, automated modeling may be helpful and surveying is essential. In multitenant environments, an examination of the interference environment, due to both other access points and clients and non-LAN interferers, may be needed to understand what quality of service is achievable. Once the network is installed, ongoing monitoring may be helpful both to ensure that coverage remains acceptable despite the inevitable changes in the propagation environment as people, equipment, and partition walls move and change and to protect against new interferers and "rogue" access points. (Rogue access points are access points connected to the wired network without approval of the corporate or facility authority. Because of the Ethernet-based transparency of the 802.11 standard and the fact that most organizations assume physical security for their internal network and impose little additional access control on its ports, it is easy for an employee or occupant to simply plug a consumer access point into the internal Ethernet network, possibly without even activating Wired Equivalent Privacy encryption. Such an access point provides an open opportunity for any nearby person equipped with a client card to obtain Internet access through the wired network, possibly without the approval of the owners of the network. Many wired networks will also allow a client to see the list of servers available on the LAN with no authentication of any kind, an invitation for industrial espionage if the client is capable of sniffing passwords or other attacks. Thus, rogue access points represent significant security concerns.)

Commercial organizations and software change rapidly, particularly in developing fields like wireless networks, and any specific listings and advice provided in a book of this nature are likely to be out of date by the time the reader could take advantage of them. Therefore, with respect to such tools, it is the intention of this section merely to provide examples of equipment and software capabilities that are available at the time of this writing, so that the reader may be apprised of the sort of resources that might be helpful. Nothing herein should be taken as either a review or an endorsement of any particular hardware tool or software package.

8.2. Surveying

The purpose of surveying is to assess received signal strength as a function of location within (and perhaps beyond) the intended coverage area. Many standards, including the 802.11 standards, require that compliant radios provide some measure of the average power of each received packet (typically known as the *received signal strength indication* [RSSI]); in consequence, any compliant 802.11 radio can be used as a survey tool. The author has verified that RSSI results for at least a couple of commercial client cards are well correlated with received power of continuous-wave signals measured using a network analyzer; RSSI provides a useful semiquantitative indication of signal strength.

The simplest approach to surveying is thus to use any portable device equipped with a client card and software that displays RSSI. Many card vendors provide management utilities that offer access to received signal strength. There are also numerous public-domain and shareware utilities that provide lists of received access point and client signals, including recent and average signal power: Netstumbler (Windows OS, www.netstumbler.org), Kismet (Linux, www.kismetwireless.net), and Kismac (Mac OS, www.binaervarianz.de/projekte/programmieren/kismac/) are some examples. A screen shot of a Kismac scan output is depicted in Figure 7-28. Simple surveying can be accomplished by manually recording received power on a facility map. For small facilities and simple installations, such an approach is quite adequate (and low in cost).

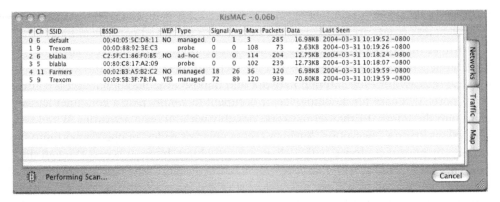

Figure 7-28: Example of Stumbler Application Scan Result, Showing Most Recent and Average Signal Strength Indicators for Each Access Point Received

For more complex installations, involving multiple access points and large multifloor buildings, manual surveys become impractically laborious. Fortunately, software tools are available to automate the process. Current tools allow the import of a graphic background (typically a building plan) on which the user can indicate a starting and stopping point by mouse clicks; data are then acquired at regular intervals as the user walks from the starting to the stopping point, and with the presumption that the walk was performed at a constant rate each data point may be assigned to a location. A thorough survey of a large facility can thus be acquired rapidly without sacrificing the flexibility of handheld portable devices for data acquisition. Surveying tools of this type are available from (at least) ABP Systems, Ekahau Inc., AirMagnet, and Berkeley Varitronics as of this writing. Berkeley Varitronics provides a custom personal digital assistant with calibrated Wi-Fi receiver, so that the absolute signal power is available, whereas the others run as software programs on a Wi-Fi–equipped portable device and are limited to RSSI inputs. Figure 7-29 shows a map of signal strength obtained using Ekahau Site Survey 2.0, superimposed on a building plan and record of the walking path used in acquiring the survey. Maps of other parameters of interest, such as data rate, coverage, S/N or S/I, and strongest access point in each location, are also available. Similar maps and features are available from the other vendors. These tools are much faster than manual surveying for a large facility!

In addition to the hardware and software needed on the receiving end, when surveying before installation the toolkit should include at least two portable easily installed access points so that candidate locations can be readily tested; two are recommended to allow for real-time examination of the overlap region between each pair. Another useful addition to the toolkit for large facilities is a tall portable support pole for raising a candidate access point close to the facility ceiling when installation there is likely.

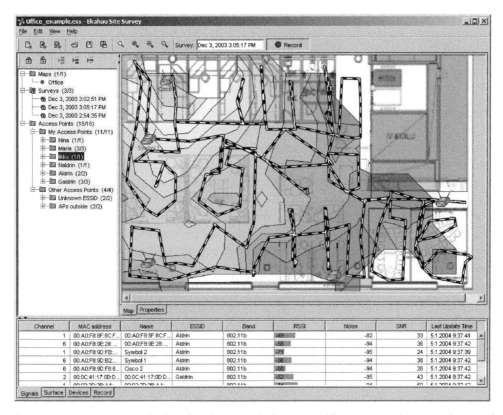

Figure 7-29: Survey Result for Signal Strength for a Specific Access Point Superimposed on Facility Map; Dashed Lines Indicate Walking Paths Used in Survey; File Management and Access Point List Windows Are Also Shown (Image Courtesy Ekahau Inc.)

8.3. Indoor Propagation Modeling

From the discussion presented in this chapter, the reader will have gathered that the first step in understanding propagation in a building is to examine the building layout and understand the obstacles to propagation that are present. For any given proposed access point location, a ruler placed over the map will allow a quick if rough evaluation of the likely signal strength, given that the approximate nature of the walls and fixed obstacles is known.

Automated modeling allows the planner to deal with larger more complex installations. Modeling uses ray tracing techniques; rays are launched at a variety of angles, and the obstacles the rays encounter result in the launch of new rays (e.g., due to reflection) and changes in the ray amplitude (due to absorption and partial transmission). The net phase accumulation along the ray can be tracked to attempt to reconstruct the actual received signal amplitude, though as we have noted it is very difficult in practice to know geometries to such precision as to make relative phase data meaningful. In most cases, the relative amplitude of the important rays is sufficient to establish average signal strength within a region. Phase accumulation is of course equivalent to propagation delay, so that power/delay profiles like Figure 7-19 can also be estimated to ensure that they do not exceed the tolerance of the relevant modulation for multipath delay.

The algorithmic art of a successful ray-tracing problem for complex environments is in the handling of the interaction of obstacles and rays to minimize the computational burden. The problems encountered are similar to those dealt with in rendering software used to produce computer-graphics images. Considerable ingenuity goes into establishing which obstacles and portions of obstacles are visible after a certain number of reflections and avoiding computation along paths that will never contribute to the received signal. A key choice in determining the size of the problem is the number of obstacle interactions to be evaluated; if one chooses (as we have for the manual approach) to model only absorption along the direct ray, the computational burden is much reduced, and qualitative features are still correctly reproduced in most cases. As with most computational modeling, the problem size increases drastically in progressing from two-dimensional to three-dimensional simulations, and many of the tricks used in other fields to reduce computational time cannot be applied here, because buildings generally do not have axes of symmetry.

Diffraction around obstacles is often treated in ray-tracing contexts using the *Uniform Theory of Diffraction* or the *Generalized Theory of Diffraction*. These methods attempt to account for the effect of an obstacle through the definition of effective rays emitted at the edges of the obstacle, where the amplitude and phase of these rays are determined by a fit to Fresnel integrals of the sort we examined in Chapter 6 (albeit conventionally using the unobstructed regions as sources rather than the induced current on the obstacle, as we have done). Such look-up–based approaches are much more computationally efficient than a complex integration when a large number of irregularly shaped obstacles must be handled.

Commercial software tools specific to indoor propagation for WLAN systems are available from such vendors as Wireless Valley, Wibhu Technologies, ABP Systems, and Ekahau, among others. Most systems can take a building plan graphic as input but require some manual intervention to mark features as walls of a particular type, so that the appropriate RF properties can be assigned. Multifloor buildings can be simulated. Results generally include maps of signal strength and related properties such as signal to noise and interference. Some vendors have constructed databases of commercial access points and clients and thus can provide expected data rate and throughput for specified clients as a function of location.

9. Capsule Summary: Chapter 7

Buildings in the modern world use a modest repertoire of structural materials, of which steel, concrete, wood, glass, gypsum, and bricks and masonry are most frequently encountered. Small buildings use load-bearing walls of concrete, reinforced masonry, or (for residences) plywood and non–load-bearing partitions of stud-reinforced gypsum. Large buildings rely on steel or reinforced concrete framing and do not use load-bearing walls. Floors may be made of wood or concrete on either wood or steel.

At microwave frequencies, typical thicknesses of gypsum, glass, and wood induce only modest absorption. Brick and masonry absorb noticeably. Concrete properties vary depending on mixture and cure state, but thick concrete walls will usually represent significant absorbers. Most common structural materials have refractive indices between 2 and 3 and so reflect significantly, though consequent transmission loss is modest except at glancing angles. Large metal obstacles, including both elements of structures such as columns or ducts and interior objects such as cabinets or server racks, cast shadows 10–20 dB deep lying within their geometric shadow. Floors of concrete on steel act as shields. With these observations in hand, one can construct a first approximation to the average received power at an indoor location by examining the absorption of the direct ray path from transmitter to receiver as it traverses walls, floors, and windows.

Real indoor structures show widely varying propagation characteristics that depend critically on what is in the way. Long RMS propagation delays, averaging around 100 nsec and reaching as high as 1000 nsec when exterior reflections are significant, occur because of reflections from interior structures, even when the direct-line distance from transmitter to receiver is small. Open areas have average signal strength close to that obtained in free space (though local fading occurs because of interactions with reflected rays); a different location at the same distance from the transmitter may have 40 dB less signal power due to the interposition of interior and exterior walls. For typical clients, average signal strength of −75 dBm at the receiver is sufficient to provide reliable high-rate service. If access points can be positioned to avoid most obstacles, interior areas approaching 100 m in span can be covered with a single access point; in the presence of multiple walls, the same access point can be limited to 10 or 15 m.

Common indoor interferers include microwave ovens, cordless phones, and rival Wi-Fi or Bluetooth devices. In each case, the time-dependent nature of the interference ensures that some traffic is likely to get through even in the worst cases, but physical configuration and channel planning can also help to achieve reliable high data rates in the presence of interferers.

Software and hardware tools are available for modeling and measuring propagation characteristics in indoor environments; use of these tools facilitates installation of complex networks and improves the utility of the result.

10. Further Reading

Building Construction Techniques

Building Construction Illustrated (3rd Edition), Francis Ching with Cassandra Adams, Wiley, 2000: *Mr. Perry's favorite pictorial reference; a beautifully illustrated book.*

Microwave Properties of Construction Materials

NISTIR 6055: NIST Construction Automation Program Report No. 3: "Electromagnetic Signal Attenuation in Construction Materials," William C. Stone, 1997; available from NIST or by web download

"Effect of Admixtures, Chlorides, and Moisture on Dielectric Properties of Portland Cement Concrete in the Low Microwave Frequency Range," K. Pokkuluri (thesis), Virginia Polytechnic Institute, October, 1998

"Different Kinds of Walls and Their Effect on the Attenuation of Radiowaves Indoors," P. Ali-Rantala, L. Ukkonen, L. Sydanheimo, M. Keskilammi, and M. Kivikoski, 2003 Antennas and Propagation Society International Symposium, vol. 3, p. 1020

"Measurement and Monitoring of Microwave Reflection and Transmission Properties of Cement-Based Specimens," Kharkovsky et al., IEEE Instrumentation and Meas Tech Conf, Budapest, Hungary, May 21–23, 2001 p. 513

"Measurement and Modeling of Propagation Losses in Brick and Concrete Walls for the 900 MHz band," D. Peña, R. Feick, H. Hristov, and W. Grote, IEEE Trans Antennas and Propagation 51, p. 31, 2003

"A Comparison of Theoretical and Empirical Reflection Coefficients for Typical Exterior Wall Surfaces in a Mobile Radio Environment," O. Landron, M. Feuerstein, and T. Rappaport, IEEE Trans Ant Prop 44, p. 341, 1996

"Reflection and Transmission Losses through Common Building Materials", Robert Wilson, available from Magis Networks, www.magisnetworks.com

The Chemistry of Silica, Ralph Iler, Wiley, 1979

Indoor Propagation Studies

"The Indoor Propagation Channel," H. Hashemi, Proceedings of the IEEE, vol. 81, no. 7, p. 943, 1993

"The Impact of Surrounding Buildings on Propagation for Wireless In-Building Personal Communications System Design," S. Seidel, T. Rappaport, M. Feuerstein, K. Blackard, and L. Grindstaff, 1992 IEEE Vehicular Technology Conference, p. 814

"Indoor Throughput and Range Improvements using Standard Compliant AP Antenna Diversity in IEEE 802.11a and ETSI HIPERLAN/2," M. Abdul Aziz, M. Butler, A. Doufexi, A. Nix, and P. Fletcher, 54th IEEE Vehicular Technology Conference, October 2001, vol. 4, p. 2294

"Outdoor/Indoor Propagation Modeling for Wireless Communications Systems," M. Iskander, Z. Yun, and Z. Zhang (U Utah), IEEE Antennas and Propagation Society, AP-S International Symposium (Digest), vol. 2, 2001, pp. 150–153

"Effective Models in Evaluating Radio Coverage in Single Floors of Multifloor Buildings," J. Tarng and T. Liu, IEEE Trans Vehicular Tech, vol. 48, no. 3, May 1999

"Correlating Link Loss with Data Throughput," D. Dobkin, High Frequency Electronics, January 2003, p. 22

Interference: General

"Literature Search and Review of Radio Noise and Its Impact on Wireless Communications," Richard Adler, Wilbur Vincent, George Munsch, and David Middleton, U.S. Naval Postgraduate School, for FCC TAC, 2002

Interference: Microwave Ovens

"A Novel Prediction Tool for Indoor Wireless LAN under the Microwave Oven Interference," W. Chang, Y. Lee, C. Ko, and C. Chen, available at http://www.cert.org/research/isw/isw2000/papers/2.pd

"Effects of Microwave Oven Interference on the Performance of ISM-Band DS/SS System," S. Unawong, S. Miyamoto, and N. Morinaga, 1998 IEEE International Symposium on Electromagnetic Compatibility, vol. 1, pp. 51-56

Interference: WLAN Self-interference

"The Impact of Power Limitations and Adjacent Residence Interference on the Performance of WLANs for Home Networking Applications," S. Armour, A. Doufexi, B. Lee, A. Nix, and D. Bull, IEEE Trans. Consumer Electronics, vol. 47, p. 502, 2001

Interference: Bluetooth and WLAN

"Coexistence of IEEE 802.11b WLAN and Bluetooth WPAN," Stephen Shelhammer, Chapter 14 in **Wireless Local Area Networks**, Bing (op. cit.), Wiley, 2002

"An Investigation of the Impact of Bluetooth Interference on the Performance of 802.11g Wireless Local Area Networks," A. Doufexi, A. Arumugam, S. Armour, and A. Nix, 57th IEEE Vehicular Technology Conference, 2003, vol. 1, p. 680

"The Performance of Bluetooth System in the Presence of WLAN Interference in an Office Environment," O, Larjalainen, S. Rantala, and M. Kivikoski, 8th International Conference on Communication Systems (ICCS), 2002, vol. 2, p. 628

"Rejection of Bluetooth Interference in 802.11 WLANs," A. Soltanian, R. Van Dyck, and O. Rebala, Proceedings 56th IEEE Vehicular Technology Conference, 2002, vol. 2, p. 932

Modeling

"Propagation Modelling for Indoor Wireless Communications," W. Tam and V. Tran, Electronics and Communications Engineering Journal, October, 1995, p. 221

"Wideband Propagation Modeling for Indoor Environments and for Radio Transmission into Buildings," R. Hoppe, P. Wertz, G. Wolfle, and F. Landstorfer, 11th International Symposium on Personal Indoor and Mobile Radio Communications, vol. 1, p. 282

"Efficient Ray-Tracing Acceleration Techniques for Radio Propagation Modeling," F. Agelet et al., IEEE Trans. Vehicular Tech 49, no. 6 November, 2000

"Improving the Accuracy of Ray-Tracing Techniques for Indoor Propagation Modeling," K. Remley, H. Anderson, and A. Weisshaar, IEEE Trans Vehicular Tech, vol. 49, no. 6, p. 2350, 2000

Indoor Setup Guidelines

"Wireless LAN Design, Part 1: Fundamentals of the Physical Layer," Jesse Frankel, WiFi Planet, San Jose, CA, fall 2003

Outdoor Networks

1. Neither Snow Nor Rain Nor Heat Nor Gloom of Night . . .

. . . ought to keep an outdoor network from transporting its appointed packets—but then there are lightning strikes too. It's a dangerous world out there in the wild outdoors for an access point as well as a very different sort of environment for propagation of radio waves. In this chapter, after examining some fundamental issues having to do with radio signals, we proceed from small to large and near to far, examining local coverage networks first, then point-to-point bridges and point-to-multipoint networks, and finally long line-of-sight links. We conclude with a few important notes on safety in design, construction, and working practices for outdoor facilities.

Before we plunge into details of implementation, it is appropriate to set the stage with a few cautions. Institute of Electrical and Electronic Engineers (IEEE) 802.11 is a *wireless local area network* (WLAN) standard set. Although wireless bridges from one building's wired network to a neighbor are quite sensible within the confines of the standard, most other outdoor uses stretch the intended functionality of the Wi-Fi protocols. The 802.11 medium access control (MAC) has proven remarkably adaptable to these alternative applications, but consideration of other approaches may often be appropriate, depending on the scope and scale of services to be provided. In particular, the recent (2003) approval and upcoming additions to the IEEE 802.16 *metropolitan area network* standards provide a standards-based solution to outdoor broadband networks that is being supported by many vendors, both with existing proprietary systems and newcomers to the field. Also known as *WiMax*, the 802.16 standard provides many capabilities particularly useful for point-to-multipoint long-reach networks and should be kept in mind if high data rates and demanding quality-of-service specifications are contemplated.

Furthermore, unlike indoor networks where most of the signal is confined in many cases within a single owner's geographic domain, outdoor signals will often cross over public and private property not owned by the user. Both enlightened self-interest and ethical standards demand that consideration be given to other actual and potential users of the same wireless medium, so that disputes are whenever possible avoided rather than litigated.

If interception of indoor signals must be considered possible, interception of outdoor signals must be presumed inevitable. This is not a concern if the main use is for access to the public Internet, but if private traffic of any kind is envisioned, security must be attended to. The same recommendations discussed in Chapter 3 are applicable here. Native Wired Equivalent Privacy, or the enhanced WPA version if available, should be supported when public access is not desired, both to signal casual users that the network is closed and to provide a first layer of link security. Real security for critical data should be implemented end to end, using virtual private network technology for LAN access and secure sockets layer for web and Internet traffic. Wired connections to outdoor access points should be treated as insecure ports and isolated from a trusted network by a firewall or equivalent security precautions.

2. Line-of-Sight Sites

The first question one should ask about an outdoor installation is how far can you go? Let's assume we're trying to evaluate the performance of a network meant to cover a large open area. The access point can use an "isotropic" antenna (section 6.8) that broadcasts in all horizontal directions but suppresses transmission straight up and straight down, directions unlikely to be of much help in most outdoor applications. Such an antenna will typically provide around 6 dBi (dB relative to an isotropic antenna) of directive gain in the horizontal plane. Because the client antenna might be oriented in some arbitrary direction, we'll conservatively give it no directive benefit over a dipole (a directivity of 0 dBi). Assuming either that cable loss is negligible or that it has been corrected for, we'll allow the maximum legal transmission of 1 W at 2.45 GHz. The Friis equation [5.40] for the received power becomes

$$P_{RX, dBm} = 30 + 6 - 0 + 20 \log \left(\frac{0.122}{4(3.14)} \right) - 20 \log (r) \tag{8.1}$$

The resulting received power is plotted in Figure 8-1. If we use the same rule of thumb for reliable high-rate 802.11 transport stated in Chapter 7—a minimum of −75 dB from a milliwatt (dBm) received power—we find that our little 1-W transmitter should easily reach 3500 m. That's 3.5 km or a bit over 2 miles: a respectable performance for a modest amount of input power. If we were to add an isotropic antenna at the receiver, at the cost of a dorky-looking client we'd get an additional factor of 2 in range, bringing us up to 7 km: quite comparable with the range achieved in practice by very expensive high-power cellular telephony base stations. So far it doesn't look very hard to run a radio outdoors. Unfortunately, that's because we haven't looked very hard yet. All sorts of obstacles, literal and figurative, restrict the achievable performance of microwave radios outdoors.

The literal obstacles play a more important and rather different role than was the case indoors, because of the change in scale. Recall from Chapters 6 and 7 that the proper way to characterize the size of an obstacle from the point of view of radio propagation is by the Fresnel zones it subtends: Equivalently, we can scale the size of the obstacle by the characteristic length for diffraction. From equation [6.34] we find

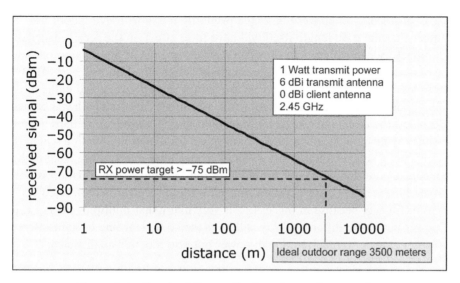

Figure 8-1: Received Power for Free-Space Propagation

$$u = \frac{x}{x_F} \quad x_F = \sqrt{\frac{R_{av}\lambda}{2}} \quad R_{av} = \frac{1}{\frac{1}{z_{src}} + \frac{1}{z_{obs}}} \qquad [8.2]$$

In an urban or suburban environment, the characteristic distance in a random direction in which an obstacle is encountered is on the order of, for example, 20 m, so at 2.45 GHz, the characteristic length scale for diffraction $x_F \approx 1$ m. A typical obstacle (a building, for example) might be 10–30 m in all dimensions, with a large building approaching 50 m, so the normalized size u is on the order of 10–30 or greater. Recall that except very close to an obstacle, normalized sizes of one to four were more typical in indoor environments (see Figure 6-36). Outdoor objects cast much deeper shadows at microwave frequencies than most indoor objects do. Outdoor objects, being composed of many indoor obstacles, are normally quite opaque to transmission as well: An empty open building with a steel frame and glass external walls might provide little impediment to propagation, but in a more typical case the direct ray through the building is likely to encounter external and internal load-bearing walls, partition walls, and people and their tools and toys, each imposing its share of attenuation. To really get 3500 m, we are going to need something approaching a clear line of sight between the transmitter and receiver.

The long distances encountered outdoors also mean that multipath takes on a whole different meaning. If we could really achieve our ideal case of 3500 m of separation, the propagation delay attendant upon this separation is 10,000 nsec or 10 µsec, significantly longer than even an 802.11a/g orthogonal frequency-division multiplexing (OFDM) symbol (see Figure 3-20). A reflected path with a significantly different path length, if of comparable magnitude with the direct signal, will overwhelm the multipath tolerance of the 802.11 protocols, which were designed primarily for short-range indoor use. Remember that the path length itself will have little effect on the received signal: An additional 1500 m will impose only $(3500/5000)^2 \equiv 3$ dB of attenuation but adds 5000 nsec of delay. A singly reflected path might suffer an additional 3–20 dB of reflection loss (see Chapter 6, section 3), though for multiply-reflected paths the loss attendant upon reflection becomes significant. The singly reflected path will be perhaps 6–20 dB below the direct path, okay for binary phase-shift keying or perhaps quaternary phase-shift keying modulation but a major problem for higher modulation schemes with more demanding signal-to-noise ratios. Thus, intersymbol interference from multipath components is potentially a major problem for outdoor links. The situation is exacerbated if the direct path between transmitter and receiver is obstructed by buildings or foliage, because in this case it is very likely that multiple paths of roughly equal power but vastly differing delay will be present. A clear line of sight offers not only a better chance to receive enough power but also a better-quality signal.

(There are technological solutions to the problem of large multipath components that work very well for cellphones, which require only around 10 Kbps for a compressed voice channel. Code-division multiple access phones can use a *rake receiver*, which we described briefly in Chapter 3. By searching for delayed copies of the code assigned to a particular logical channel in a particular phone and adding them together with the necessary delay correction, a rake receiver can combine the power from several paths. This technique can also be used for 802.11 classic signals, which use a Barker code, but the 11 Barker-code chips provide only about a microsecond of multipath tolerance, and of course the Barker code is not used at higher data rates. *Equalizers* use a similar trick, reconstructing the transmitted signal from a weighted sum of delayed versions of the received signal, but equalizers for high-rate data become complex when the delay is much longer than the symbol times.)

Outdoor links are also subject to many time-dependent phenomena that are generally not encountered indoors: wind, rain, snow, seasonal and time-dependent variations in foliage, and (for long links) refractive misdirection of the direct path. Even once the known obstacles are dealt with, the prudent planner must provide for additional *link margin* to allow for all these new potential failure modes. How much margin is required depends on what system reliability the user is trying to achieve and how much they are willing to pay to achieve it. Recall that our -75 dBm guideline already includes about 8 dB of margin over the typical 11-megabits per second (Mbps) sensitivity for an 802.11b client (see Table 7-4) to account for Raleigh-type fading; in an outdoor environment it would not be at all unusual to tack 10 or even 20 dB on top of this to achieve the sort of 97+% availability that one might expect at a minimum for a usable network. Resources are available, discussed later in the chapter, to approximate the effects of precipitation and other common impairments on link performance, but to get to 99% and start tacking nines on after the decimal point, it is indispensable to monitor the specific link of interest, because each geographic location has its own unique issues. Outdoor microwave links, like politics, are local.

3. Outdoor Coverage Networks

3.1. Propagation

We already know we can go a few tens of meters indoors, depending on the building we're in. We've established that we could go a few kilometers out of doors if only things didn't get in the way. What happens when they do? What does propagation look like in an urban or suburban environment? Most urban environments are organized around streets, often lined with and sometimes containing trees, and the regions between the streets, occupied by buildings and various sorts of open spaces. We need to understand how the mechanisms of reflection, diffraction, absorption, and refraction act in such an environment.

Let's examine reflection first. Figure 8-2 shows a cartoon of reflected rays along a street whose margin is populated by buildings of varying size and configuration. Near the transmitter, the direct ray will likely dominate, because reflection coefficients are usually modest for high angles (unless large metallic walls are present). As the receiver moves down the street, the reflected rays will arise from higher angles and thus become more significant due to the increase in reflection coefficient. The average power is likely to be higher than that expected from unobstructed propagation, though the instantaneous power at any location will exhibit Rayleigh-like fading because the relative phase of the various reflected rays is unpredictable. At very large distances, the reflectivity of truly smooth surfaces like windows will continue to increase, but random scattering will decrease the reflection from rougher surfaces like brick, cut stone, or poured concrete. Scattering will be more noticeable at 5 GHz than at 2.45, as the surface roughness is scaled by the wavelength.

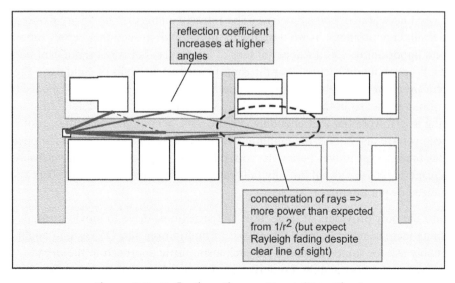

Figure 8-2: Reflection Along a Street (Top View)

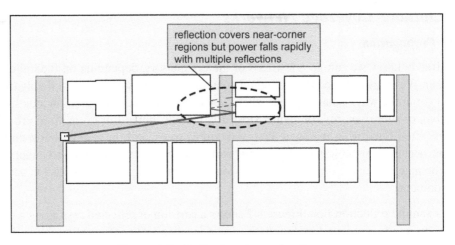

reflection covers near-corner regions but power falls rapidly with multiple reflections

Figure 8-3: Reflection Around a Corner

Reflection from buildings will also result in some propagation around corners (Figure 8-3), but once the intersection is far from the transmitter, multiple reflections would be needed to get very far (unless a specular surface at an odd angle happens to be present on one of the buildings at the corner). Each reflection incurs additional losses of at least several decibels; multiple reflections will provide coverage within a few street widths of a corner, but not much farther.

We've already noted that propagation through buildings is not likely to play a large role, except in areas of economic distress like that sadly visible in many locations in Silicon Valley at the time of this writing. (But business parks filled with empty buildings don't need network coverage anyway.) The refractive index of air is very close to 1 under all circumstances; as we describe in more detail in section 4 below, refraction plays an important role at distances of tens of kilometers but is not significant between streets. What about diffraction over the roofs or around the walls of buildings? The only relevant problem that we've examined in detail is that of diffraction at the edge of a semiinfinite plane, so in the spirit of looking where the light is good, let's treat a building as if it were flat and unending (Figure 8-4). When the transmitter, roof edge, and receiver are just in line, the amplitude is reduced by a factor of 2 and the power by 6 dB; as the receiver moves below the roof edge (the displacement being measured in normalized units), the signal initially falls rapidly, being attenuated 30 dB for normalized displacements > 10.

Because the only providers of large, thin, flat structures, drive-in movie theaters, are becoming increasingly rare with the advent of multiplexes and DVDs, one might reasonably ask for some validation for such a simplistic approach to the urban diffractive jungle. Fortunately, Frederiksen and coworkers performed some of the requisite measurements. Their measurement arrangement is depicted in Figure 8-5.

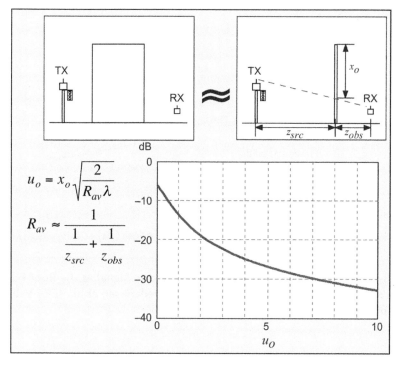

Figure 8-4: Diffraction Over the Roof of a Building Treated as Diffraction by the Edge of a Semiinfinite Plane

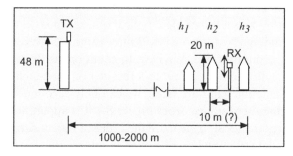

Figure 8-5: Measurement Arrangement for Rooftop Diffraction: Transmitter at 1.7 GHz; Receiver Height Varied in 5-m Steps From Roof to Ground

(They measured locations with various relationships between the heights $h_{1,2,3}$; here we only report the data for $h_1 < h_2$.) Figure 8-6 shows that the simple edge-diffraction model is within a few decibels of the measured data. More sophisticated models (see Frederiksen et al. for citations) that attempt to account for multiple edges and complex shapes exist, but for our purposes are unnecessarily complex.

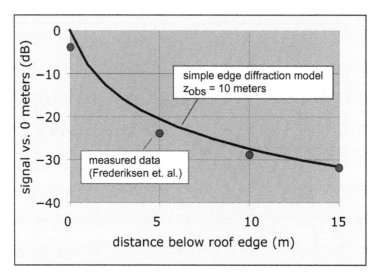

Figure 8-6: Measured Signal Strength, Normalized to Signal at the Roof Edge, Compared With Edge-Diffraction Model (From Frederiksen et al., IEEE Vehicular Tech. Conf. 2000, p. 896)

Having validated the simple model, what scaling length should we use to apply it? In the worse case, the receiver is fairly close to the building (e.g., 5 m), so if the transmitter is distant, we have $R_{av} \approx 5$ m and the scaling length $x_F \approx 0.5$ m at 2.45 GHz. A normalized displacement of 10 is around $10(0.5) = 5$ m (15 feet) below the roof edge. Because the receiver is probably a meter or two off the ground for convenient human use, this would correspond to a rooftop about 6–7 m high, typical of a ranch-style residence or a two-story commercial building. To get even 20 dB attenuation, the receiver needs to be within a few meters of the roof line. We have to count on 30 dB of attenuation in hopping from a transmitter on one street to a receiver on the neighboring street. If the transmit power and antenna gains are fixed, we can only accommodate this increased path loss by decreasing the distance: from Figure 8-1 it is apparent that 30 dB of diffraction loss reduces the ideal range to on the order of 100 m. A transmitter can get to the neighboring parallel street by hopping over an intervening roof if it isn't too high, but the signal won't go much farther.

Can the signal turn a corner? That is, what happens along perpendicular streets that intersect that on which the transmitter lives? Near the corner, as we've noted, the received signal can benefit from rays reflected once or twice off the neighboring buildings. Farther away, the only path available is the diffracted path (assuming that the buildings are high and opaque). We'd expect again to encounter attenuations of 30 dB and more. Measured data for tests performed in New York and Boston by Erceg and colleagues is shown in Figure 8-7, along with theoretical predictions they made using the uniform theory of diffraction we briefly mentioned above. At a location 100 m down the

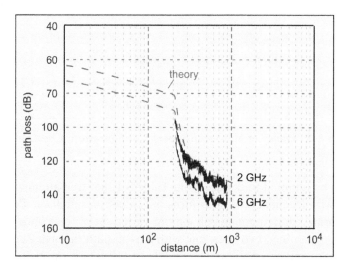

Figure 8-7: Measured Path Loss Along a Perpendicular Street (Corner at 200 m From Transmitter), With Prediction Accounting for Reflection and Diffraction (After Erceg et al., IEEE Trans. Vehicular Tech. 43, 1994)

perpendicular street, they found path losses of 35, 40, and 50 dB at 0.9, 2, and 6 GHz, respectively. By reference to Figure 8-1, these losses will limit the range of our 1-W link to only a few tens of meters: low-power unlicensed WLAN links cannot turn corners far from the transmitter.

Finally, like a Sierra Club member we must also ask what about the trees? Recall from Chapter 6 that liquid water is a very effective absorber of microwaves at a few gigahertz (see Figure 6-7) as well as having a high refractive index. Because tree branches and leaves contain lots of water, we might expect strong absorption and scattering to result when microwaves and trees meet. A closer look engenders even more concern: Trees and their leaves are comparable in size with the wavelengths of interest, and their shapes and configurations vary with the type of tree, wind, precipitation, and the seasons of the year. It shouldn't surprise us to find that trees are significant and highly variable obstacles to microwave propagation. Let's take a look at the results of some measurements and see.

Dalley and coworkers measured absorption due to a single chestnut tree in leaf; their measurement configuration is shown in Figure 8-8. In the frequency range of interest for WLAN technologies, attenuation averages 11–12 dB with excursions to around 20 dB. Although 13 GHz is a bit high for WLAN, it is worth noting that a wet tree at this frequency causes an additional 7 dB of attenuation; attenuation at lower frequencies will almost certainly increase by several decibels when it rains. The authors note that the tree causes some rotation of the axis of polarization, evident as a component orthogonal to the original polarization at about 12 dB below the main signal. Changes in polarization of the transmitted signal result in minor (1 dB) changes in attenuation.

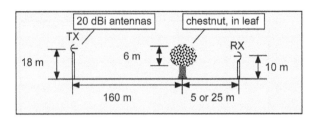

Figure 8-8: Dalley et al. Measurement Configuration, Chestnut Tree

One might also be tempted to deduce from Table 8-1 that tree attenuation increases monotonically with frequency, but this turns out to be too simplistic. Perras and Bouchard performed measurements on a more varied set of trees under more varied conditions. They examined two sites, one with a 64-m path through 21 m of deciduous trees (three maples and one crab) and a second site 110 m long through 25 m of conifers (spruce and pine). They made measurements at 2.45, 5.25, 29, and 60 GHz, using a continuous-wave source, for a total of more than 2 billion data points. They measured at the first (deciduous) site in both summer, when the trees were in leaf, and winter, when they were bare. Some of their results are shown in Figures 8-9 and 8-10.

Figure 8-9 shows the probability distribution of attenuation values for both sites, including in-leaf and winter data. The results thoroughly demolish any illusion that trees might be simple. Coniferous trees absorb more strongly at 30 GHz, but 60-GHz behavior resembles lower frequencies. Deciduous trees with leaves block 5 GHz and pass 2.45 GHz; after the loss of their leaves, high frequencies penetrate readily, whereas attenuation at 2.45 GHz is little changed.

These results can be qualitatively understood by reflecting on the sorts of features present in each case. Conifers are marked by the presence of needles, on the order of 3–15 cm long and a few millimeters in diameter. At 2 or 5 GHz the needle diameter is much less than the wavelength, so the needles mainly affect propagation when they happen to be aligned with the polarization of the incoming radiation. If the foliage is dense, unobstructed paths

Table 8-1: Attenuation Due to Chestnut Tree

Frequency (GHz)	Average Attenuation (dB)	90% Link Margin (dB)
3.5	11.2	16.2
5.85	12	20
13	13.5	21.3
13 (wet)	20.5	
26.5	17.7	25.3

90% Link Margin, 90% of measured attenuations were less than this value. After Dalley et al., 1999 IEEE Conference on Antennas and Propagation.

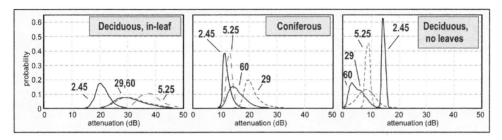

Figure 8-9: Probability Distribution of Attenuation Values for Different Trees in Differing Seasons; Flagged Are Measurement Frequencies in GHz (After Perras and Bouchard, 5th International Symposium on Wireless Multimedia, p. 267, 2000)

through the trees large compared with the wavelength are unlikely. At 30 or 60 GHz (wavelengths of 1 cm and 5 mm, respectively), interaction with the needles is stronger and less influenced by orientation, but the smaller wavelengths can also take advantage of accidental clear paths only a few centimeters wide.

Maple leaves are characteristically on the order of 10 cm wide, with edge serrations of around 3–4 cm characteristic dimension. When the trees are in leaf, the strong attenuation at 5.25 GHz might plausibly be accounted for by strong scattering off the leaf edges; the dense foliage leaves few gaps for millimeter waves to penetrate. When the trees lose their leaves, higher frequencies do a better job of sneaking through the open gaps between the bare branches, and attenuation decreases monotonically with frequency.

Wind moves trees and thus induces additional time-dependent variations in signal strength (Figure 8-10). In strong winds, Perras and Bouchard found strong (50 dB) local fades. The effects of wind are worse at higher frequencies, as one might expect (because smaller displacements of the trees are needed to induce significant changes in phase). The authors note that gentle winds produce shallower but longer lasting fading. Note

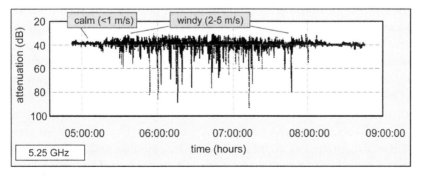

Figure 8-10: Attenuation by Deciduous Trees in Leaf vs. Time; Wind Velocity Data Were Collected Contemporaneously (After Perras and Bouchard, Op. Cit.)

Table 8-2: Mean and Variance (in dB) of Single-Frequency Attenuation for Various Tree Sites

		Frequency (GHz)			
		2.45	*5.25*	*29*	*60*
Deciduous, in leaf	Mean	−21.8	−39.1	−32.8	−32.1
	Variance	7.9	19.6	37.4	34.8
Coniferous	Mean	−12.6	−13.5	−21.7	−16.2
	Variance	1.9	1.1	10	9.7
Deciduous, no leaves	Mean	−14.9	−9.2	−8.7	−5.5
	Variance	0.2	0.5	10.9	6

After Perras and Bouchard, op. cit.

that these 40–50 dB fades are in part the result of the use of single-frequency measurements; a broadband signal like those used in WLAN technologies would encounter time-dependent frequency-selective fading, showing up as distortion in an 802.11 classic or b signal or knocking out particular subcarriers in an 802.11a/g OFDM signal, rather than the loss of communication that would result from a 50-dB reduction in total signal power.

Quantitative summaries of the data are provided in Table 8-2. At WLAN frequencies, deciduous trees without leaves and conifers display modest attenuations around 10 dB and small variances therein, but deciduous trees with leaves are strong attenuators with large variance. Although these results allow for simple and reliable installations in mountainous regions in the southwestern United States, where conifers dominate, networks in urban and forested regions elsewhere must contend with large and fluctuating attenuation if links that penetrate stands of trees are contemplated.

Like a chess position, every installation is unique and must be studied individually, but we can draw a few general conclusions from the data and arguments presented in this section. In urban areas with large buildings, coverage will generally follow streets (if they are straight). Ground-level transmitters will provide only such coverage; elevated transmitters have a better chance of reaching across the roof ridge to a neighboring street, but the signal can't turn corners well enough to help coverage much. Elevation may also help reduce the impact of trees, when present, if the transmitter can be placed the above the tops of the trees, though as the receiver moves down the street, the direct path thereto must inevitably penetrate a larger number of trees with increased attenuation of the signal resulting. The range along the street the transmitter is located on is likely to limited by foliage (if it is present) long before the free-space limits are reached; a range of around 800 m can be obtained with a 10-m transmitter elevation in a street with trees typically a few meters shorter. These guidelines are qualitatively summarized in Figure 8-11. In more rural areas, foliage will represent the main limit to coverage and range.

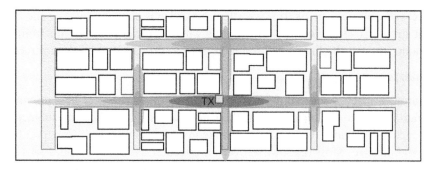

Figure 8-11: Cartoon of Urban Area, Depicting General Areas of Coverage for an Elevated Transmitter as Darkened Regions

A result for a suburban installation, using a 1-W transmitter suspended about 10 m above ground level, is shown in Figure 8-12. The reported measurement is the percentage of uplink "ping" packet that is successfully received by the access point from a mobile client. The environment in this location is composed of numerous trees, virtually all in leaf, surrounding one to two story wood-frame houses with two-story wood-frame apartment buildings in some locations. The access point is located on a light post at the intersection of two streets.

As expected, excellent coverage is obtained along the straight portions of the streets intersecting the access point location; high-quality links are available at 500-m range,

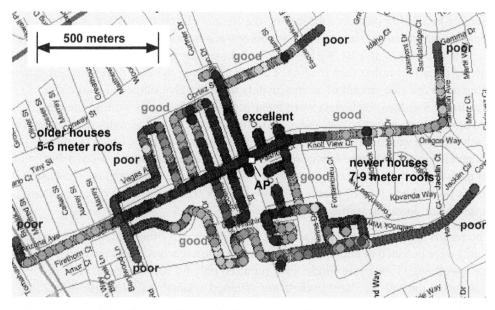

Figure 8-12: Coverage Map in a Suburban Area Using 1-W 802.11b Access Point and Laptop-Based Client (Image Courtesy Tropos Networks, Inc.)

and coverage continues to 1000 m. Excellent coverage is also obtained on the next-adjacent street. Coverage quality falls rapidly when a corner is turned far from the access point, though the link is maintained out to as far as 800 m in some directions.

3.2. Interference

All the usual suspects discussed in Chapter 7—microwave ovens, cordless phones, other 802.11 radios, Bluetooth devices—rear their ugly heads again when the subject of interference is broached. However, the change in circumstances results in modifications in their significance. Most microwave ovens are located indoors. An outdoor user is likely to have some separation and at least one external (absorbing) wall between their radio and the oven. Cordless phones are also mostly confined to indoor use, but the handset is mobile and may be found near windows as well as entryways or balconies; cordless phones are a challenge for outdoor users. The larger ranges involved in outdoor networks tend to reduce the likelihood of Bluetooth devices (other than those carried by the user of the outdoor network) being close enough to the client to either interfere or suffer interference relative to an indoor environment.

In the outdoor environment, one must also deal with some interferers that weren't often apparent indoors. In the United States, many gasoline filling stations maintain data links to a local office of the American Automobile Association, often in the 2.4-GHz ISM band. Bridges from one building to another and long point-to-point or point-to-multipoint links (discussed in the next sections) may also be encountered.

Results of a brief survey of other access points as interferers are summarized below. The survey was conducted in April 2004 using a commercial 802.11b client card in monitor mode with packets collected by the Kismac stumbler software package. In a separate measurement the card and stumbler were shown to be capable of capturing packets at a rate of at least 800 Kbps. (The data stream used was probably limited in rate by the server rather than the link. There is no reason to believe the stumbler cannot intercept a full-rate stream of approximately 6 Mbps.) Five outdoor locations in downtown San Jose, California were monitored; these locations were in the close proximity of museums, hotels, and a convention center as well as numerous businesses. Each location was monitored continuously for 300 consecutive seconds (5 minutes). Channels 1, 6, and 11 were monitored in continuous sequence with 1-second dwell time on each channel. Only access points or probe requests are shown below; the busiest access points had up to 10 clients, but client uplink traffic was in all cases less than access point downlink traffic.

The results are summarized in Table 8-3. An average of 17 individual 802.11 access points were detected at each site. The busiest location received 10,445 packets, which at the lowest 802.11 data rate would have required only 6.2 seconds or 2.1% of the monitoring time. If the client packets are assumed symmetrical, the busiest location would consume only about 4% of the available time. The average access point–offered load was only 1% of the total available, so that the total offered load is likely to be less

Table 8-3: Outdoor Interference by Existing 802.11b Access Points/Probes

Location	No. of MACs	Packets	Data (kilobyte)	Packet Size (bytes)	Packet Time at 1 Mbps (s)	Offered Load (%)
Chavez at Fairmont	23	10,445	778	74	6.2	2.1
Convention Center	15	1717	102	59	0.8	0.3
First and San Carlos	15	3295	246	75	2.0	0.7
First and San Fernando	18	4408	437	99	3.5	1.2
Tech Museum	16	3648	269	74	2.2	0.7
Averages	17	4703	366	76	2.9	1.0

than 2%. The average packet length was only 60–70 bytes in most locations, suggesting that many packets are beacons or management frames.

A histogram of the data in kilobytes per access point is shown in Figure 8-13. Over 80% of the access points transmitted less than 50 kilobytes during the monitoring time, compared with a presumed maximum capacity for a symmetric traffic stream of about 900,000 kilobytes.

There are various remaining questions, including whether other interferers were present that were limiting the ability to receive 802.11 packets successfully and how many networks were operating in 802.11g-only mode. However, the consistency of the results suggests that, at least at the present time, most 802.11 access points are lightly loaded at this time and do not represent significant interferers in the outdoor environment.

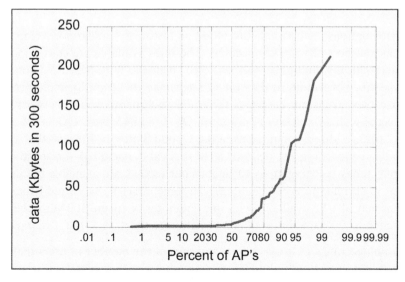

Figure 8-13: Cumulative Probability Plot of Data Received

3.3. Backhaul

An outdoor coverage network has a big potential problem: getting a connection to the global Internet to the access point. Access point indoors will generally be connected to an existing wired network, possibly through a firewall to protect the integrity of the wired network, so that an existing Internet connection can be exploited. An outdoor coverage network with no such connection isn't very useful for most applications.

Traditional backhaul to the global Internet is provided by various legacy wired networks. *Digital subscriber lines* (DSLs) use existing copper twisted-pair wiring to deliver around 1 Mbps of capacity to locations served by a wired phone connection and reasonably close to a telephone provider's central office; *T-1* lines provide a fixed 1.5-Mbps symmetric service that can be run over twisted copper wire but is also often provided using optical fiber. Both technologies require that a location be served by existing fixed wiring, and pricing varies from about U.S.$30/month for residential/ commercial single-use DSL to U.S.$500/month for T-1 service. *Cable modems* exploit the coaxial cable infrastructure originally installed to provide television service in addition to or in competition with broadcast television; the service is traditionally entertainment based and thus primarily residential, with little coverage of commercial areas (at least in the United States). Single modems are normally limited to downlink speeds of around 1–3 Mbps, and uplinks are much slower because of the limited allocation of uplink bandwidth. (The cable-modem downlink uses 6-MHz channels in the range of 550–850 MHz also used for digital television signals; the uplink has so far been relegated to narrower channels in the 5- to 45-MHz region, which was historically unused for cable television because of the prevalence of interference sources at the lower frequencies.) All these cited options provide much lower data rates than the native capacity of even an 802.11b access point.

Higher speed options for backhaul are based on historical standards developed for hierarchical transport of voice connections. The rates available are constrained by the need to hierarchically multiplex 64-Kbps voice channels. Wire-based *T-3* connections provide 44.7 Mbps of symmetric data rate. Faster links are generally implemented using fiber-optic cables. These are known as *SONET* (synchronous optical network) or *SDH* links and come with a fixed set of data rates: OC-1 (51.84 Mbps), OC-3 (155.52 Mbps), and OC-12 (622.08 Mbps) connections in the United States; in Europe, the corresponding connections are denoted STM-0, STM-1, and STM-4 respectively. (Faster links, today reaching up to 40 gigabits per second, are available but unlikely to be relevant to current WLAN technology.) In some areas, native wired Ethernet connections may also be available, with raw data rates of 10 and 100 Mbps that provide a better match to the capabilities of 802.11 networks than the SONET hierarchy.

Given the limited range of access points and thus of the number of users that can be served, spending a few hundred dollars per month for each access point to provide backhaul is likely to result in costs that exceed revenues that can reasonably be expected for any outdoor data service. The high-speed links that could actually serve the access point at full rate are much more expensive and are normally only used for connections between peering sites and data centers or for the wired networks of large corporate facilities employing hundreds of employees or having special high-density data requirements. (Open plains or deserts with minimal obstacles will allow a much larger coverage area but generally don't have any customers to pay the bills.) Inexpensive WLAN technology can become prohibitively costly when it is time to move the data.

There are three economical solutions to the backhaul problem for outdoor networks. The simplest is the one used indoors: exploit an existing backhaul connection. This is the approach taken in public Wi-Fi networks created informally by residential users who intentionally leave encryption off or by neighborhood organizations seeking to provide local coverage. In general, the owner of the access point in this case has already purchased a broadband connection through a DSL or cable modem and thus incurs no additional cost from public use of the access point. (The service provider company is arguably not so fortunate: their network is sized assuming an average data utilization for a residential customer, and expensive reengineering may be needed if average use increases due to sharing of the connection. In response, some service providers seek to impose restrictions on wireless sharing as a condition of service. However, one may plausibly argue that this is a cavil without much force, because the same effects on their networks would result from increased internet usage by the residents to whom their services is sold. A more sensible practice, already used by some providers, is to sell tiered services based on the total bits transferred per month; it is hoped that this model will in the end win out, providing freedom for users to experiment and bear the true costs when relevant.) Existing backhaul can also be exploited for commercial, industrial, and retail establishments with their own wired networks and such locations as telephone booths and transport centers that are equipped with data access for other reasons.

A different and more innovative solution is to use outdoor access points to provide wireless backhaul as well as access: a *mesh network*. In a mesh network, each access point also acts as a bridge to neighboring access points, and the whole group of access points together acts to route packets between internal clients and an external or Internet connection at one or more of the access points. Mesh networks can use conventional 802.11 radios and MAC layers, with a network-aware operating system at the TCP/IP layer where network forwarding takes place (see Figure 3-1). Mesh networks are able to substitute software for wireless capability, by intelligently optimizing routes to use the highest quality links and avoid interference. Network capacity can be increased gracefully by adding network nodes with wired backhaul connections. Network coverage grows when an additional node in radio range is turned on, using automated discovery and route optimization to take advantage of additional coverage. Conventional Internet routing protocols must be modified to avoid excessive routing traffic in large networks. Intelligent routing protocols can also allow a client to maintain its Internet protocol address as it roams from the coverage area of one node to that of another, thus minimizing or avoiding interruption in data streams and address-sensitive applications like virtual private networks.

Mesh networks allow flexible placement of the wired-backhaul nodes to take advantage of existing or inexpensive additional backhaul capacity, thus optimizing the overall installation and operating cost of a coverage network. Open-source implementations of many of the elements of a mesh network are available for the Linux operating system. The MIT Roofnet project (http://www.pdos.lcs.mit.edu/grid/) has demonstrated useful fixed-wireless coverage over an area of roughly 1 square mile (3 square kilometers), using approximately 60 simple nodes with current cost of about U.S.$650. Each node has a rooftop-mounted 8-dBi omni antenna. (The authors reported that higher gain omni antennas tend to become unusable after storms; the 8-dBi units, with a broad 20-degree beam, are robust to small displacements.) Commercial Senao 200-mW 802.11b cards provide radio connectivity. Internet throughputs of a few hundred kilobytes per second are reported for most nodes. The network is self-configuring and grows whenever a node is added, though the current routing protocols are not expected to scale to very large networks, and enough wired backhaul nodes need to be incorporated in the network to avoid swamping any given wired node. Routing messages in a network with links whose quality is time varying and often asymmetric is different from routing within a substantially static network and is an area of active research.

Proprietary commercial implementations from such companies as Tropos Networks and MeshNetworks were in early commercial deployment at the time of this writing. Good coverage is obtained in suburban and residential areas using on the order of 10 nodes per square mile (3–4 nodes per square kilometer). Symmetric rates (uplink and downlink) of as high as 2 Mbps are claimed. The commercial nodes have 1-W output power and low-noise receivers and achieve useful ranges of up to 800 m (along a street) versus typical range of 100 m for the Roofnet off-the-shelf equipment. Because the node cost for commercial units is on the order of $3000 and would come down considerably in large volumes, it is practical to cover significant urban or suburban areas at modest capital cost.

Finally, a third solution is to use an array of higher gain antennas to provide increased coverage and range from a single access location, minimizing the cost of backhaul per area covered. An example of a commercial implementation of this type of outdoor network is the Vivato array, shown in Figure 8-14. The structure shown is actually 16 vertically oriented slot antenna arrays laid side by side to create a horizontal phased array. The horizontal phased array is in turn driven by a fixed phase shifter, such as the Butler matrix described in Chapter 5 (see Figure 5-42). Thirteen 802.11 radios are connected to 13 nodes of the phase shifter to provide 13 separately directed beams; the whole collection provides coverage over a lateral angle of about 100 degrees.

Each vertical slot array has about 17-dBi gain, constraining radiation into the horizontal plane, and the array of 16 antennas adds about 12 dBi (10 log 16) for a total of 29 dBi.

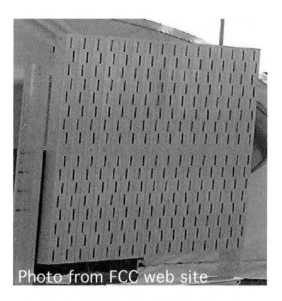

Photo from FCC web site

Figure 8-14: Vivato Slot-Array Antenna

A receiver listens to each active port of the phase shift array at all times to ensure that uplink packets are received; transmitters can be switched from one port to another depending on the target client. Beacons and association are performed separately for each pointing direction and sequentially within those beams that are on the same channel. Beam area is about 0.016 sr, implying a beam width of about 7 degrees.

At the time of this writing, the Vivato equipment was approved by the FCC under the point-to-point rules due to the fixed nature of each pointing direction. These rules allow higher effective isotropic radiated power (see Figure 5-13 if you've forgotten what that is!), because it is assumed that fixed point-to-point links represent less objectionable sources of interference. Specifically, FCC 15.247(b)3(i) requires 1/3 dB backoff for each dB of additional antenna gain over 6 dB (see Appendix 1). The Vivato system uses radios with output power of about 20 dBm, for an equivalent isotropically radiated power (EIRP) of $(20 + 29) = 49$ dBm, which is greatly in excess of the 36 dBm allowed for mobile radiators. This represents about a fivefold increase in line-of-sight range or penetration through several absorbing walls. In a field trial performed by the Horwitz International consultancy, a single unit at 12 m height was able to provide reasonable coverage to the interior and exterior of seven two-story wood-frame buildings spread over a campus area about 180 by 90 m. Similar coverage with conventional access points would have required about 7 units, with additional backhaul requirements.

4. Point-to-Multipoint Networks

A point-to-multipoint network is in some sense a generalization of the Vivato example discussed in the previous section: a single fixed-location wireless node, equipped with a wired backhaul connection, serving a number of (generally fixed) clients, providing data services. In view of the propagation issues discussed in section 3.1, it is apparent that for such a scheme to work, the serving node (often known as a *base station*) must have a clear or nearly clear line-of-sight to each client. This usually means a base station is placed on a tall building or tower in flat geographies or on a mountaintop in rougher terrain. Figure 8-15 shows a representative arrangement. Both approaches inevitably involve higher cost than simply sticking an access point on a pole, but because the base station is meant to serve many clients, it can support such added site costs as well as the cost of backhaul. Mountaintop locations have the further disadvantage that the easiest region to serve, within 2–3 km of the base station, is generally devoid of customers, because few people tend to live on the top of a mountain.

The great advantage of implementing such a network using 802.11-based technology is cost, particularly that of the *customer premises equipment*. Wi-Fi clients, even high-performance units, are much cheaper than most other data radios. In addition, operation in unlicensed bands means that there are no costs associated with bandwidth acquisition. However, there are some corresponding disadvantages. Operation in the unlicensed bands means one must accept any interference other users generate: such uncontrolled intrusions make it difficult to guarantee any particular quality of service to clients and make constant network monitoring and management necessary.

There are also some fundamental limitations related to the use of the 802.11 MAC outside of its designed application space. The MAC is based on collision sensing (carrier-sense multiple access with collision avoidance—see Chapter 3, section 3.2), which works best when most participants in a given basic service set can hear each other's broadcasts and defer to one another, but in a point-to-multipoint geometry this will rarely be the case. In general, all clients can hear the base station and the base station can hear all clients, but they can't hear each other. The normal response to such a condition is to implement request-to-send/clear-to-send (RTS/CTS) packets, but

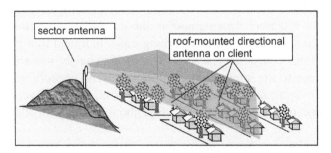

Figure 8-15: Point-to-Multipoint Configuration Serving a Residential Area

outdoors the efficacy of this approach is limited, because of the long propagation delays. The short interframe space (SIFS, see Figure 3-6) is normally on the order of 10 μsec long. In a small network, propagation delays are minimal, and all stations will hear an access point packet and defer for the packet and SIFS time. Recall, however, that to propagate a kilometer requires 3.3 μsec; a network with a reach of more than 2–3 km will have round-trip propagation delays considerably longer than the usual SIFS, so that medium reservations may not be heard in time and timer settings may be inappropriate. If the SIFS is adapted to the long distances, short packet exchanges will occur slowly due to the long interpacket waits. Furthermore, the long propagation delays represent a considerable overhead on RTS/CTS handshakes, reducing total data rate. When clients are at disparate ranges from the base station, a nearby client will capture the channel from a faraway client it cannot hear, even in the middle of a transmission, causing packet loss for the faraway client.

Commercial implementations (*wireless Internet service providers* [WISPs]) can use slightly modified MAC layers to adapt to the exigencies of the long-reach environment as well as to allocate bandwidth fairly to the multiple users. An alternative is to "layer" clients of a given access point at roughly equal distances, so that all clients come in at about the same power and contend on an equal basis for the channel. The threshold for RTS/CTS can be set to a very small packet size, though anecdotal reports indicate that the expected benefits in overall throughput don't always materialize. Interference from sources not in the WISP network is likely to play an important role in limiting throughput at long distances: unlike a local coverage network, the base station antenna is susceptible to interference from a large physical area.

Base stations in these networks generally use sector antennas (i.e., antennas with a narrow more-or-less horizontal beam that split the 360-degree view from the tower site into three or four azimuthal sectors). Three sectors are typical, though up to eight are possible. Antennas must be selected with care: they are exposed to the outdoors, often on mountaintop towers even more demanding than an ordinary outdoor site, and must survive temperature cycling, wind, rain, and possibly ice and snow. RF planning is needed to make such a system work. It is difficult to provide enough isolation from one antenna to another on the same tower to enable reuse of a channel. Recall that a distant client's signal, perhaps at the sensitivity limit of −85 dBm, is competing with the 1-W (30-dBm) transmission of the neighboring antenna a couple of meters away. An isotropic link would produce an interferer at −16 dBm, annihilating the wanted signal. The directive gain of the two antennas along the unwanted antenna–antenna link—the antenna–antenna isolation—must exceed 70 dB to avoid interference. That sort of isolation is achievable (barely) with parabolic antennas but not with broader beamed sector antennas, and such a demanding setup is vulnerable to reflection or scattering from objects on or near the tower. Thus, to provide high total service capacity, it is necessary to partition service to multiple towers or use the Unlicensed National Information Infrastructure band where more channels are available. Even once channels are partitioned one to a sector, it may be helpful to provide additional filtering to minimize adjacent-channel interference; commercial filters for 802.11 channels 1, 6, and 11 are available for this purpose.

Coordination between service providers sharing the same or nearby sites is needed to minimize frequency reuse and consequent interference. This coordination is informal today, but as service requirements evolve, providers may either find themselves in litigious conflict or be forced seek regulated bandwidth.

Out-of-band interferers, not normally a threat in indoor environments, are of some importance to base stations because they may be collocated with other broadcasting equipment. In the United States, the FCC has allocated a band at 2.5 GHz for the *multichannel multipoint distribution service*. The service was intended to allow cable television–like redistribution of broadcast and satellite signals. It is not widely used, but where present the transmitters may operate at up to 100 W. Few unmodified Wi-Fi radios have enough front-end performance to reject such large out-of-band signals (recall from Chapter 4 that these signals may exploit third-order distortion in the low-noise amplifier or RF mixer to create spurious signals in band). External filtering may be needed to enable operation in close proximity to multichannel multipoint distribution service transmitters. High-power ultrahigh frequency television broadcasting may also product harmonics that are within FCC regulations but are challenging for collocated microwave radios.

One of the main motivations for a point-to-multipoint architecture is the control of backhaul costs. The use of a small number of base stations means that larger backhaul cost can be amortized over many users. Many cellular telephone base stations are provided with T-1 lines served by the operator or local telephone provider; WISPs that share the base station thus have easy access to wired backhaul, albeit at high prices and with limited capacity. Point-to-point millimeter-wave links are also popular for cellular base station backhaul and may be available for collocated services or for independent WISP installations.

A customer with a fortunate ground-level line of sight to a nearby base station may be able to use the service with an unmodified desktop or laptop computer, perhaps exploiting a universal serial bus adaptor with movable antenna. Most installations, however, will require a directional antenna mounted on a rooftop to achieve at least near–line-of-sight conditions. In residential districts, trees are a major challenge; in commercial areas, other buildings may be the prime impediment to service. Yagi-Uda antennas (typically with around 12 dBi of power gain) or parabolic mesh antennas are often appropriate for the customer's site. When clear line-of-sight is not achievable, multipath distortion may make high-rate links impossible even though adequate total signal power is available for communication. In such circumstances, it may be more practical to turn to OFDM schemes with a large number of carriers (such as 802.16-compliant equipment) rather than struggle with 802.11 clients.

5. Point-to-Point Bridges

Bridges in wired Ethernet networks connect different LAN segments together so that packets can be routed to the appropriate receiving node. Wireless bridging is a straightforward extension of this function to the wireless environment, in which the access point establishes a link with a second access point to which MAC packets can be addressed based on appropriate forwarding tables. Many ordinary commercial access points can be configured as bridges; some can even be used simultaneously for communicating with local clients and bridging to a neighboring access point.

Access points configured as bridges provide a convenient method of connecting networks in neighboring buildings to create a single larger network. Directional antennas and rooftop mounting may be used to achieve unobstructed link ranges of several hundred meters. In conjunction with coverage access points, they can create a networked region with local coverage in and around buildings and wireless backhaul to possibly limited wide-area-network access.

An example of such a network implementation is shown schematically in Figure 8-16. The site, the Hidden Villa farm in Los Altos Hills, California, is in a valley surrounded by wooded rising terrain. The network links buildings separated by up to 270 m.

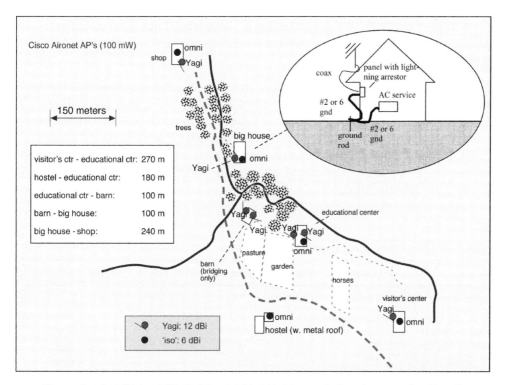

Figure 8-16: Hidden Villa Bridged 802.11b Network (Implementation Details Courtesy Greg DesBrisay and Pat McCaffrey)

Coverage access points provide in-building network access. Yagi-Uda directional antennas are used for the longer links. One site (the "barn") has no need for coverage, because the animals housed therein are not currently users of high-speed data services, but serves only as an intermediate bridge to ensure reliable penetration of the relatively heavy foliage surrounding the "big house" site, where DSL backhaul to the wired network is located. All the access points are Cisco Aironet access points with 100-mW transmit power (donated to this nonprofit site).

A simplified link analysis for the site is shown in Table 8-4. The received power in free space with no obstructions would greatly exceed that needed for reliable 802.11b service; this large margin is provided to ensure reliable service in the presence of foliage and to allow for pointing errors, wind-driven antenna displacements, and precipitation.

6. Long Unlicensed Links

Longer links can be implemented in unlicensed spectrum. In the United States, the FCC limits mobile transmitters in the ISM 2.45-GHz band to 36 dBm EIRP or, equivalently, to 1 W of transmit power in an antenna with no more than 6-dBi power gain (see Appendix 1). A user may employ an antenna with higher gain but must reduce transmit power to maintain the same EIRP. Such constraints reward the community by limiting interference except in the pointing direction of the antenna but hold no benefit for link integrity because the power density in the direction of maximum antenna gain is fixed.

A fixed point-to-point link is treated more leniently, because it is presumed that such links will interfere only with receivers in the direction in which they are pointed. Such transmitters are required to back off 1 dB for each additional 3 dB of antenna directive gain in the 2.45-GHz band and need not back off at all in the 5.725- to 5.850-GHz band. For high-gain antennas that's a huge advantage: a 21-dBi parabolic antenna can use up to 26-dBm output, allowing an EIRP of 47 dBm. This extra power density in the pointing direction allows line-of-sight links at ranges of up to tens of kilometers. Recall that using a 6-dBi transmit antenna and 0-dBi client antenna, we

Table 8-4: Link Analysis for Layout of Figure 8-17

Link	Antenna 1	Antenna 2	Distance	Friis Power	Margin	Remarks
Visitor ctr ⇔ edu ctr	12 dBi	12 dBi	270 m	−45 dBm	30 dB	Line of sight
Hostel ⇔ edu ctr	6	12	180	−47	28	Line of sight
Edu ctr ⇔ barn	12	12	100	−36	39	Trees
Barn ⇔ big house	12	6	100	−42	33	Trees
Big house ⇔ shop	12	12	240	−44	31	Trees

Transmit power = 100 mW. "Friis power" = power calculated assuming free space propagation; margin is relative to rule-of-thumb high-rate power of −75 dBm.

expected a range of 3.5 km in an unobstructed link using 1 W of transmit power (Figure 8-1). If we substitute 21-dBi directional antennas on both ends and reduce the transmit power as per regulations, we obtain an additional link margin of

$$(D_{TX} - 6) + (D_{RX} - 0) - \frac{1}{3}(D_{TX} - 6) = 15 + 21 - 5 = 31 \text{ dB} \qquad [8.3]$$

(where D is the antenna directivities in dB). This corresponds to an increase of $10^{31/20} \approx 35$ in the range of the link: from 3.5 km to 125 km! If available power were the only limitation, very long links could be constructed using only unlicensed radios.

Unsurprisingly, lots of practical obstacles stand in the way of such very long reach. Real links can reach 10–20 km; ranges >50 km are very difficult to achieve reliably even using licensed bands and higher transmit power. The obstacles are essentially the same as those encountered in licensed links (with the exception of interference), so much of the discussion of this section follows Trevor Manning's excellent book, *Microwave Radio Transmission Design Guide* (see section 9).

First, reliable long links require a clear line of sight; remember that two or three trees in the path will use up all the 31 dB of extra link budget gained from point-to-point status. Because the earth's surface is curved, this means that at the minimum, the transmitter and/or receiver must be raised above ground level by larger distances as the link length increases. (Raised transmit and receive sites also help clear local obstacles such as trees and buildings.) The effective height of an ideal averaged earth surface is

$$h = \frac{d_{TX} d_{RX}}{12.75} \qquad [8.4]$$

where the distances d are in kilometers and the height h in meters. For "flat" local conditions, the highest point is midway between the transmitter and receiver; for a 20-km link that's 8 m. The height is quadratic in distance, so a 40-km link would require 32 m of clearance: a tower, tall building, or the aid of a mountain is required for such a long link.

How clear is clear? Recall from our discussion of straight-edge diffraction that when the edge is right on the line-of-sight path, the signal is reduced by 6 dB from the unobstructed signal (see Figure 6-33). To get close to unobstructed path loss, the line of sight should be above the highest obstruction edge by at least a normalized length $u_o \approx 0.6$, that is, we'd like to have at least most of one Fresnel zone clear of obstructions. The Fresnel zone side varies with distance from the transmitter and receiver; at the midpoint between the two, the required clearance is approximately

$$h_{clear} \geq 0.6\sqrt{\lambda \frac{d[\text{meters}]}{2}} = 13.4\sqrt{\lambda d[\text{km}]} \qquad [8.5]$$

where d is the distance to either node from the midpoint and wavelength λ and height h are in meters. For a 20-km link the clearance required is about 14 m, which needs to be added to the height of the line of sight. Note that equation [8.5] should be taken as a useful but approximate guideline; a perfectly straight edge on a half-plane is unlikely to be a very accurate representation of the real obstacle environment for any given link. One could interpolate between the straight-line model and an aperture model (equation [6.30]) to try to estimate the consequences of a path constrained both by ground and vertical obstacles such as neighboring buildings or a path through a gap in mountainous terrain, but the accuracy of such approximations is unlikely to be good enough to be worth a great deal of effort. For the reader determined to be quantitative, a collection of appropriate estimates is available from the International Telecommunications Union as their recommendation ITU-R P.526-5.

To establish a clear line of sight is not so simple in the real bumpy world. If you've already determined the locations of your prospective endpoints, you can establish an optical line of sight using a telescope at one intended endpoint of the link; if the other end is not prominently visible, it will be helpful to have a colleague at the other end armed with a hand mirror to direct sunlight toward the observer. (Two-way communications by, e.g., cellular telephone are indispensable in such an effort.) Note, however, that the average optical path is not identical to the average microwave path because of the gradients in refractive index of the atmosphere: the optical path correction factor k (which is explained in detail below) is around 1.5 versus a typical microwave value of 1.3. If you just happen to have a helicopter with a radar altimeter and global positioning satellite (GPS) available, you can overfly the route and map the heights of obstructions (but these resources are unlikely to be justified for an unlicensed link!).

To establish a line of sight for prospective sites, it is first necessary to know their exact locations to within about 20 m. Low-precision GPS may not be accurate enough for this purpose; reference to surveyed points or differential GPS from a known location may be needed. The proposed link path may then be reviewed against a topographical map, taking into account the earth's curvature and refraction (discussed below). The proper approach using a printed map is to draw the path with a ruler and extract the height of each contour line intersected; these heights are then corrected for the curvature of the earth and plotted versus distance to create a side view (profile) of the path. If a digital terrain model is available, the process can be automated. Note that typical low-resolution models, with 200-m pixel sizes, are not sufficient for final verification of the path. Digital terrain data are available from numerous commercial vendors as well as government sources; a nice introduction to these tools can be found (as of this writing) at www.terrainmap.com.

We have intimated several times already that for long links refraction must be considered. The density of the earth's atmosphere falls with increasing altitude, providing a background gradient in refractive index upon which one must superimpose local and global variations in temperature and humidity. Because the refractive index n is very close to 1, it is customary in treating atmospheric refraction to define the differential index N as

$$(n-1) \cdot 10^6 \equiv N \qquad [8.6]$$

The differential index is mostly determined by the ratio of total pressure to absolute temperature (which by the ideal gas law is proportional to the total density) and the humidity. (The water molecule is highly polar relative to other major constituents of the atmosphere; recall from Figure 6-7 that liquid water has a large refractive index.) To a good approximation,

$$N \approx 77.6 \left(\frac{P}{T}\right) + 3.73 \cdot 10^5 \left(\frac{P_{H_2O}}{T^2}\right) \qquad [8.7]$$

where the pressure is measured in mbars and the temperature in Kelvin. (For those who love unit conversions, 1 bar = 105,000 Pa = 0.987 standard atm = 750 Torr = 29.5 in Hg, and 0°C = 273 K). The partial pressure of water can be obtained from Figure 6-6 for a given humidity and temperature using the molar weight of water (18 g/mol) and the ideal gas law $PV = NRT$, where $R \approx 0.083$ mbar m^3/mol K. A typical value of N for room-temperature ground level conditions is around 300–330. Because of the generally exponential decrease in density with increasing altitude in the atmosphere, on average the refractive index falls as well:

$$N(h) \approx N_o e^{-\frac{h}{h_o}}; \quad h_o \approx 7.35 \text{ km} \qquad [8.8]$$

The vertical gradient of N, dN/dh, is often denoted G. The gradient solely due to the average pressure change with altitude may be obtained once we recall that the derivative of an exponential is that exponential multiplied by the derivative of the argument; we get a gradient of about −43 differential index units per kilometer of altitude. The average measured value of the gradient in clear weather, in data acquired in France over a number of locations and time periods, is about −39, suggesting that although humidity plays a strong local role, on average water vapor occupies the lower levels of the atmosphere in a fairly uniform fashion, at least away from clouds.

Recall from Chapter 6 that a gradient in refractive index causes a ray to curve, with radius of curvature R related to the magnitude of the gradient:

$$R = \frac{n_o}{\partial n/\partial h} \approx \frac{1}{\partial n/\partial h} = \frac{10^6}{G} \quad \text{where } G \equiv \partial N/\partial h = 10^6(\partial n/\partial h) \qquad [8.9]$$

Thus, the experimental value of G corresponds to a radius of curvature of about 23,000 km, compared with the actual radius of curvature of the earth's surface of about 6370 km: Refractive deflection in typical conditions is quite small, though not completely negligible. However, values of G as large as +300 and as small as −300 are observed in temperate regions about 10% of the time over the course of a year because of the local vagaries of temperature and humidity. Such values correspond to radii of curvature around 3000 km, smaller than that of the earth. For long links refraction must be considered if high availability is an important goal.

In the literature on this topic it is common (if confusing to the uninitiated) to use an *effective radius of curvature* for the earth instead of explicitly considering the effects of the refractive index gradient. The definition is as follows:

$$R_{eff} = kR_{earth} \qquad [8.10]$$

In this view, the microwave (or optical) ray is considered to travel in a perfectly straight line over a surface whose curvature is modified to account for refraction. Values of $k > 1$ correspond to a flatter effective earth surface; $k < 1$ implies more curvature than the real earth. After a bit of algebra, one can derive an expression for k in terms of G from equation [8.9]:

$$k = \frac{1}{1 + R_{earth}\partial n/\partial h} = \frac{1}{1 + R_{earth}G \cdot 10^{-6}} \approx \frac{1}{1 + 0.00637G} \approx \frac{157}{157 + G} \qquad [8.11]$$

With this equation we find that the typical range of observed gradients, from about $G = 0$ to $G = -157$, corresponds to values of k from 1 to 8. (An infinite value of k simply denotes a flat effective earth.) Larger positive values of G give rise to *subrefraction*, where the beam is bent away from the earth's surface; larger negative values of G cause *ducting*, redirection of the beam into the ground (Figure 8-17).

Subrefraction occurs when n increases with altitude despite the decrease in pressure. This sort of behavior can result from a strong decrease in temperature with altitude or a large increase in moisture (or both). Subrefraction is characteristically the result of cold moist air moving over warm ground. Such a condition is unstable to natural convection, in which the cool air sinks into the lighter warm air, explaining the rarity of subrefraction; unusual conditions are required to make $k < 1$.

Ducting occurs when the refractive index decreases rapidly with altitude. Ducting can result from, for example, dry air flowing over air humidified by evaporation from a lake or ocean. In some ocean areas, semipermanent ducts form within about 20 m of the water surface. In the continental United States, warm fronts tend to slide on top of existing cool air, creating appropriate conditions for ducting, particularly over bodies of

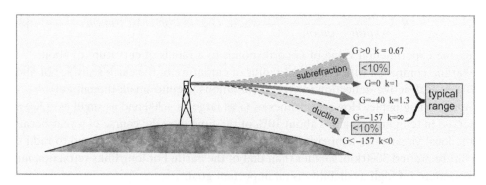

Figure 8-17: Ranges of G And k Depicted vs. Ray Path Relative to Earth

water. Elevated ducts sometimes form below stratocumulus clouds. Most ducting in temperate regions is short-lived, except for those over the ocean. Long-lived elevated ducts are more frequently observed in the tropics.

Rain is water, and water has both a high refractive index and absorbs strongly at WLAN frequencies (see Figure 6-7), so one would expect additional attenuation during precipitation. However, the actual density of rainwater in the air is fairly small, and the droplets are very small compared with the wavelength of ISM or Unlicensed National Information Infrastructure radiation, so they don't scatter very strongly. The result is that rain attenuation is only significant for very heavy rains and then only in the 5-GHz band (Table 8-5).

Just as in the indoor environment, rays following multiple paths from transmitter to receiver can interfere destructively, leading to multipath fading. However, for long links and highly directional antennas, only glancing reflection from the ground or from rare elevated ducts can lead to multipath. Multipath fading can result from specular reflections from calm water surfaces. It can also be a problem if the main beam is not properly centered on the receiving antenna or is defocused by refraction so that nonspecular reflection from the ground can compete with the direct ray. A pair of diversity antennas, separated as usual by at least a half a wavelength (which is unavoidable using large directional antennas), is generally sufficient to ameliorate the worst effects of fading.

Finally, the use of highly directional antennas, typical of long outdoor links, brings with it demanding requirements for pointing accuracy. Recall that the beam width is inversely proportional to the square root of the antenna directivity (equation [5.18]). An antenna with directivity of 20 dBi has a beam width of 20 degrees; by the time 30 dBi is reached, the beam is only about 6 degrees wide. To avoid significant changes in received intensity, displacement of the beam from the intended target at the transmitter or receiver should be significantly less than the beam width, so for a 30-dBi antenna, displacement should not exceed about 2 degrees. An antenna of this type at 2.45 GHz would be around 150 cm in diameter, so a 2-degree rotation corresponds to a displacement of about 2.5 cm (1 inch) of the antenna rim. Such displacements greatly

Table 8-5: Attenuation vs. Rainfall for Frequencies Relevant to Outdoor WLAN Operation

f (GHz)	Rainfall (mm/h)	Attenuation (dB/km)
2.45	25	<<0.01
	100	0.02
5.5	25	0.04
	100	0.3

After Antenna Engineering Handbook, Johnson, pp. 45–49.

exceed the imprecision expected of any antenna mounting hardware in calm conditions, but in high winds both the antenna mount and support structure are exposed to significant mechanical stress, and if care is not taken, displacements of this magnitude are quite likely. A commercial-grade radio link is normally expected to have only 10-dB variations in signal strength during a "10-year" wind and survive without damage in a "50-year" wind. To achieve such a standard, accurate initial pointing adjustments and well-designed rigid mountings are indispensable. Stresses induced on the antenna mount are greatly reduced if grating antennas are used instead of, for example, solid parabolic reflectors, at the cost of slight reductions in directivity and a significant penalty in front-to-back ratio. Grating antennas should be considered when only one link is present at a given site, though if multiple links are placed in a single location on the same or neighboring frequencies, the degradation in isolation may not be tolerable.

7. Safety Tips

Any network installation requires that some thought be given to safety of both installers and users, but outdoor networks involve numerous potential dangers that are rarely encountered indoors. Some of the most important precautions are cited below for convenience.

Because of the requirement of ground clearance, outdoor networks of all kinds often involve working high above the ground. In some cases the heights involved are sufficient to kill you if you fall. Some general guidelines for working safely are as follows:

- If > 2 m (7 feet) from ground level, use guardrails or a safety belt with a shock-absorbing harness.

- Working on a high roof, within 2 m (6 feet) of any edge without a guard rail, use a retracting safety line (known as a "yo-yo"). The line attachment must be able to support a 1500-kg (3000-pound) load, because this is the expected impulse load from a fall.

- Always wear a safety line on any roof whose slope exceeds 3:12.

- Don't use a conductive ladder (steel or aluminum) close to high-voltage power lines.

- Don't work on outdoor installations in stormy weather.

Outdoor installations (antennas and towers) should always be carefully grounded. It is important that there is only one ground potential for any given site. Multiple independent "ground" networks should be avoided, because they can give rise to differing potentials that can cause equipment problems and hazardous conditions. Thus, grounds for the AC power system, phone system, towers, antennas, and any underground metal piping that is present in the area should be tied together.

For light-duty sites, such as a single access point mounted in or on a residence, a ground resistance of around 25 Ω is sufficient. Industrial/commercial sites, such as cellular base station collocations or sites containing networking equipment and multiple access points or other radio equipment, a system resistance of less than 5 Ω to true ground is appropriate.

During site selection, soil resistivity should be measured. For those familiar with semiconductor characterization, the procedure is a bit nostalgic: one constructs a giant-sized four-point probe using grounding rods. Four rods are driven at least 20 cm into the ground, evenly spaced along a line. Resistance is measured by forcing current into the outer pair and measuring voltage across the inner pair. The soil resistivity in Ω-cm is then approximately $\rho = 633S\,R$, where S is the spacing in meters and R the measured four-point resistance (ratio of voltage to current). This information can be used to ensure that the system ground resistance meets the requirements cited above.

Grounding should be performed using special-purpose ground rods; don't rely on underground piping to provide low-resistance grounding. Typical grounding rods are 15 mm (0.6 inch) diameter and around 2.4 m (8 feet) long. Rods should not have any paint or other nonconductive finishing. Normally, it is safe to use steel rods with copper cladding. If the soil pH < 5 (i.e., in strongly acidic soil), use solid copper rods, larger diameter clad rods, or special electrolytic rods to minimize corrosion. In acidic soil, grounding integrity should be verified yearly. Grounding rods may be driven into the ground at an angle but should be at least 75 cm deep.

Connections from the ground rods to the network should use solid copper wire of AWG no. 2 or larger. The wires should not be bent sharply to avoid cold working and fatigue. Large towers or buildings should be surrounded by a ring of ground rods. All metal objects within 3 m of the external ground system should be connected to it, including metal fences, cable runways, roof-mounted antennas, and supports or guy wires attached to any tower or antenna. All internal systems and equipment racks should also be grounded. In no case should braided grounding straps be used. Existing towers should not be welded to or drilled for ground connections without a review of the structure by the tower manufacturer or a professional engineer. Existing ground connections should always be checked for current flow, using an inductive current meter, before being disconnected. Disconnecting a live ground line may result in large inductive surges and dangerous overvoltages. Lightning arrestors connected to ground should be placed wherever outdoor RF cables (or other electrical connections) transition through a wall into an indoor environment.

Some outdoor sites will use backup batteries to ensure power in the case of a power grid disruption. These batteries are typically lead–acid types and thus contain sulfuric acid and release hydrogen when charging. Therefore, there is danger of explosions and exposure to corrosive liquids. When working with backup batteries, use insulated tools and eye protection.

Whenever working with high voltages, work with a buddy who can provide assistance in case of an accident. Don't use a grounded wrist strap. Don't mix aluminum and copper conductors. Consumer-grade electrical strips should never serve as a permanent power source for industrial equipment. Cabling used in conduits should be plenum rated even if local regulations allow otherwise; nonplenum cabling may release toxic compounds in the case of a fire. If cabling or ducting penetrates a firewall (a wall that is required by building codes to withstand a fire for a certain period of time), precautions, such as a fireblock within the duct, must be taken to ensure the integrity of the firewall.

Battery racks, cabinets, and other heavy objects should be solidly braced if the site has any likelihood of seismic activity (that's pretty much anyplace in California!). Two folks should work together to lift any object heavier than 18 kilograms (40 pounds). Check with local utilities or utility locating services before digging or trenching at the site.

If developing a new outdoor site, be aware that in addition to local permits and zoning approval, U.S. Federal Aviation Administration approval is required for any tall free-standing structure. If you're merely attaching additional antennas to a tower, it may be advisable to get an analysis by a civil engineer to ensure that the extra loading is acceptable. Transmission lines from a tower-mounted antenna should include a drip loop to prevent water from running along the line into a building. Cables should be routed using cable runways, never hanging J-bolts for support. Cables should not be mounted to the rungs of a ladder.

Finally, if working on a collocated site with, for example, a cellular phone provider, be aware that radiated power for licensed sites can be as high as several hundred watts. If you have to work close to a high-power antenna, make sure it is not active when you're there!

The reader who needs to perform extensive installation work outdoors should consider reviewing (at least) the Motorola publication Standards and Guidelines for Communications Sites cited in section 9.

8. Capsule Summary: Chapter 8

Use of WLAN technology outdoors is possible and sometimes very useful. If unobstructed, WLAN radios can easily reach several kilometers, but in the real world propagation is limited by many obstacles. Coverage networks in urban areas are limited by the presence of buildings and foliage; as a result, effective ranges of hundreds of meters can be obtained along a straight street with a well-placed access point, but the signal will not turn distant corners and may not serve adjacent streets very well. The expense of providing wired backhaul to outdoor coverage networks can greatly exceed the cost of the network hardware; solutions to this conundrum include exploiting existing backhaul connections, mesh networking, and use of a small number of high-performance access points.

WLAN technology can be used to provide point-to-multipoint data services at very low equipment cost. Base station siting with good line-of-sight coverage is necessary to provide enough potential customers for a viable business. The 802.11 MAC can be used with minor modifications, though such use stretches the capabilities of a system designed for short-range indoor applications.

Point-to-point bridges to link building networks can be readily implemented over several hundred meters using commercially available antennas and access points. For longer links, refraction, absorption, and pointing accuracy issues that are negligible at shorter ranges must be considered. Long point-to-point links of tens of kilometers are achievable with high-gain antennas and good siting, although more appropriate technologies should be considered when high throughput and guaranteed quality of service are desired.

Working outdoors is potentially much more dangerous than working indoors. Network installers must attend to the hazards of elevated workplaces, inclement weather, high-voltage electric power, and high-power RF radiators.

9. Further Reading

Propagation in Urban Environments

"Prediction of Path Loss in Environments with High-Raised Buildings,"
F. Frederiksen, P. Mogensen, and J. Berg; IEEE Vehicular Technology Conference 2000, p. 898

"Diffraction Around Corners and Its Effects on the Microcell Coverage Area in Urban and Suburban Environments at 900 MHz, 2 GHz, and 6 GHz," V. Erceg, A. Rustako, and R. Roman, IEEE Trans Vehicular Technology 43, no. 3, August 1994

"A Comparison of Theoretical and Empirical Reflection Coefficients for Typical Exterior Wall Surfaces in a Mobile Radio Environment," O. Landron, M. Feuerstein, and T. Rappaport, IEEE Trans Antennas & Propagation, vol. 44, no. 3, p. 341, 1996

"Propagation Loss due to Foliage at Various Frequencies," J. Dalley, M. Smith, and D. Adams, 1999 IEEE Conf. Antennas & Propagation

"Fading Characteristics of RF Signals due to Foliage in Frequency Bands from 2 to 60 GHz," S. Perras and L. Bouchard, 5th International Symp. Wireless Multimedia, p. 267, 2002

High-Speed Wired/Optical Backhaul

Understanding SONET/SDH and ATM, S. Kartalopoulos, IEEE Press, 1999

Mesh Networks

http://wire.less.dk/wiki/index.php/MeshLinks

http://www.pdos.lcs.mit.edu/grid/

"Opportunistic Routing in Multi-Hop Wireless Networks," Sanjit Biswas and Robert Morris, Proceedings of the Second Workshop on Hot Topics in Networking (HotNets-II), Cambridge, Massachusetts, November 2003

"A High-Throughput Path Metric for Multi-Hop Wireless Routing," Douglas S. J. De Couto, Daniel Aguayo, John Bicket, and Robert Morris, Proceedings of the 9th ACM International Conference on Mobile Computing and Networking (MobiCom '03), San Diego, California, September 2003

Array Network Coverage

Comprehensive Site Survey, Sterling University Greens Apartment Housing Complex, Horwitz International April 2003; available from Vivato web site www.vivato.net or Horwitz International http://www.wi-fi-at-home.com/

Long Links

Microwave Radio Transmission Guide, Trevor Manning, Artech 1999

Recommended formulae for diffraction loss: ITU-R P.526-5, 1997

Digital terrain models: http://www.terrainmap.com/

Safety

http://www.osha.gov/SLTC/constructionfallprotection/index.html

Standards and Guidelines for Communication Sites, report no. 68P81089E50-A, 2000; available from Motorola Broadband Communications

Afterword

I hope you've found this book informative and perhaps on occasion as entertaining as a technical tome can hope to be. I'd like to offer thanks again to the many people who helped and accept responsibility for the inevitable errors of commission and omission. Comments and criticism may be addressed to me at enigmatics@batnet.com.

Daniel M. Dobkin
Sunnyvale, CA
April 27, 2004

Regulatory Issues

1. A Piece of History

Anarchy degenerates into government. If you wait long enough, you get revolution and anarchy, and the cycle starts again. So it is (hopefully only in a metaphorical sense) with radio. After early experiences with anarchy in the use of wireless at sea, culminating in the highly publicized sinking of the *Titanic* in April 1912, a U.S. law passed in August of that year gave the Secretary of Commerce the responsibility of administering licensing of radio stations within the United States. Broadcast radio in the United States arose around 1921, and the initial legal framework recognized de facto possession of spectrum as constituting a property right: the first broadcaster in any jurisdiction to broadcast in a given frequency band owned the spectrum they transmitted in. However, confusion and contrary court decisions caused this simple guideline to falter, leading to broadcast anarchy and interference with popular stations in major broadcast markets. In 1927 the U.S. Congress passed the Radio Act, which established a Federal Radio Agency with responsibility for stewardship of spectrum; 7 years later the Congress replaced the Federal Radio Agency with the Federal Communications Commission (FCC). The FCC through its early years followed a strict command-and-control regulatory model, in which licenses for spectrum were granted only for specific uses in specific locations, with no rights to trade, modify, or assign the license.

Almost all other nations in the world followed suit in establishing a national agency to regulate the use of the radio and microwave spectrum. Coordination between these agencies is today provided by the *International Telecommunications Union* (ITU), which operates under the auspices of the United Nations. The ITU gathers the national regulatory agencies once every 3 years at the *World Radiocommunications Conference*, which like most such exercises generates only slow and painful progress toward world consensus on the regulation of radio technology.

Fortunately for wireless local area networks (WLANs), things began to change in the 1980s, as the mantra of deregulation slowly spread through American government and politics. In 1985, the FCC approved the use of *Industrial, Scientific, and Medical* (ISM) spectrum at 900 MHz and 2.4 GHz (and a bit around 5 GHz) for unlicensed use, with limitations on power, allowed modulation techniques, and antenna gain to minimize consequent interference. Additional spectrum around 5 GHz, known as the *Unlicensed National Information Infrastructure* (UNII) band, was made available in 1997. Similar actions were being taken by other regulatory agencies around the world. The manifest success of these actions in promoting innovative use of spectrum induced the FCC to make additional 5-GHz spectrum available in 2003 (see Figure 3-15) as well as to allow ultrawideband (UWB) radios to operate without licenses in spectrum previously reserved for other uses (see Figure 3-27).

The proper model for regulation of spectrum usage is a topic of active debate within and outside of governments. The success of unlicensed spectrum has demonstrated the feasibility of a "common" model in which neither detailed use regulations nor property rights are applied; the success of the auctioned spectrum used for cellular telephony has demonstrated the utility of a market in spectrum rights. Both models are arguably superior to the command-and-control model practiced for most of the twentieth century.

It is important to recall the physics that frames the changing debate. The 1912 Act in the United States dealt with radio wavelengths in the range of tens of meters to over 1600 m (i.e., from tens of kilohertz to around 10 MHz). Buildings and even small mountains represent sub-wavelength obstacles at these frequencies. Diffraction is easy, absorption is small, and very high powers were often used. Direct ranges of tens to hundreds of kilometers are typical. The lower frequencies are below the plasma frequency of the ionosphere (the layer of partially ionized gas that lies above the stratosphere) and thus are reflected by it, so that they can hop across hundreds or thousands of kilometers. That is, early radio stations broadcast everywhere, limited only by radiated power, and interference was intrinsically a public issue.

By contrast, the WLAN emissions we've studied are much easier to corral. For indoor networks, low radiated power, combined with good antenna management and a helpful building (thick concrete or wooden walls and conductive-coated windows), can keep most of the radiated power within a single structure. In many locations, foliage provides a strong limitation on outdoor propagation distance except for line-of-sight applications. Management of interference becomes much more akin to a local property right, administered by property owners for indoor networks and perhaps by local authorities (city or county governments) for outdoor networks. There is less need and less justification for government intervention in radio operation in these circumstances.

As communications evolves toward the use of still higher frequencies, with more intelligent radios and antennas, it can be expected that the laws and codes regulating the use of radio will continue to evolve as well. It is to be hoped that further innovation will produce more ways of sending data at lower cost as a result. That's the fun part; the remainder of this appendix is necessarily best suited for curing insomnia, though still less so than the original documents. Only the brave need continue.

2. FCC Part 15

Use of unlicensed spectrum in the United States and other countries is not unregulated. Limitations are placed on total power, modulation approaches, antennas, and conditions of operation in an attempt to minimize interference within the intended band. Further limitation of radiated and conducted emissions (i.e., stuff that sneaks back into the AC power system through the equipment power cord) is imposed to minimize interference outside of the intended bands. Equipment must generally be approved before marketing; the main thrust of the approval process is to ensure that the regulations regarding interference are met by the equipment. Users are responsible to see that the limitations in the regulations are met during operation.

The regulations of the FCC are contained in Title 47 of the U.S. code of Federal regulations. Part 15 deals with operation of unlicensed radio transmitters. Table A1-1 summarizes some of the key aspects relevant to WLAN applications. Some of the curious choices in modulation and architecture made in the various protocols may become clearer on review of the relevant regulations. Of particular interest are ISM-band regulations (15.247), UNII band regulations (15.401), and UWB regulations (15.501). The most recent changes in UNII regulation have not yet made it into the code, which was updated in October 2003. Table A1-1 is meant to help the reader understand what the law is so as to remain within it; when in doubt, review the original text (available at the Government Printing Office web site, www.access.gpo.gov) or seek legal counsel if appropriate.

For the tinkerer and network installer, it is important to consider specific restrictions and loopholes. The intent of the regulatory apparatus for unlicensed radios is stated in 15.5, paragraph (b): an unlicensed radio must not create harmful interference, and must accept any interference it encounters. The basic framework followed to ensure this result is to require certification of all unlicensed equipment before sale (15.201). Furthermore, the rules seek to allow the use of only those antennas and power amplifiers that have been certified with a particular transmitter (15.203, 15.204). However, several loopholes are provided in this regulatory fence. First, 15.23 specifically allows "home-built" radios that are produced in small quantities and are not sold. In addition, 15.203 implies noncertified antennas can be used when systems are "professionally installed."

How should a responsible citizen who also wants a network that works act in view of the regulations? (Note that the following is the author's personal view and not a legal opinion.) First, do no harm. Remember that the legal obligation of 15.5(b) is the same as the ethical obligation to your fellow spectrum users: avoid harmful interference. Proper consideration of the interests of your neighbors in setting up your network will avoid most disputes. The language of 15.23 provides the second important guideline: "Good engineering practices should be used to ensure that devices operate within the restrictions of the regulations." Good engineering practice in this context means that (for example) before swapping out an antenna on a certificated radio, the user is responsible to at least demonstrate by calculation, using the nominal output power of the transmitter and the manufacturer-specified power gain of the antenna, that the regulatory requirements will still be met. Other changes to any certificated system

Table A1-1: Summary of Code of Federal Regulations Title 47 Part 15 as of April 2004

Section	Topic	Summary, Excerpts, Remarks
15.1	Scope	This section sets out the conditions under which unlicensed operation is permitted
15.5	General unlicensed operation	"Operation of an intentional, unintentional, or incidental radiator is subject to the conditions that no harmful interference is caused and that interference must be accepted that may be caused by the operation of an authorized radio station, by another intentional or unintentional radiator, by industrial, scientific and medical (ISM) equipment, or by an incidental radiator."
15.7	Testing	Testing of noncertificated equipment is allowed to ensure compliance with regulations; shipping even prototype units to customers before certification is not.
15.23	Home-built devices	Home-built devices are allowed if in small quantities and not for sale. Good engineering practices should be used to ensure that devices operate within the restrictions of the regulations.
15.31	Compliance measurement procedures	Read 'em and weep. This is why you pay a testing laboratory.
15.33	Measured frequency range	Frequency range for radiated measurements. Start at the lowest frequency generated by the device and (for devices operating at <10 GHz) and continue to the tenth harmonic or 40 GHz, whichever is lower. Thus, for 2.45-GHz devices the limit will be 24.5 GHz and for UNII-band 40 GHz.
15.35	Pulsed transmitters	Specifies how to measure compliance of pulsed transmitters with requirements on average emission.
15.201	Certification	All intentional radiators (except home-builts as in 15.23) should be certified under part 2 subpart J (a separate chapter of the regulations, not covered in this table) before marketing.
15.203	External antennas	All devices must either use built-in antennas or nonstandard connectors not easily available to the public (maybe true years ago but hardly today), except those that are required to be "professionally installed." In the latter case, the installer is responsible for ensuring that the limits of radiation are not exceeded.
15.204	External power amplifiers	External power amplifiers and antennas may not be marketed except as part of an approved system. Only the antenna that a system is authorized with may be used with the intentional radiator.

Table A1-1: (*Continues*)

Section	Topic	Summary, Excerpts, Remarks
15.205	Spurious emissions only	Lists bands in which only spurious emissions are allowed. Bands of interest for WLAN are 2310–2390, 2483.5–2500, 4500–5150, and 5350–5460. (Note: This band is still excluded under the 11/03 revision; see Figure 3-15.) Limits for emission are stated in 15.209. Note there are limits on conducted emissions to the AC power line and low-frequency (<1 GHz) emissions that are not specifically cited in this summary, though emissions may exist due, e.g., to local oscillator frequencies, and must be measured in certification.
15.209	Spur limits	For $f > 900$ MHz, field strength from an intentional radiator measured at 3 m shall be less than 500 μV/m (except as provided in 15.217 through 15.255 for specific frequency bands) Note that this field strength corresponds to an EIRP of about −41 dBm as per the conversion cited above. Spurs must be less than the intended fundamental. Measurements should use an averaging detector above 1 GHz. Measurements at distances other than 3 m are converted using procedures from 15.31,33,35.
15.247	ISM band regulations	This one is important and long. Summary presented separately after the table.
15.249	ISM point-to-point links	This appears to specifically apply to fixed point-to-point operations. Field strengths are not to exceed 50 mV/m (fundamental) and 500 μV/m (harmonics) in the 2400–2483 and 5725–5850 bands, measured at 3 m distance. Out-of-band emissions other than harmonics shall be attenuated at least 50 dB from the fundamental or less than the requirements of 15.209, whichever is less demanding. Average detection is permitted but the peak shall not exceed the specs by more than 20 dB under any conditions.
15.401	UNII general	
15.407	UNII band regulations	The equivalent of 15.247 for the UNII band; summarized after table below.
15.501	UWB general	
15.503	UWB band definitions	fH and fL are the 10-dB-down upper and lower frequencies. fC is average of fH and fL. Fractional bandwidth = (fH − fL)/fC.

(*Continued*)

Table A1-1: (*Continues*)

Section	Topic	Summary, Excerpts, Remarks
		UWB = fractional bandwidth > 0.2 *or* bandwidth > 500 MHz.
		Several sections cover requirements for special applications of UWB (e.g., through-wall imaging); they are not summarized here.
15.517	Indoor UWB technical requirements	Must be only capable of operation indoors; e.g., mains-powered, not intentionally directed outside the building, no outdoor-mounted antennas.
		UWB bandwidth must be contained between 3100 and 10,600 MHz; radiated emissions at or below 960 MHz must obey 15.209.
		Above 960 MHz (using a 1-MHz resolution bandwidth): EIRP limits in dBm:
		960–1610 MHz −75.3
		1610–1990 MHz −53.3
		1990–3100 MHz −51.3
		3100–10,600 MHz −41.3
		>10,600 MHz −51.3
		Additional restrictions using a measurement bandwidth "not less than 1 kHz" (does that mean equal to 1 kHz? I don't know.)
		1164–1240 MHz −85.3
		1559–1610 MHz −85.3
		Additionally, in 50 MHz centered on frequency of maximum radiation, less than 0 dBm EIRP.
15.519	Handheld devices	Technical requirements are similar, though these devices may of course be used outdoors.
15.521	Other UWB uses	UWB may not be used for toys, on board aircraft, ship, or satellites. The frequency of maximum radiated emission must be within the UWB bandwidth.

Unquoted text is the author's brief summary; quoted text is taken from the regulations.

should be similarly validated. The author's limited personal and broader vicarious experience is that the FCC is a relatively open-minded and technologically astute bureaucracy. If you've done your technical homework and talked with your neighbors, you should be okay. Note, however, that 15.204 is very specific in forbidding external power amplifiers without certification: Proceed at your own risk, even if calculations and measurements show that you are within regulatory limits!

Designers and manufacturers of commercial radio equipment are specifically permitted to develop and test their hardware but are specifically forbidden to offer it for sale on the open market before certification. Certification is a complex rule-bound process, best relegated to the many testing laboratories that exist to perform this function.

Table A1-1 provides a summary of the sections of part 15 that seem most relevant to WLANs. Some of the regulations are framed in terms of field strength. Field strength in dBμV/m at 3 m can be converted to equivalent isotropically radiated power (EIRP) in decibels from a milliwatt (dBm) by subtracting 95.2 dB.

Section 15.247: ISM band operation

Frequencies covered: 2400–2483.5 MHz, 5725–5850 MHz

"Frequency hopping systems shall have hopping channel carrier frequencies separated by a minimum of 25 kHz or the 20 dB bandwidth of the hopping channel, whichever is greater. The system shall hop to channel frequencies that are selected at the system hopping rate from a pseudorandomly ordered list of hopping frequencies. Each frequency must be used equally on the average by each transmitter. The system receivers shall have input bandwidths that match the hopping channel bandwidths of their corresponding transmitters and shall shift frequencies in synchronization with the transmitted signals."

Frequency-hopping (FH) systems in the 2400- to 2483-MHz band: At least 15 nonoverlapping channels must be used. Occupancy in any channel less than 0.4 seconds in a period of (0.4 × (no. of channels)). Systems that use 75 channels are allowed 1-W transmit power; any other FH system is limited to 1/8 W.

Digital modulation techniques are allowed in 2400–2483 MHZ and 5725–5850 MHz; 6-dB bandwidth shall be >500 kHz, with 1-W maximum transmit power. (Note that there is no longer a "spread-spectrum" requirement, which was present in the original incarnation of the regulations. Thus, 802.11b complementary code keying and packet binary convolutional coding modulations, which have essentially no processing gain, are formally legal.)

2400–2483 MHz: If antenna gain is >6 dBi (dB relative to an isotropic antenna), power must be reduced by 1 dB for each extra dB of antenna gain, that is, EIRP is limited to 36 dBm. However, if the system is for fixed point-to-point use only, power must be reduced by 1 dB for every 3 dB of extra gain over 6 dBi.

5725–5850 MHz: Fixed point-to-point links are allowed with any antenna gain, and no requirement to reduce power. A professional installer who exploits these options must ensure the system is used only for fixed point-to-point applications.

All systems must protect the public from harmful RF energy; see section 1.1307(b)(1).

Out of Band: In any 100-kHz slice outside of the target band, the radiated energy must be at least 20 dB less than that found in the 100-kHz slice with the highest intentional power (typically center of the radiated band or near the center).

Radiations in the restricted bands defined in 15.205 must be within the limits of 15.209.

Digitally Modulated Systems: The conducted emission to the antenna in any 3-kHz band shall not exceed 8 dBm at any time during continuous transmission.

Coordination: An individual transmitter can adjust hops to avoid interference, but coordination between multiple transmitters is not allowed. (This appears to be a precaution to avoid people simply having a bunch of transmitters all hopping, but the net effect of which is to fill all the band and exceed the limits on radiated power.)

Section 15.407: UNII Band Operation

The old regulations were as follows:

5.15–5.25 GHz: Power must be less than the lesser of 50 mW or (4 dBm +10 log BW) (BW is the 26-dB bandwidth in MHz).

5.25–5.35 GHz: Power must be less than the lesser of 250 mW or (11 dBm +10 log BW); not to exceed 11 dBm in any 1-MHz band. If antenna gain > 6 dBi, back off 1 dB for each extra dB of antenna gain—that is, maintain a constant EIRP.

5.725–5.825 GHz: Power must be less than the lesser of 1 W or (17 dBm +10 log BW); <17 dBm in any 1-MHz band. Antenna gain > 6 dBi: back off to maintain EIRP. However, systems used only for fixed point-to-point links may use any gain up to 23 dBi without reducing power. After that gain limit, decrease power to maintain the EIRP corresponding to 23 dBi of antenna gain. The operator or professional installer is responsible for ensuring only point-to-point use of the system.

Peak power must be measured with instruments calibrated in RMS voltage, adjusted for detector response time and so on to obtain a true peak measurement. The peak of the envelope should be less than 13 dB over the peak transmit power.

Spurious Emissions

5.15- to 5.25-GHz equipment: Less than −27 dBm/MHz EIRP for all emissions outside band.

5.25–5.35 GHz: Same as above, except that the system can emit in 5.15–5.25 within the regulations set above.

5.725–5.825: Band edge to 10 MHz beyond the band edge: less than −17 dBm/MHz. Outside of 10 MHz beyond edges, less than −27 dBm/MHz.

For frequencies less than 1 GHz, radiation must comply with 15.209, and conducted emissions on AC power must meet 15.207; all intentional radiators here are subject to 15.205.

Operation Restrictions

5.15–5.25 GHz: The antenna must be an integral part of the device, and only indoor operation is allowed.

Recall that the UNII band requirements have recently been changed (FCC 03-287, November 2003). These changes are not yet reflected in the Government Printing Office rules. In the future the regulations will be changed as follows:

5.25–5.35 GHz and 5.47–5.725 GHz: Peak power must be less than the lesser of 250 mW or (11 dBm + 10* log (BW)), where BW is the 26-dB bandwidth in MHz. The power spectral density must be less than 11 dBm per MHz in any 1-MHz band. If antenna gain > 6 dBi is used, power must be backed off to maintain EIRP and spectral density. This is the same requirement previously used only for the 5.25-to 5.35-GHz band.

5.47–5.75 GHz: Spurious emissions outside the intended band must be less than −27 dBm/MHz EIRP.

Transmit power control: Devices operating in the 5.25-to 5.35- and 5.47-to 5.725-GHz bands must be able to decrease their power to no higher than an EIRP of 24 dBm (vs. the allowed maximum EIRP of 30 dBm). Devices that transmit no more than 500 mW EIRP at peak power don't need to control power.

Dynamic frequency selection: Devices in the 5.25- to 5.35-and 5.47- to 5.725-GHz bands must be able to detect radar pulses with a sensitivity of −64 dBm (for high-power devices; devices with EIRP < 200 mW have a slightly relaxed −62-dBm requirement). Before a UNII device starts using a new channel, it must check for 60 seconds for any radar-like transmission above the sensitivity level. If a radar signal is found, the devices have 10 seconds to vacate the channel (consisting of 200 msec of normal data transmission, followed by some additional time for the management frames required to coordinate a channel move). A channel on which a radar signal was detected must be left vacant for at least 30 minutes thereafter.

3. European Standards

Individual national regulatory bodies continue to operate within European states, but the nations of the European Union generally seek to harmonize their individual regulatory actions following standards promulgated by the *European Telecommunications Standards Institute* (ETSI) and the *European Radiocommunications Committee* (ERC).

At the time of this writing, ETSI documents were available for free download (after registration) from the ETSI web site, www.etsi.org. Brief summaries of some of the key portions of ETSI documents governing unlicensed band utilization are provided below.

ISM Bands: ETSI EN 300 328 V1.5.1

The original ETSI ISM band document, EN 300 328, was published in November 1994. The most recent document, v1.5.1, was promulgated in March 2004 and unsurprisingly (since I'm writing this is April) is awaiting final approval. The key technical specifications are given below. This document provides a great deal of information on approved test procedures and is thus a useful reference if you are involved in testing.

Section 1:   The ISM band extends from 2.4 to 2.4835 GHz.

Section 4:   EN300 defines two differing standards for power. Frequency-hopping (FHSS) modulations must use at least 15 channels, separated by at least the 20-dB bandwidth of the signal, with the dwell time per channel not to exceed 0.4 second. Any modulation that doesn't satisfy the above requirements is considered to be direct-sequence spread spectrum.

For all systems, the EIRP is limited to 100 mW (20 dBm), for any combination of antenna and transmit power. Further limitations are placed on power spectral density. For FH systems, an EIRP of 100 mW/100 kHz is allowed, whereas for direct-sequence systems only 10 mW/1 MHz is permitted. (Note that, for example, a 16-MHz wide orthogonal frequency-division multiplexing [OFDM] signal can't even use this power density: at 10 mW per MHz, the total power would be 160 mW, exceeding the total EIRP limit.)

Separate limits are provided for "narrowband" spurious emission and "wideband" spurious emission. This sort of approach is followed to allow for some leakage from local oscillators or their harmonics, where the total power is contained in a narrow spectral line and must be specified while separately limiting broadband emissions, which should be characterized in terms of power spectral density. In the 1- to 12-GHz range, narrowband transmitter emissions are limited to less than −30 dBm and broadband emissions to less than −80 dBm/Hz (that's −20 dBm/MHz), when the radio is operating. Lower power levels are required when the radio is in "standby" mode, and more stringent standards are also set for receiver spurious outputs.

Section 5:   Extensive discussion of the tests required to demonstrate compliance of a particular equipment set with the standards.

HiperLAN Bands ERC/DEC(99)23 and EN 301 893

The original regulatory basis for 5-GHz operation in Europe seems to be a document from the European Radiocommunications Committee, ERC/DEC(99)23, "ERC Decision of 29 November 1999 on the harmonised frequency bands to be designated for the introduction of High Performance Radio Local Area Networks," available from http://www.ero.dk/documentation/docs.

Some key points are as follows:

- 5150–5350 and 5470–5725 MHz are designated as HiperLAN bands.

- The lower band is only for indoor use, with a maximum EIRP of 200 mW allowed.

- The upper band is allowed for indoor or outdoor use with a maximum EIRP of 1 W.

- The following features are mandatory:

  1. transmit power control

  2. dynamic frequency selection to provide a uniform spread of channel loading across 330 MHz, or 255 MHz if only used in the upper band

The recent ETSI document EN 301 893 provides more detailed specifications. A few key points are as follows:

- Nominal channel center frequencies start at 5180 MHz and continue every 20 MHz to 5320; then we skip to 5500 MHz and again have a channel every 20 MHz up to 5700 MHz (thus there are 30-MHz guard bands at the band edges).

- Center frequencies are to be held to ±20 parts per million (ppm).

- EIRP should be less than 23 dBm for the low band and less than 30 dBm for the high band.

- Power spectral density (EIRP) should be less than 11 dBm/MHz for the low band and 18 dBm/MHz for the high band.

- Transmit power control must provide a range of at least 6 dB below these limits.

- Out-of-band spurious emissions: a rather complex regimen applies below 1 GHz, though above 1 GHz the requirement is simply −30 dBm EIRP in 1 MHz.

- Dynamic frequency selection shall ensure uniform spectral loading across the operative bands. Operation of the dynamic frequency selection function is generally similar to that described above in FCC 03 287 (which was derived from the ETSI document): the access point must check for radar interference during startup and usage of a channel and abandon the channel for 30 minutes if radar is detected. Distinctions are made between clients and access points (master and slave devices here), which are not explicit in FCC 03 287.

An extensive discussion of the compliance test suite is also provided.

In a pleasantly surprising bit of good news, it appears that U.S. and European regulations in this area will end up reasonably harmonized.

4. Japan

In Japan, the *Ministry of Posts and Telecommunications* regulates the use of the radio spectrum. The Ministry of Posts and Telecommunications appears to delegate the creation of "voluntary" Japanese standards to the *Association of Radio Industries and Businesses* (ARIB). Equipment certification is performed by the *Telecom Engineering Center*.

Below we provide a brief summary of regulations for operation in the 2.4-GHz ISM band, as summarized in the ARIB standard STD-T66, version 2.1, "Second Generation Low-Power Data Communication System/Wireless LAN System," March 2003. The regulations appear to have undergone significant changes in the last few years, as operation over the whole ISM range is allowed, which was not contemplated in the original release of the 802.11 standards.

- Band addressed: 2400–2483.5 MHz

- Direct sequence or OFDM systems:

  - Transmit power ≤ 10 dBm/MHz, EIRP ≤ 12.14 dBm/MHz (? see below)

  - Minimum bandwidth 500 kHz; minimum 5:1 processing gain

- FH or "hybrid" FH/DS or FH/OFDM systems

  - Transmit power ≤ 3 dBm/MHz (note that the 10-dBm limit appears to apply if only the bands 2400–2427 MHz and 2470.75–2483.5 MHz are used.)

  - < 0.4 second dwell time, < 26-MHz "necessary" bandwidth

- 99% bandwidth ≥ 500 kHz

- Carrier frequency ± 50 ppm

- Transmitter spurious output:

  - 10–2387 MHz: ≤ -26 dBm/MHz

  - 2387–2400 MHz: -16 dBm/MHz

  - 2483.5–2496.5 MHz: -16 dBm/MHz

  - 2496.5–8000 MHz: -26 dBm/MHz

- Receiver spurious outputs less than -54 dBm below 1 GHz, less than -47 dBm for 1–8 GHz

- Antennas:

  - Maximum of 12.14 dBi power gain

  - This is followed by a very confusing paragraph (3.6 (2)a) that appears to indicate that a higher gain antenna can be used as long as EIRP is less than

22.14 dBm for spread spectrum or 15.14 dBm for FH using the center of the ISM band.

○ For spread-spectrum systems, the half-power beam width must be less than 360(16 mW/EIRP per MHz) (in degrees) for both vertical and horizontal planes through the antenna pattern; for FH systems operating in the center 2427–2470 region, the corresponding standard appears to be 360(3.3 mW / EIRP). This appears to be an attempt to ensure that the region subject to interference is minimized for higher EIRP.

- Interference prevention:

  ○ User shall be able to change operating frequencies and turn emission off

  ○ A pair of discussions on coexistence with RFID and amateur operators is provided; the basic guidelines appear to be as follows:

    ○ Preexisting stations have preference

    ○ An "expert" should survey before activating a new station and resolve interference issues proactively

I have not been able to identify any corresponding information for the 5-GHz band at this time.

5. China

The Ministry of Information Industry regulates radio operations in the People's Republic of China (the mainland). The regulations do not appear to be readily available in English; the documents cited here were obtained in Chinese with the assistance of my wife, Nina (a native of Hong Kong), and translated by Franz Chen of RIQ Technology.

The most recent regulatory pronouncement on the 2.4- to 2.4835-GHz ISM band appears to be MII Notification no. 253, from August 23, 2002. This notice provides for the following conditions:

- EIRP $\leq$ 20 dBm for antenna gain $\leq$ 10 dBi

- EIRP $\leq$ 27 dBm for antenna gain $\geq$ 10 dBi

- Direct-sequence power spectral density $\leq$ 10 dBm/MHz for antenna gain $\leq$ 10 dBi

- Direct-sequence power spectral density $\leq$ 17 dBm/MHz for antenna gain $\geq$ 10 dBi

- FH PSD [PT5] $\leq$ 20 dBm/MHz for antenna gain $\leq$ 10 dBi

- FH PSD $\leq$ 27 dBm/MHz for antenna gain $\geq$ 10 dBi

- Carrier frequency $\pm$20 ppm

- Out-of-band "transmission" power $\leq -80\,\text{dBm/Hz}$ (i.e., $\leq -20\,\text{dBm/MHz}$)
- Radiated power beyond ± 2.5(carrier frequency):
  - 30–1000 MHz: $\leq -36\,\text{dBm/100\,kHz}$
  - 2.4–2.4835 GHz: $\leq -33\,\text{dBm/MHz}$ (?)
  - 3.4–3.53 GHz: $\leq -40\,\text{dBm/MHz}$
  - 5.725–5.85 GHz: $\leq -40\,\text{dBm/MHz}$
  - Other 1–12.75 GHz: $\leq -30\,\text{dBm/MHz}$

The requirements on out-of-band radiation are somewhat confusing. The reference to multiples of the carrier frequency might perhaps be intended to refer to the transmission bandwidth; otherwise, it is difficult to reconcile with the frequency bands specified.

The notification specifically cites WLANs, wireless "access" systems, Bluetooth, point-to-point, and point-to-multipoint systems as users, with preference granted to compliant businesses and industries assigned with a fixed frequency. The transmitter is to use an integral antenna, and swapping of antennas or addition of a power amplifier is prohibited. Equipment must be certified under a previous notification, no. 178 (1998). Users appear to be required to register with the local administrative authority, although there is no rule applied to interference other than that the contending parties ought to work things out. Point-to-point spread spectrum is allowed in rural areas with approval of the local administration, but prohibited in urban areas.

The 5.8-GHz band regulations are described in MII Notification no. 277 (July 2002). They apply to the 5725- to 5850-MHz band. Applications specifically allowed are point-to-point and point-to-multipoint spread spectrum, WLAN, wireless broadband access, Bluetooth, and autoidentification for automobiles. Again, preference is granted to businesses and industries operating on assigned fixed frequencies.

Technical specifications are as follows:

- Transmit power $< 27\,\text{dBm}$
- EIRP $\leq 33\,\text{dBm}$
- Power spectral density $\leq 13\,\text{dBm/MHz}$, EIRP PSD $\leq 19\,\text{dBm/MHz}$
- Carrier frequency $\pm 20\,\text{ppm}$
- Out-of-band "transmission" power $\leq -80\,\text{dBm/Hz}$ (i.e., $\leq -20\,\text{dBm/MHz}$)
- Radiated power:
  - 30–1000 MHz: $\leq -36\,\text{dBm/100\,kHz}$
  - 2.4–2.4835 GHz: $\leq -40\,\text{dBm/MHz}$

- ○ 3.4–3.53 GHz: ≤ -40 dBm/MHz

- ○ 5.725–5.85 GHz: ≤ -33 dBm/MHz "beyond 2.5 times carrier"

- ○ Other 1–40 GHz: ≤ -30 dBm/MHz

Again the puzzling reference to "2.5 times the carrier frequency" may actually be directed to the signal bandwidth. As in notification no. 253, equipment must use an integral antenna, and swapping antennas or adding power amplifiers is prohibited. Users pay a bandwidth occupancy fee based on notification no. 226 and must gain approval from local authorities. Permits are required for point-to-point or point-to-multipoint communications systems, wireless base stations, and fixed-location stations. Connection to the public network requires a telecom business permit. Local governments can apparently also use the band for automobile identification and traffic control.

Legally approved fixed-bandwidth stations are protected from interference, but other wireless stations have to work things out among themselves. All equipment must receive a "model" permit from the national wireless administration office, and marketing may not represent the equipment as being usable without a permit.

6. Concluding Remarks

The regulations for unlicensed operation worldwide continue to evolve with the technology. By the time this book gets into your hands, some of what is described here will have changed (even when I got it right in the first place!). There's no substitute for checking the source materials. The APEC Telecommunications and Information Working Group web page, www.apectelwg.org/apec/alos/osite_1.html, provides a very useful listing of telecommunications regulatory links and sites for many countries in Asia and elsewhere. Other relevant web sites are listed below:

International Telecommunications Union, Radio sector: www.itu.int/ITU-R/

ITU-R World Radio Conference site: www.itu.int/ITU-R/conferences/wrc

U.S. Federal Communications Commission: www.fcc.org

U.S. Code of Federal Regulations, Title 47:

http://www.access.gpo.gov/nara/cfr/waisidx_03/47cfr15_03.html

European Telecommunications Standards Institute: www.etsi.org;

European Radiocommunications Committee: http://www.ero.dk/documentation/docs

Japan Ministry of Posts and Telecommunications radio-related laws: www.soumu.go.jp/joho_tsusin/eng/laws.html

Japan Association of Radio Industries and Businesses (ARIB): www.arib.or.jp/english/index.html

Japan Telecommunications Engineering Center (TELEC): www.telec.or.jp/ENG/index_e.htm

National Radio Spectrum Management Center, People's Republic of China: http://www.srrc.gov.cn/ (in Chinese)

7. Further Reading

"The Federal Communications Commission," R. Coase, The Journal of Law and Economics, vol. II, p. 1, 1959

"Spectrum Management: Property Rights, Markets, and the Commons," G. Faulhaber and D. Farber, from "Spectrum Policy: Property or Commons?" Stanford University, Stanford, CA, USA, March 2 and 3, 2003

Appendix 2

Measurement Tools

1. The RF Toolbox

There are four tools that are somewhere between useful and indispensable for working with RF signals in communications applications: spectrum analyzers, network analyzers, signal generators, and vector signal analyzers.

A *spectrum analyzer* is just a radio with a screen. It has a local oscillator to convert the incoming RF signal to an intermediate frequency signal, which is filtered and then applied to a detector to generate a voltage corresponding to the power in the received bandwidth (Figure A2-1). By ramping the frequency in time, the screen displays a picture of the power spectral density (power in a given frequency range) of the signal received by the analyzer. Spectrum analyzers are used to examine in-band power, distortion (through looking for changes in the shape of the emitted spectrum), and spurious output.

Modern spectrum analyzers are remarkably versatile: in top-of-the-line commercial tools, the frequency of operation extends over multiple octaves: 10 kHz–26 GHz tools are nothing special. Frequency can be fixed or swept over any range (the frequency *span* of the sweep) extending up to the whole range of the instrument, and sweep time can be varied from milliseconds to tens or hundreds of seconds. The intermediate frequency bandwidth of the receiver (usually known as the *resolution bandwidth* [RBW]) can be varied over a similarly huge range, typically from a few hundred hertz to 1–5 MHz.

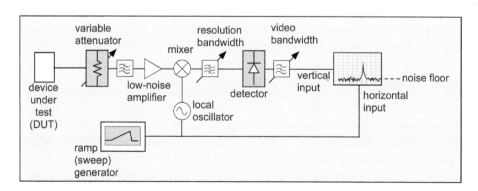

Figure A2-1: Spectrum Analyzer Block Diagram

In most cases, the frequency is swept over a range of interest by a ramp generator driving the voltage-controlled oscillator. The voltage from the detector is filtered before being applied to the vertical deflection control on the display; the inverse response time of this filter is known as the *video bandwidth*. The video bandwidth needs to be large enough to respond to the fastest change in detected power that is expected from the sweep: for a perfect CW signal, this time is approximately [sweep time(RBW/span)]. Because by the properties of Fourier transforms a signal with a bandwidth of RBW can't change any faster than about (1/RBW), the most typical approach to setting up a measurement is to ensure that the video bandwidth is about equal to the RBW. The shortest sweep time that makes sense is about (span/RBW^2). For a 1-MHz bandwidth and an 80-MHz sweep (covering the Industrial, Scientific, and Medical band), the minimum sweep time would be 80μsec; typically sweep times of a few milliseconds are fine.

Spectrum analyzers, like other radios, have a finite dynamic range. However, unlike most radios, this range is adjustable on both ends, allowing the user great flexibility in capturing both large and small signals. The noise floor in the analyzer changes when the resolution bandwidth is changed: recall from Chapter 4 that the amount of thermal noise entering a radio is proportional to bandwidth, so shrinking the bandwidth reduces noise. Because the RBW typically can vary over a range of at least 1 MHz to 1 kHz, this is a large effect. When a small signal needs to be extracted from the noise, a narrow RBW must be used. Most analyzers also have a video averaging function, which will help to clean up noise if the signal is the same on every sweep. (Unfortunately, for a bursty signal like an 802.11 packet, very narrow RBW corresponds to slow instrument response: packets shorter than (1/RBW) may be invisible to the analyzer.)

When large signals are introduced into the analyzer, distortion within the analyzer may contribute spurious harmonics or odd-order distortion products, misrepresenting the true signal from the device under test. The origin of any given signal component can be empirically established by changing the attenuator setting (typically in steps of 10 dB). Because the gain of the spectrum analyzer is adjusted to compensate for the attenuator, any true external signal will be unaffected (though the noise floor must rise by the amount of attenuation introduced). If a signal amplitude is affected by the attenuator setting, the signal is contaminated by internal distortion; the attenuator must be increased until no further change is noted. External attenuators can be added (with calibration of their effects) to allow smaller steps if desired.

Most spectrum analyzers provide a number of additional capabilities. One of the most useful for bursty signals is the peak hold display, which retains the highest recorded power in any given frequency bin. Some analyzers offer time-gated analysis, which is very useful for examining the spectrum of a bursty packet-based source (as long as the signal power is large enough to allow the analyzer to identify the beginning of the packet).

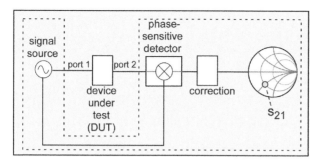

Figure A2-2: Vector Network Analyzer

A *network analyzer* consists of a signal source that can be swept in frequency and a detector. A *scalar network analyzer* uses a detector that is only sensitive to amplitude; a *vector network analyzer* can also detect the phase of the signal relative to that of the source (Figure A2-2). Network analyzers can generally characterize both transmitted and reflected signals. Network analyzers have various applications in microwave circuit and system characterization. For example, a network analyzer can be connected to an antenna and the reflected signal from the antenna measured: the frequencies where the reflected signal is small will generally be those frequencies at which most of the input power is being radiated and thus constitute the range over which the antenna is likely to be useful. Scalar and vector network analyzers are used extensively in circuit design and test. Vector network analyzers can display the results in many formats, including conventional rectangular plots of power or reflection coefficient versus frequency and *Smith chart* images that provide a useful view of complex reflection or transmission coefficients (see Appendix 3).

Calibration is an important issue for network analyzers. A scalar network analyzer is readily calibrated and can even be used without calibration in many cases, because the loss of cables and connectors is normally small if the cables are kept short and connectors snugged properly. A vector network analyzer must be carefully calibrated, because a typical measurement setup has a huge phase shift. (At 2.4 GHz, two pieces of cable each 1/2-m long represent a combined phase of about 77 radians or 4430 degrees. A tiny change in system phase completely changes the relative phase due to the object under test.) Modern network analyzers include sophisticated digital correction capability, but good calibration standards and careful measurement are still needed to get accurate results.

A *signal generator* provides a test signal. Older units, often known as *sweepers*, were capable of generating a single frequency, which could be swept across a frequency band. Simple amplitude modulation could be applied to the signal. Modern *arbitrary signal generators* (also known as *vector signal generators*) include sophisticated digital-to-analog converters and can thus generate signals appropriate to various communications protocols associated with wireless local area networks, cellular telephony, or just about

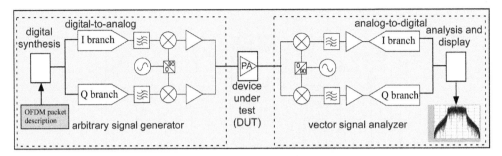

Figure A2-3: Schematic Depiction of Arbitrary Signal Generator and Vector Signal Analyzer

any other signal source (Figure A2-3). By providing the characteristics of a given type of signal (such as an orthogonal frequency-division multiplexing packet signal) in software, the arbitrary generator can digitally synthesize the desired analog signal at baseband. The arbitrary signal generator can be used as a source to test the performance of transmitter up-conversion and amplification with realistic input signals or up-converted to validate power amplifier or receiver performance.

Vector signal analyzers are the fancy digital version of an oscilloscope. They down-convert and demodulate a received signal into I and Q baseband components and then display the signals in a number of useful views. Signal voltages can be depicted in the complex plane, where by gating at sampling times compliance to the desired modulation (e.g., quaternary phase-shift keying or 16 quadrature-amplitude-modulation) can be visually established. Error vector magnitude can also be quantitatively extracted by software to complement the visual display. Depiction of amplitude in time and the probability distribution of amplitudes can be used to measure the peak-to-average ratio of a signal. Fast Fourier transforms of the signal can be used to extract the spectrum of the packetized signal.

Commercial vendors of microwave measurement equipment include Agilent, Wiltron, Rohde-Schwartz, Anritsu, and Tektronix. Modern instruments are remarkably capable, but this capability comes at a price: new ones will set you back tens of thousands of dollars (U.S.$). Even decade-old Hewlett-Packard spectrum analyzers with dial controls for frequency ranges are still selling on E-Bay for several thousand dollars.

A properly equipped measurement facility also requires innumerable small items. A selection of good-quality cables (see Chapter 5) is indispensable to get things hooked up. Be sure to characterize cable loss before performing critical measurements. A power meter (a simple broadband tool that measures total power) is very nice to have for verifying absolute power levels, though a well-maintained spectrum analyzer will generally give the same answer to better than 1 dB. Adaptors are a big help: no matter what cable you buy, it always has the wrong connectors on the end for the device you need to test. Connectorized attenuators are commercially available and are useful for throttling back signals from, for example, power amplifiers to levels that test instruments can tolerate; they can be hooked up in series to get to large attenuation values. Connectorized low-pass filters are helpful to remove harmonics from signal sources before applying them to devices when performing distortion measurements. If you don't have a network analyzer available, you can make scalar measurements of reflection using a directional coupler, which sends the forward and backward waves on a line to separate ports.

Reflection and Matching

1. Reflection Coefficients

In most of our discussion, we have assumed that radios, transmission lines, and antennas are all well matched, so that any power coming from one goes into the other. What happens when this is not the case? How do we measure the deviation from ideality, and what can we do about it? In this appendix we provide a very brief introduction into reflection coefficients and impedance matching.

In microwave land, a *port* is a connection from one microwave environment to another—for example, from a cable to an antenna. The signal traveling along the cable may be partially reflected by the antenna if the impedances do not match (Figure A3-1).

The ratio of the reflected signal to the incident signal is the reflection coefficient:

$$\Gamma \equiv \frac{v_{ref}}{v_{inc}} \qquad [A3.1]$$

The reflection coefficient is in general complex, because the phase of the incident and reflected waves may not be the same. It can be shown that the reflection coefficient is related to the impedance seen by a wave at port 1:

$$\Gamma = \frac{Z_1 - Z_o}{Z_1 + Z_o} \qquad [A3.2]$$

where Z_1 is the (generally complex) impedance of the port and Z_o is the impedance of the transmission line, typically 50 Ω. Recall that the impedance of a capacitor C with frequency f is $Z_c = 1/(i\omega C)$ and an inductor has an impedance of $Z_L = i\omega C$, with $\omega = 2\pi f$.

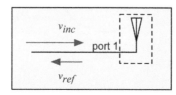

Figure A3-1: A Reflected Signal

The reflection coefficient must always have a magnitude less than 1 (if port 1 has only passive circuits inside of it) and varies from +1 for an open (infinite load impedance) to −1 for a short (zero load impedance), as can be easily verified from [A3.2]. The magnitude of the reflection coefficient in decibels is often known as the *return loss*; the terminology implies that this is the loss suffered by the incident signal making a return trip to the sending instrument.

The phase of the reflection coefficient changes if we measure it at a different location along the cable, because the incident wave gains phase as we move to the right and the reflected wave gains phase as we move to the left. Because the ratio of the voltages takes the difference of the phases (see Chapter 2), the phase of Γ changes by 4π each time the measurement plane moves one wavelength. Because the total voltage at any location is the sum of the incident and reflected voltages, the total voltage will also vary with position along the cable. The ratio between the largest and smallest magnitude of voltage is known as the *voltage standing wave ratio.* (The importance of this somewhat funky parameter is partially historical; in the days when phases of microwaves were very difficult to measure, it was relatively easy to move a pickup along a waveguide and measure the difference in power received.) By reference to equation [A3.2] it is easy to see that voltage standing wave ratio can be expressed in terms of the reflection coefficient:

$$VSWR = \frac{1 + |\Gamma|}{1 - |\Gamma|}$$ [A3.3]

If we display the reflection coefficient corresponding to a particular complex impedance in the complex plane, with scales showing the corresponding impedance, we get an extremely useful graphical tool for matching and other microwave circuit operations, the *Smith chart.* (Such an operation is formally known as a *conformal map*; circles map to circles and angles are preserved.) A simplified chart is shown in Figure A3-2.

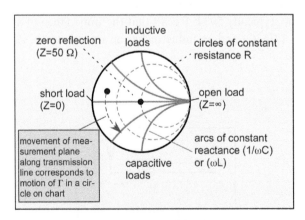

Figure A3-2: Smith Chart

The Smith chart maps the infinite impedance half-plane into a finite region (a circle of radius 1). It provides a very useful visual summary of what adding any element to the circuit of port 1 does to the consequent reflection. Adding some transmission line moves the reflection coefficient on a circle around the point $\Gamma = 0$ (as shown in Figure A3-2). Adding a capacitance or inductance moves the reflection coefficient on a circle of constant resistance; adding a (series) resistor moves Γ on an arc of constant reactance. Another nifty property of the Smith chart is that the picture for admittances (the reciprocal of an impedance, corresponding to elements added in parallel instead of in series) is just the same chart but reflected through the y-axis (Figure A3-3).

2. A Simple Matching Example

Let's look at a particularly simple example of matching. Recall that a square patch of copper used as a microstrip (patch) antenna presents a real input impedance of about 120 Ω at its resonant frequency (see Chapter 5, section 6.6). From equation [A3.2] the reflection coefficient of this patch, if attached directly to a 50 Ω line, is

$$\Gamma = \frac{120 - 50}{120 + 50} \approx 0.4 \equiv -7.7 \; dB \tag{A3.4}$$

Having a large reflection coefficient causes two problems. The first is that about 20% of the power ends up reflected rather than reaching the antenna, corresponding to a reduced transmitted power. (For a receiver, the same remark applies to the received power, by reciprocity.) This power is simply wasted, forcing an increase in the power amplifier rating and DC power consumption to achieve the same transmitted power. The second problem is that the reflected wave is now bouncing around inside the system: if not properly terminated, it can reflect again and cause variations in the transmitted power (and other performance characteristics). Because these effects will depend on the relative phase of the reflected wave, they are very sensitive to effective propagation distance and thus could vary with cable length, temperature, and other not-necessarily-controlled factors. Even in a single patch, such a large reflected signal is a

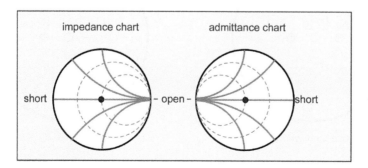

Figure A3-3: Smith Chart For Admittance Is The Mirror Image Of Smith Chart For Impedance

problem; in an array of patches used to make a directional antenna, reflections from every patch could add in phase at some frequencies and completely bollix up the antenna performance. We need to reduce this reflected wave by *matching* the impedance of the antenna to that of the transmission line.

A particularly simple way to manage this feat is to use a *quarter-wave transformer*. This is simply a length of line of differing impedance inserted into the connection between the cable or transmission line and the antenna. To see why it works, we first have to imagine ourselves standing at the end of the inserted line, looking into the antenna impedance. Construct a Smith chart appropriate to the impedance of the inserted line; that is, if the line has $Z_o = 77\Omega$, the center of the chart corresponds to 77 instead of 50 Ω (Figure A3-4). If the load impedance is larger than 77 Ω, the reflection coefficient will appear on the real axis to the right of $\Gamma = 0$. Now recall that as we move along the line, the reflection coefficient traces out a circle (clockwise toward the radio, counterclockwise toward the load). A quarter-wave brings us half-way around the chart, so the real load of 120 Ω is transformed to—what a coincidence!—a real load of 50 Ω. This load will produce no reflection from the incoming 50 Ω line, that is, the quarter-wave transformer has matched the patch.

The image of the structure shown in Figure A3-4 makes it clear that this sort of matching is very easy to realize using lines formed on a circuit board, and so it is commonly used for patch antennas. Analogous tricks using short lengths of cable of differing diameter can be used for matching to wire antennas or Yagi-Uda structures.

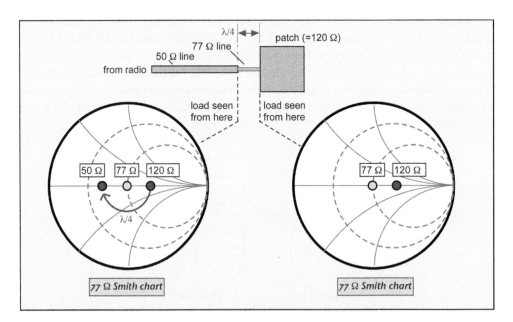

Figure A3-4: Matching a 50 Ω Line to a 120 Ω Patch Using a Quarter-Wave 77 Ω Line as a Transformer

The astute reader may have noticed that Figure A3-4 looks different from Figure 5-29 (showing a patch array). The patch array in Figure 5-29 is designed using lines with a characteristic impedance of about 120 Ω; the short stubs applied to the patches there are merely acting to remove a bit of reactance and place the load on the real axis rather than to transform the real impedance value. A quarter-wave transformer is then used at the input of the antenna to match to the 50 Ω external world. This type of design allows more patches to fit into a given space, because the high impedance lines are narrower than a 50 Ω line would have been, but incurs a bit more loss in getting to the patches.

Note that a quarter-wave transformer match is fairly narrowband: at a different frequency, the length of the transformer is no longer exactly a quarter-wave, so the transformed impedance point will either come up short of the real axis or overshoot it. This is a minor issue if the original impedance is not too far from the impedance to be matched, but as the load gets closer to the edges of the chart, small changes in frequency will cause large displacements of the final impedance. The match becomes more and more sensitive to frequency changes (or other errors, such as line length variation). An example is shown in Figure A3-5 for a low-impedance load such as that we might find in a short dipole structure or other very small antenna. Here a 10% variation in frequency (or line length or phase velocity along the line) gives rise to an unacceptably large reflection coefficient. Matching to antennas or other loads with impedances far from 50 Ω is practically very difficult because of the sensitivity of the result to small variations in just about everything; this is one of the reasons why the most popular antennas are those that need little or no matching to connect to a cable or transmission line.

Many other kinds of matching structures can be constructed; for example, discrete surface-mount capacitors and inductors on a printed circuit board can be used if they

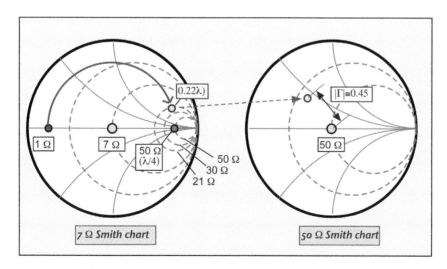

Figure A3-5: Matching to a Load Far From 50 Ω

are small enough (much smaller than a wavelength). These structures can also be analyzed using the Smith chart: each element moves the load along a circle of constant resistance, and each length of line turns on a circle around the center. In this fashion, matching circuitry can be designed for any load. Many software packages are available to automate the process, but as usual we note that software in the computer is a tool rather than a substitute for the software in your head.

The techniques introduced here can also be applied in a much more general fashion for devices with both input and output ports by defining *scattering parameters* (generally known as S-parameters) describing the transmitted and reflected waves resulting from signals at any of the ports.

3. Further Reading?

There are so many books on microwave design and engineering that it's hard to decide on any particular one to recommend. Johnson's **Antenna Handbook**, previously cited, has a very thorough if not very beginner-friendly reference chapter dealing specifically with antenna matching (Bowman, chapter 43).

For system and circuit design, the two most popular software packages specifically oriented toward microwave applications are Microwave Office (my personal favorite) by Applied Wave Research (www.appwave.com) and Agilent's Advanced Design System (ADS) (www.agilent.com), though many other general-purpose electronic design tools now support the requirements of RF and microwave design.

The Lorentz Gauge

1. No Royal Road

In Chapter 5 we showed that ideal dipoles did not couple along their axes. In this appendix we demonstrate that this result is not peculiar to the ideal dipole but is an inevitable consequence of the conservation of charge. To simplify the derivation, we consider only plane waves far from a spatially limited clump of charges and currents. We first demonstrate that the scalar and vector potentials are related to one another: the Lorentz gauge. We then show that as a consequence of the Lorentz gauge, only the transverse part of the vector potential **A** couples to a distant receiver.

This appendix assumes the reader is familiar with typical vector notations, Gauss' law (the divergence theorem), the chain rule, and the other jargon normally associated with electromagnetism. If you don't know this stuff, you'll just have to accept the conclusions on faith, at least until I find a simpler way to derive the results.

2. Lorentz Gauge Derivation

Recall the source equations for the potentials were

$$A_z(\vec{r}) = \frac{\mu_o}{4\pi} \int J_z \frac{e^{-ikr}}{r} dv \quad \phi(\vec{r}) = \frac{\mu_o c^2}{4\pi} \int \rho \frac{e^{-ikr}}{r} dv \qquad [\text{A4.1}]$$

where we have assumed only z-directed currents for simplicity. (The demonstration can be carried out for the other directions in the same fashion.) We begin by noting the following important fact: *the currents and charges that act as sources for the potentials are not independent.* They are coupled by the conservation of charge. Any change (divergence) in the current flow must result in an accumulation or depletion of charge density. Mathematically speaking we have, recalling that everything is assumed proportional to $e^{i\omega t}$,

$$\nabla \cdot \vec{J} = -\frac{\partial \rho}{\partial t} = -i\omega\hat{\rho} \Rightarrow \omega\hat{\rho} = i\frac{\nabla \cdot \vec{J}}{\omega} \qquad [\text{A4.2}]$$

where the caret denotes the time-independent amplitude of the charge density.

The goal of the remainder of the derivation is to exploit this relationship to derive an expression for the scalar potential in terms of the vector potential. We begin by substituting the expression for the charge density into the equation for the scalar potential:

$$\hat{\phi}(\vec{r}) = \frac{\mu_o c^2}{4\pi} \int \hat{\rho} \frac{e^{-ikr}}{r} dv = \frac{\mu_o c^2}{4\pi} \int i \frac{\nabla \cdot \vec{J}}{\omega} \frac{e^{-ikr}}{r} dv$$

$$= i \frac{\mu_o c^2}{4\pi\omega} \int \nabla \cdot \vec{J} \frac{e^{-ikr}}{r} dv$$

[A4.3]

Note that the divergence of the current here is taken locally in the volume *v* where the sources are, but the measurement point is presumed to be far away at the end of some vector **r**.

The argument of the new integral for the scalar potential is one piece of the divergence of the current source term in the vector potential, as we can see by applying the chain rule:

$$\nabla \cdot \left(\vec{J} \frac{e^{-ikr}}{r} \right) = \frac{e^{-ikr}}{r} \nabla \cdot \vec{J} + \nabla \left(\frac{e^{-ikr}}{r} \right) \cdot \vec{J}$$

[A4.4]

We need to work out what the second term in [A4.4] looks like to establish a relationship between the scalar and vector potentials. Terms that look like a gradient of (1/*r*) shrink quite rapidly (about quadratically) as *r* grows large, so for distances much larger than the extent of the source currents and charges, we can treat the (1/**r**) term as a constant and pull it out of the gradient. We are left with the gradient of an exponential, which just multiplies the exponential by the derivative of its argument. We get

$$\nabla \cdot \left(\vec{J} \frac{e^{-ikr}}{r} \right) \approx \frac{e^{-ikr}}{r} \nabla_v \cdot \vec{J} + \frac{1}{r} \nabla_v (e^{-ikr}) \cdot \vec{J} = \frac{e^{-ikr}}{r} [\nabla_v \cdot \vec{J} + ik\hat{r} \cdot \vec{J}]$$

[A4.5]

where the second term is the projection of the current onto the unit vector in the direction of **r**. A subtle but important point to note here is that the gradient of the exponential is being taken in the source coordinates. When we move the source of the vector **r** to the right by *dr*, we *decrease* the length *r* by *dr*, that is, the gradient on the origin introduces an additional factor of −1, which cancels that from the argument of the exponential, so that both terms in the brackets are positive.

Now we integrate over a region large compared with the source and note that we can use Gauss' law to convert the integral of the divergence over a volume to the integral of the vector argument over the enclosing surface:

$$\int \nabla \cdot \left(\vec{J} \frac{e^{-ikr}}{r} \right) dv \approx \int \frac{e^{-ikr}}{r} [\nabla_v \cdot \vec{J} + ik\hat{r} \cdot \vec{J}] dv$$

$$\updownarrow \text{ Gauss' law}$$

[A4.6]

$$= \int \left(\vec{J} \frac{e^{-ikr}}{r} \right) \cdot \hat{n} dS$$

Because the volume is taken to be outside of all the sources (but still much smaller than the distance to the observation point), the surface integral must vanish, as $\mathbf{J} = 0$ on the surface. Therefore, we can write

$$\int \frac{e^{-ikr}}{r}[\nabla_v \cdot \vec{J} + ik\hat{r} \cdot \vec{J}]dv = 0 \Rightarrow \int \frac{e^{-ikr}}{r}[\nabla_v \cdot \vec{J}]dv = -ik \int \frac{e^{-ikr}}{r}[\hat{r} \cdot \vec{J}]dv \quad [A4.7]$$

Equation [A4.7] permits us to write the scalar potential in terms of the source current only:

$$\hat{\phi}(\vec{r}) = i\frac{\mu_o c^2}{4\pi\omega} \int \nabla \cdot \vec{J} \frac{e^{-ikr}}{r}dv = i\frac{\mu_o c^2}{4\pi\omega}\left(-ik \int \frac{e^{-ikr}}{r}[\hat{r} \cdot \vec{J}]dv\right)$$

$$= i\frac{\mu_o c^2}{4\pi\omega}\left(-ik\hat{r} \cdot \int \frac{e^{-ikr}}{r}[\vec{J}]dv\right) = i\frac{c^2}{\omega}\left(-ik\hat{r} \cdot \frac{\mu_o}{4\pi} \int \frac{e^{-ikr}}{r}[\vec{J}]dv\right) \quad [A4.8]$$

$$= c\frac{1}{k}\left(k\hat{r} \cdot \frac{\mu_o}{4\pi} \int \frac{e^{-ikr}}{r}[\vec{J}]dv\right) = c\vec{A}_{long}$$

In the last steps, we used $\omega = ck$ and equation [A4.1]. The dot product of $\mathbf{A}$ and the unit vector in the r-direction is the longitudinal portion of the vector potential.

Now, the divergence of the vector potential is also proportional to the longitudinal component:

$$\nabla_r \cdot \vec{A}(\vec{r}) = \frac{\mu_o}{4\pi r} \int \vec{J} \cdot (-ik\hat{r}e^{-ikr})dv = -i\frac{\mu_o k}{4\pi r}\hat{r} \cdot \int \vec{J}e^{-ikr}dv = -ik\hat{r} \cdot \vec{A}(\vec{r}) = -ikA_{long}$$

$$[A4.9]$$

Thus, we can also write this expression as

$$\frac{-i\omega}{c^2}\hat{\phi}(\vec{r}) = (\nabla_r \cdot \vec{A}(\vec{r})) = -\frac{1}{c^2}\frac{\partial}{\partial t}\hat{\phi}(\vec{r}) \quad [A4.10]$$

which is the more conventional way of writing the Lorentz gauge condition.

3. Coupling of the Potentials

We're almost done. We just need to examine what happens to the induced voltage when we impose our new relationship between the vector and scalar potentials.

First, let's calculate the voltage induced along a small displacement parallel to the vector **r**. The line integral in the vector potential simply becomes the dot product of **A** and the displacement if the displacement is very small. If the displacement is in the direction of **r**, the dot product is just the longitudinal component of **A**. The scalar potential contributes a voltage due to the integral of its gradient over the displacement (which again becomes just the product for small displacements). Plugging in the value of the scalar potential from [A4.9], we get

$$V = v + \delta\phi = i\omega A_{long}\delta r + \nabla\phi \cdot \delta r = i\omega A_{long}\delta r - ikcA_{long}\delta r = 0 \quad [A4.11]$$

In the direction perpendicular to **r**, the scalar potential must be constant. (This is easy to see directly from [A4.1]; when one is far from the source charges, an infinitesimal displacement perpendicular to **r** has a quadratically small effect on the distance and thus makes no change at all in the integral.) Thus, the voltage arises only from the line integral of the vector potential and only the transverse portion makes a contribution:

$$V = v + \delta\phi = i\omega\vec{A} \cdot \delta\hat{\theta} + 0 = i\omega A_{trans}|\delta\hat{\theta}| \qquad [\text{A4.12}]$$

We have thus justified the assertion that only the transverse part of the vector potential contributes to coupling to a receive antenna (or any other structure) at long distances from the sources.

Appendix 5
Power Density

1. Rederiving P

In Chapter 5 we provided a heuristic argument to derive an equation for the power density that could be delivered by a propagating potential. In this appendix we provide a somewhat more exhaustive (or perhaps just exhausting) justification for this formula.

The question we wish to address is how much power is carried in a plane wave potential? This is the same as asking how much power can such a potential deliver to a physical object? One way of getting to the answer is to imagine a plane wave impinging upon a region in which the potential can induce a current (Figure A5-1).

In Figure A5-1 we have asserted a particular relationship between the potential and current and a resulting behavior of the potential. Before continuing, let us justify these assertions. It will be understood that in the following we deal only with the transverse part of the impinging potential, because any longitudinal part induces no voltages and does not couple (see Appendix 4).

To find the solution for the potential in the region of the absorber, we introduce the differential form of the propagation equation (the reader can insert the integral forms,

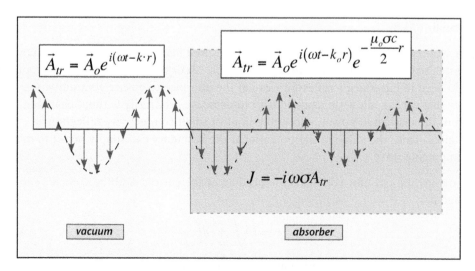

Figure A5-1: Plane Wave Impinging on an Absorber

equations [A4.1] or equivalently [5.1], of the source equations and verify that they satisfy the differential form in vacuum):

$$\frac{d^2 A}{dr^2} - \frac{1}{c^2}\frac{d^2 A}{dt^2} = -\mu_o J \quad J = -i\omega\sigma A \qquad \text{[A5.1]}$$

Assuming that the potential is of the form of an exponential as usual, we can substitute for **A** and obtain

$$-k^2 + \frac{\omega^2}{c^2} = i\omega\mu_o\sigma \quad k^2 = \frac{\omega^2}{c^2} - i\omega\mu_o\sigma \qquad \text{[A5.2]}$$

Clearly, to use the exponential form of the solution, we must allow for a complex value of the wavenumber k. We can approximate this value if the conductivity σ is small:

$$k = \sqrt{\frac{\omega^2}{c^2} - i\omega\mu_o\sigma} = \sqrt{\frac{\omega^2}{c^2}\left(1 - \frac{i\omega\mu_o\sigma c^2}{\omega^2}\right)} \approx \frac{\omega}{c}\left(1 - \frac{i\omega\mu_o\sigma c^2}{2\omega^2}\right) = \frac{\omega}{c} - \frac{i\omega\mu_o\sigma c}{2\omega} = \frac{\omega}{c} - \frac{i\mu_o\sigma c}{2}$$

$$\text{[A5.3]}$$

Substituting this value of k in the equation for **A** and multiplying out imaginary part, we obtain

$$\vec{A}_{tr} = \vec{A}_o e^{i(\omega t - k_o r)} e^{-\frac{\mu_o \sigma c}{2} r} \qquad \text{[A5.4]}$$

as we had asserted in Figure A5-1. Equation [A5.4] shows why we named the material an absorber; the potential decreases in amplitude as r increases. If we assume the absorbing region is arbitrarily large in extent, the potential must become completely negligible at large distances within it. The power carried by the potential must have been converted to currents within the material. Therefore, by integrating the dissipation in the absorber, we should be able to find the power that was transported by the potential.

(The astute reader will note that we have not ensured that the potential is continuous at the interface between the vacuum region and the absorbing region. In general, one would need to introduce a reflected wave at the interface to ensure continuity. We find below that we can allow the conductivity to decrease arbitrarily without changing the form of the dissipated power, and in the limit of small conductivity reflection is negligible. Thus, the value of the potential at the boundary can simply be set to that of the impinging wave.)

The dissipation per unit volume is the product of the current and the voltage. The voltage is

$$v = i\omega \int \vec{A} \cdot \vec{dl} \qquad \text{[A5.5]}$$

Substituting for the current and assuming an infinitesimal region, we find

$$P = \frac{1}{2}i\omega \int \vec{A} \cdot \vec{dl} \, * -i\sigma ck\vec{A} = \omega A dz * \sigma ck A (da)^2$$

$$= \frac{1}{2}\omega\sigma ckA^2(d^3r)$$

$$[A5.6]$$

where da is the area element perpendicular to the propagation vector **r**. Integrating this dissipation over the distance r from the edge of the absorber (taken to $r = 0$ for convenience) to ∞ and assuming unit area in the direction perpendicular to propagation, we obtain the power dissipated per unit perpendicular area:

$$P = \frac{1}{2}\omega\sigma ckA_o^2 \int_0^\infty e^{-\mu_o\sigma cr}dr = \frac{1}{2}\omega\sigma ckA_o^2 \frac{1}{\mu_o\sigma c} = \frac{\omega k A_o^2}{2\mu_o} \qquad [A5.7]$$

which is the same as equation [5.15], obtained from the heuristic argument. (We remind the reader again that only the transverse part of the potential is to be considered here.) Note that the dissipated power is *independent of the conductivity* σ. This means that we can allow the conductivity to be arbitrarily small without changing the dissipated power. It is straightforward to show that the reflected potential wave at the interface is proportional to the conductivity (when the latter is small), so that the reflected power may be made arbitrarily small without changing the form of the dissipated power. Thus, in the limit of very small σ, the vector potential in expression [A5.7] is in fact the same as the impinging potential, and therefore all the impinging power is dissipated in the absorber. This justifies identifying the quantity P as the power density associated with the propagating potential **A**.

The impinging power per unit area can be reexpressed in terms of the angular frequency only as

$$P = \frac{\omega^2 A_o^2}{2\mu_o c} \qquad [A5.8]$$

which has the explicit form of the square of a voltage per length $(\omega A)^2$ divided by the impedance of free space $\mu_o c$.

Conventional E & M

1. Speaking Their Language

This book has applied an approach to electromagnetic theory that is not (yet) commonly used. We have formulated the theory in terms of the magnetic permeability of free space, μ_0, the velocity of light in vacuum c, and the refractive index n. Most of the literature the reader will encounter uses different nomenclature and defines different fundamental quantities, though the underlying physics is not changed. In this appendix, we briefly describe the conventional quantities to assist in translation.

The conventional approach is based on the electric field

$$\vec{E} = -\nabla \phi - \frac{d\vec{A}}{dt} \qquad [A6.1]$$

and the magnetic field

$$\vec{B} = \nabla \times \vec{A} \qquad [A6.2]$$

which together impose a force on a particle or other object with net charge q of

$$\vec{F} = q(\vec{E} + \vec{v} \times \vec{B}) \qquad [A6.3]$$

Here the symbol "×" represents the vector cross-product and the cross-product of a gradient operator with a vector field is the curl of the vector field. These concepts, which have been (thankfully) absent from our development of electromagnetism, are integral to the conventional approach.

The effect on free charges within a metal or other conductive material is phenomenologically expressed in a relation between the current and electric field:

$$\vec{j} = \sigma \vec{E} \qquad [A6.4]$$

where the conductivity σ is the same as that we defined in equation [A5.1].

The ancillary quantities

$$\vec{D} = \kappa \varepsilon_0 \vec{E} \equiv \varepsilon \vec{E} \qquad [A6.5]$$

and

$$\vec{H} = \frac{\vec{B}}{\mu} \qquad\qquad [A6.6]$$

are often defined to avoid explicitly dealing with the induced charges and currents that occur when the fields interact with dielectric materials (κ) and magnetic materials (μ). In most practical cases, the magnetic susceptibility is negligible and $\mu = \mu_0$.

The propagating energy density is conventionally taken to equal the Poynting vector,

$$P = \frac{\vec{E} \times \vec{B}}{\mu} = \vec{E} \times \vec{H} \qquad\qquad [A6.7]$$

The dielectric permittivity of free space is defined in terms of the velocity of light in vacuum as

$$\mu_o \varepsilon_o = \frac{1}{c^2} \qquad\qquad [A6.8]$$

and thus the refractive index is related to the relative dielectric constant as

$$n = \sqrt{k} \qquad\qquad [A6.9]$$

The electric field can always be derived from [A6.1] by substituting for the scalar potential using the Lorentz gauge (see Appendix 4), but far from the source, the electric and magnetic fields (in vacuum) are related by

$$\vec{E} = \frac{1}{i\omega\mu_o\varepsilon_o} \nabla \times \vec{B} = -i\frac{c}{k} \nabla \times \vec{B} \qquad\qquad [A6.10]$$

For plane waves, the gradient operator simply brings down a factor of -ik in the direction of propagation. Thus, far from the source where the radiation is essentially planar, the magnetic field is the cross-product of the direction of propagation and the vector potential. It is thus perpendicular to the direction of propagation (because the cross-product of two vectors is perpendicular to both of them). The electric field is also perpendicular to the propagation vector for the same reason, from [A6.10]. Thus, both the electric field and magnetic field are transverse to the direction of propagation at long distances and perpendicular to each other. The magnitudes are related by $E = cB$ or $E = \mu_0 cH$, so that the induction field H plays the role of a current/length, which when multiplied by the impedance of free space becomes the voltage/length E.

Table of Symbols Used in the Text

Although I have tried to be consistent in use of symbols throughout the text, there aren't enough unique letters in the Latin and Greek alphabets to unambiguously name every concept, and differing fields also reuse the same letter or symbol for different meanings. The relevant alternative can be discerned from context in most cases. The table below may be of aid to the puzzled reader. The table is in alphabetical order in both alphabets (e.g., thus. σ will be found with s and ϕ with f, etc.).

Symbol	Meaning
A	Electromagnetic vector potential
A_{RX}	Effective collecting area of receiving antenna
α	Absorptive loss (Nepers/meter)
	Phase shift between adjacent antennas in an array
BW, [BW]	Bandwidth (Hz)
c	Speed of light in vacuum
dB	Decibels
dBm	Decibels from 1 mW
dBi	Decibels of directive gain relative to an isotropic antenna
δ, Δ	A small change in a quantity
D	Directivity or directive gain
E_b	Bit energy (Joules)
e	Efficiency
ε	Relative dielectric constant
ε_0	Dielectric permittivity of free space (Farad/meter)
f	Frequency (Hz)
ϕ	Electromagnetic scalar potential
G	Gain
	Refractive index gradient (1/meters)
g_m	Transconductance (change in current/controlling voltage)
h	Height (m)
i	Imaginary unit; $\sqrt{(-1)}$
I	Current (amps)
φ	Azimuthal angle (degrees or radians)
J	Vector of current density (A/m^2)
k	Wavenumber (radians/meter)

(*Continued*)

(*Continues*)

Symbol	Meaning
k	Boltzmann's constant
	Refractive correction to Earth radius of curvature
λ	Wavelength (m)
μ_0	Magnetic permeability of the vacuum (Henry/meter)
N	Noise power (Watts or dBm)
	Differential refractive index
N_0	Noise power spectral density (Watts/Hz)
	Input thermal noise to an amplifier
n	Refractive index
P	Power (Watts or dBm)
	Total pressure (Pa or Bars)
q	Magnitude of the electron charge (coulombs)
Q	Charge (coulombs)
r	Distance from source to observer
$\mathbf{r}$	Vector from source to observer
R	Resistance (Ω)
	Average distance from source to observer
ρ	Charge density (coulombs/m^3)
	Radial distance in diffraction calculations
	Resistivity (Ω-cm)
S	Signal power (Watts or dBm)
$<S>$	Average shadow depth (dB)
Σ	Sum or add function
σ	Conductivity (Siemens/meter)
	Width parameter of Rayleigh distribution
s	Incremental path length
θ	Angle (degrees or radians)
T	Time interval
	Absolute temperature (K)
u_0	Distance normalized by Fresnel length
V, v	Voltage (volts)
v_{ph}	Phase velocity (m/sec)
ω	Angular frequency (radians/sec)
Ω	Ohm
	Solid angle (steradians)
W	Watt
Z	Complex impedance (Ω)
z	Distance to diffracting obstacle

Index

Printed and bound by CPI Group (UK) Ltd, Croydon, CR0 4YY

03/10/2024

01040338-0006